D1644326

Lecture Notes in Artificial Intelligence

Subseries of Lecture Notes in Computer Science

LNAI Series Editors

Randy Goebel
  University of Alberta, Edmonton, Canada
Yuzuru Tanaka
  Hokkaido University, Sapporo, Japan
Wolfgang Wahlster
  DFKI and Saarland University, Saarbrücken, Germany

LNAI Founding Series Editor

Joerg Siekmann
  DFKI and Saarland University, Saarbrücken, Germany

Robert Buchmann   Claudiu Vasile Kifor
Jian Yu (Eds.)

Knowledge Science, Engineering and Management

7th International Conference, KSEM 2014
Sibiu, Romania, October 16-18, 2014
Proceedings

 Springer

Volume Editors

Robert Buchmann
University of Vienna
Währingerstr. 29
1090 Vienna, Austria
E-mail: rbuchmann@dke.univie.ac.at

Claudiu Vasile Kifor
Lucian Blaga University of Sibiu
10 Victoriei Blv.
550024 Sibiu, Romania
E-mail: claudiu.kifor@ulbsibiu.ro

Jian Yu
Beijing Jiaotong University
Haidian District 100044, Beijing, China,
E-mail: jianyu@bjtu.edu.cn

ISSN 0302-9743   e-ISSN 1611-3349
ISBN 978-3-319-12095-9   e-ISBN 978-3-319-12096-6
DOI 10.1007/978-3-319-12096-6
Springer Cham Heidelberg New York Dordrecht London

Library of Congress Control Number: 2014950380

LNCS Sublibrary: SL 7 – Artificial Intelligence

Typesetting: Camera-ready by author, data conversion by Scientific Publishing Services, Chennai, India

Printed on acid-free paper

Springer is part of Springer Science+Business Media (www.springer.com)

Preface

KSEM 2014 (the International Conference on Knowledge Science, Engineering and Management) followed an established tradition of long-standing conferences and was the seventh in the KSEM series, following the successful events in Guilin, China (KSEM 2006), Melbourne, Australia (KSEM 2007), Vienna, Austria (KSEM 2009), Belfast, UK (KSEM 2010), Irvine, USA (KSEM 2011), and Dalian, China (KSEM 2013).

KSEM 2014 was held in Sibiu, Romania, one of the most beautiful medieval towns in Transylvania. Due to its cultural richness along with its history, traditional values, architectural and geographical beauty, Sibiu was designated The European Capital of Culture for the year 2007 and, according to Forbes, it is one of Europe's "most idyllic places to live." The conference was hosted by Lucian Blaga University of Sibiu and was an opportunity to establish an international forum for aggregating and consolidating different technical viewpoints on the concept of "knowledge," as they emerge from its various facets – engineering, management, and science. Epistemic aspects were discussed in relation to mathematical formalisms, business goals, or information system architectures, in order to highlight and stimulate refinements on the key enabling concept for a knowledge-driven society. Integration requirements emerged from a wide array of problems and research challenges, and the problem of bridging the gap between knowledge management and knowledge engineering was tackled from multiple perspectives.

Based on the reviews by members of the Program Committee, a selection of 30 full papers and four short papers were presented and included in this year's proceedings. Additionally, three highly valued speakers provided keynote presentations, whose abstracts are also included here. We thank Acad. Prof. Ruqian Lu, Prof. Dr. Pericles Loucopoulos and Prof. Dr. Gheorghe Cosmin Silaghi for their inspiring keynote talks.

A large scientific community was involved in setting up KSEM 2014. We would like to extend our gratitude to everybody who contributed to the success of the event. First of all, we thank the authors who submitted their valuable work and the members of the Program Committee who provided their expertise for selecting and guiding the improvement of submissions. We greatly appreciate the support and advice of the conference chairs: Prof. Dr. Ioan Bondrea (Lucian Blaga University of Sibiu, Romania), Prof. Dr. Dimitris Karagiannis (University of Vienna, Austria), and Prof. Dr. Hui Xiong (Rutgers University, USA). We also thank the team at Springer led by Alfred Hofmann for the technical support in the publication of this volume.

The event would not have been possible without the extensive efforts of the Organizing Committee from Lucian Blaga University from Sibiu and of our host, Rector Prof. Dr. Ioan Bondrea.

October 2014

Robert Buchmann
Claudiu Vasile Kifor
Jian Yu

Organization

KSEM 2014 was hosted and organized by the Faculty of Engineering at Lucian Blaga University of Sibiu, Romania. The conference was held during October 16–18 2014 in Sibiu, Romania.

Organizing Committee

General Co-chairs

Ioan Bondrea	Lucian Blaga University of Sibiu, Romania
Dimitris Karagiannis	University of Vienna, Austria
Hui Xiong	Rutgers University, USA

Program and Publication Co-chairs

Robert Buchmann	University of Vienna, Austria
Claudiu Vasile Kifor	Lucian Blaga University of Sibiu, Romania
Jian Yu	Beijing Jiaotong University, China

Local Organizing Committee

Ioan Bondrea	Lucian Blaga University of Sibiu, Romania
Claudiu Vasile Kifor	Lucian Blaga University of Sibiu, Romania
Carmen Simion	Lucian Blaga University of Sibiu, Romania
Marius Cioca	Lucian Blaga University of Sibiu, Romania
Daniel Volovici	Lucian Blaga University of Sibiu, Romania
Remus Brad	Lucian Blaga University of Sibiu, Romania
Eduard Stoica	Lucian Blaga University of Sibiu, Romania
Dana Preda	Lucian Blaga University of Sibiu, Romania
Lucian Lobonţ	Lucian Blaga University of Sibiu, Romania
Radu Pascu	Lucian Blaga University of Sibiu, Romania
Eva Nicoleta Burduşel	Lucian Blaga University of Sibiu, Romania
Alina Lungu	Lucian Blaga University of Sibiu, Romania
Cristina Protopopu	Lucian Blaga University of Sibiu, Romania
Mihai Zerbeş	Lucian Blaga University of Sibiu, Romania
Radu Petruşe	Lucian Blaga University of Sibiu, Romania

Steering Committee

David Bell	Queen's University, Belfast, UK
Cungen Cao	Chinese Academy of Sciences, China
Dimitris Karagiannis	University of Vienna, Austria
Zhi Jin	Peking University, China

Jérome Lang   Université Paris-Dauphine, LAMSADE, France
Yoshiteru Nakamori   JAIST, Japan
Jorg Siekmann   DFKI, Germany
Eric Tsui   The Hong Kong Polytechnic University,
                                      SAR China
Zhongtuo Wang   Dalian University of Technology, China
Kwok Kee Wei   City University of Hong Kong, Hong Kong,
                                      SAR China
Mingsheng Ying   Tsinghua University, China
Zili Zhang   Southwest University, China
Yaxin Bi   Ulster University, Belfast, UK
Ruqian Lu (Honorary Chair)   Chinese Academy of Sciences, China
Chengqi Zhang (Chair)   University of Technology, Sydney, Australia

Program Committee

A Min Tjoa   Vienna University of Technology, Austria
Abel Usoro   University of the West of Scotland, UK
Adina Florea   Politehnica University of Bucharest, Romania
Andreas Albrecht   Middlesex University, UK
Anthony Hunter   University College London, UK
Carl Vogel   Trinity College of Dublin, Ireland
Chunxia Zhang   Beijing Institute of Technology, China
Claudiu Vasile Kifor   Lucian Blaga University of Sibiu,
                                      Romania
Costin Bădică   University of Craiova, Romania
Dan O'Leary   University of Southern California, USA
Daniel Volovici   Lucian Blaga University of Sibiu, Romania
Dimitris Karagiannis   University of Vienna, Austria
Elsa Negre   Université Paris-Dauphine, LAMSADE, France
Enhong Chen   University of Science and Technology of China
Gabriele Kern-Isberner   Technische Universität Dortmund, Germany
Gheorghe Cosmin Silaghi   Babeş-Bolyai University, Cluj Napoca
Hans-Georg Fill   University of Vienna, Austria
Heiner Stuckenschmidt   University of Mannheim, Germany
Huynh Van Nam   Japan Advanced Institute of Science and
                                      Technology
Ioan Salomie   Technical University of Cluj-Napoca, Romania
Irek Czarnowski   Gdynia Maritime University, Poland
James Lu   Emory University, USA
Jérôme Lang   Université Paris-Dauphine, LAMSADE, France
Jia-Huai You   University of Alberta, Edmonton, Canada
Jian Yu   Beijing Jiaotong University, China
Jiangning Wu   Dalian University of Technology, China
Jie Wang   Arizona State University, USA

Stewart Massie	Robert Gordon University, UK
Sven-Volker Rehm	WHU - Otto Beisheim School of Management, Germany
Takayuki Ito	Nagoya Institute of Technology, Japan
Ulrich Reimer	University of Applied Sciences St. Gallen, Switzerland
Victor Ion Munteanu	West University of Timisoara, Romania
Vladimir Creţu	Politehnica University, Timisoara, Romania
Weiru Liu	Queen's University Belfast, UK
Wenjun Zhou	University of Tennessee, USA
Xun Wang	University of Technology, Sydney, Australia
Yong Tang	South China Normal University, China
Yoshinori Hara	Kyoto University, Japan
Zhendong Ma	Austrian Institute of Technology, Austria
Zhi Jin	Peking University, China
Zhi-Hua Zhou	Nanjing University, China
Zhisheng Huang	Vrije University Amsterdam, The Netherlands
Zili Zhang	Deakin University, Australia

Additional Reviewers

Zobia Rehman
Bo Liu
Chao Qian
Alexandru Stan
Domenik Bork
Ignacio Traverso
Daniel Fleischhacker
Yi Bi
Md Solimul Chowdhury
Melinda Jiang

Li Li
Jun Wang
Guohua Liu
Mihaela Colhon
Sergio Martinez
Ivan Ruiz-Rube
Niksa Visic
Nicolas Schwind
Monsterrat Batet

Keynote Abstracts

Korchestration and the Korc Calculus

Ruqian Lu

Academy of Mathematics and Systems Science,
CAS Key Lab of Management, Decision and Information Systems,
Sino-Australian Joint Lab of Quantum Computing and Quantum Information
Processing,
rqlu@math.ac.cn

Abstract. The orchestration technique has been popular in various fields of computing science and has got different names, such as computation orchestration, service orchestration, business orchestration, cloud orchestration, etc. Most of them have similar but slightly different meanings. In accordance with this, languages programming orchestration such as Orc, BPEL and Now have been developed. However, we have noticed there are two important aspects that are still less studied in orchestration research. They are orchestration for full cycle knowledge service and big data driven orchestration. We propose the concept of korchestration, which is short for knowledge orchestration, to fill this gap. At the same time we introduce the Korc calculus as a conservative extension of Orc calculus towards application of orchestration techniques in the above mentioned two areas. The various new features of Korc include weakly open world assumption, abstract knowledge source assumption, Boolean site calls, parallel logic programming, massive parallelism, fault tolerant computing etc.

Enterprise Knowledge Modelling: Facilitating Flexible Dynamically Changing Systems

Pericles Loucopoulos

University of Manchester
pericles.loucopoulos@mbs.ac.uk

Abstract. Turbulence is in the nature of business environments. Changes brought about because of different requirements such as social, political, technical and economic, exert pressures on organisations to respond in a timely and cost effective way to these challenges. In such an unstable environment information system developers are challenged to develop systems that can meet the requirements of dynamically changing organisations in a flexible manner. Against this dynamic business backdrop, emergent application software is regarded as a key component in the service industry of tomorrow. The effective and efficient development of such systems can have a major impact on the economic value of digital companies – that is companies for which enterprise software becomes the decisive driver behind product and service innovation. Rapid organisational change, knowledge-intensity of goods and services, the growth in organisational scope, and information technology, have all intensified organisational needs for a more formal approach to dealing with enterprise knowledge. In addition virtual organisations that are made up of complementary allied entities place greater demands on knowledge sharing. This talk advances a position, based on research work and the application of this work on many industrial and commercial applications, which states that, "central to successful business evolution through the use of information technology is *Enterprise Knowledge Modelling*". Enterprise Knowledge Modelling involves many facets of the information systems domain including considerations such as technical (business processes, flow of information etc.), organisational and social (policies, structures and work roles etc.) and teleological (purposes and reasons). Conceptual modelling plays a central role in the way that one can capture, reason, represent, use for negotiation and agreement between many stakeholders and discover new knowledge from legacy systems.

Multi-Criteria Resource Negotiation and Scheduling for Hybrid Distributed Computing Infrastructures

Gheorghe Cosmin Silaghi

Business Information Systems Department,
Faculty of Economic Sciences and Business Administration,
Babeş-Bolyai University Cluj Napoca, Romania,
gheorghe.silaghi@econ.ubbcluj.ro

Abstract. Assembling and jointly using different types of computing infrastructures like grids and clouds is an increasingly met phenomenon. To achieve this goal, research communities are building bridging technologies between the various sorts of infrastructures. These infrastructures are characterized by positive attributes like cost effectiveness, reliability, high performance and greenness. With this respect, joint commercial exploitation and increased user satisfaction represent contradicting challenges. To advance towards these goals, we will discuss two aspects of the interaction between resource providers and consumers: negotiation and scheduling in a multi-criteria setup. While both types of players possess limited knowledge about the opponents, we will design two interaction mechanisms allowing for service levels establishment and jobs placement, given the mitigation between providers and consumers.

Table of Contents

Clustering and Classification

Metamodelling and Conceptual Modelling

Enterprise Knowledge

Knowledge Discovery and Retrieval

Formal Knowledge Processing

Ontology Engineering and Management

Knowledge Management and Knowledge Systems

Coming Upon the Classic Notion
of Implicit Knowledge Again

Bernhard Heinemann

Faculty of Mathematics and Computer Science,
University of Hagen,
58084 Hagen, Germany
bernhard.heinemann@fernuni-hagen.de

Abstract. Subsequently, we introduce a novel semantics for the bi-modal logic of subset spaces, denoted by LSS. This system was originally invented by Moss and Parikh for the purpose of clarifying the intrinsic relationship between the epistemic notion of knowledge and the geometric concept of topology. Focussing on the knowledge-theoretic side in this paper, we re-adjust LSS to multi-agent scenarios. As a result, a particular dynamic logic of implicit knowledge is obtained. This finds expression in the technical outcome of the paper, which covers soundness, completeness, decidability, and complexity issues regarding the arising system.

Keywords: epistemic logic, implicit knowledge, subset space semantics, topological reasoning.

1 Introduction

Reasoning about knowledge constitutes an important foundational issue in Artificial Intelligence. We concentrate on some of its logical aspects in this paper. In particular, we are concerned with the idea of *implicit* (or *distributed*) *knowledge* of a group of agents.

Meanwhile, several instructive and very readable treatises on the diverse logics of knowledge are available. While more recent publications rather stress the *dynamics* of informational perceptions, including aspects of *belief, desire,* or *intention (BDI)* (see, e.g., [3], as well for further references), the classic textbooks [7] and [13] can thoroughly serve as a common ground for the fundamentals of epistemic logic needed here. Accordingly, given a finite collection G of agents, a binary *accessibility relation* R_A connecting *possible worlds* or *conceivable states of the world,* is associated with every agent $A \in G$. The *knowledge of A* is then defined through the *validity* of the corresponding formulas at all states the agent considers possible at the actual one. Now, *collecting together* such 'locally allocated' knowledge means ruling out those worlds that are inconceivable to some of the agents in G. To put it another way, the implicit knowledge of the agents under discussion is represented exactly by *intersecting* the respective sets of accessible states; see [13], Sect. 2.3, or [7], Sect. 2.2 and Sect. 3.4. (Throughout this paper, the term *implicit knowledge* is used, as in [13]; on the other hand,

R. Buchmann et al. (Eds.): KSEM 2014, LNAI 8793, pp. 1–12, 2014.

the term *distributed knowledge* is employed in the latter reference, since the idea of *awareness* (and, therefore, that of *explicit* knowledge) enters the field there.)

Moss and Parikh's bi-modal logic of subset spaces, LSS (see [14], [5], or Ch. 6 of [1]), may be rated as a cross-disciplinary framework for dealing with topological as well as epistemic scenarios. This is exemplified in the single-agent case subsequently. The *epistemic state* of an agent in question, i.e., the set of all those states that cannot be distinguished by what the agent topically knows, can be viewed as a *neighborhood U* of the actual state x of the world. Formulas are then interpreted with respect to the resulting pairs x, U called *neighborhood situations*. Thus, both the set of all states and the set of all epistemic states constitute the relevant semantic domains as particular subset structures. The two modalities involved, K and □, quantify over all elements of U and 'downward' over all neighborhoods contained in U, respectively. This means that K captures the notion of knowledge as usual (see [7] or [13] again), and □ reflects *effort to acquire knowledge* since gaining knowledge goes hand in hand with a shrinkage of the epistemic state. In fact, knowledge acquisition is this way reminiscent of a topological procedure. The appropriate logic of 'real' topological spaces as well as that of further computationally interesting spatial structures (viz tree-like ones) were examined by Georgatos rather promptly; see [8], [9]. The ongoing research into subset and topological spaces, respectively, is reported in the handbook [1]. More recent developments include the papers [12], [2], and [15], with the last two forging links between subset spaces and *Dynamic Epistemic Logic (DEL)*; see [6].

Most papers on LSS deal with the single-agent case. Notwithstanding this, a multi-agent version was suggested in [10] (see also [11]). The key idea behind these papers is as follows: incorporate the agents in terms of additional modalities and, apart from this variation of the *logic*, let the original semantics be unchanged. However, what happens when, in contrast, the *semantics* is modified, and even in a way suggesting itself, namely to the effect that the agent structure is reflected in the *atomic* semantic entities already? – It turns out that the scope of the modality K has to be restricted then, but fortunately in a quite acceptable manner: K hereby mutates to an *implicit knowledge operator* (and, as will become apparent later, the logic remains the same in this case).

This idea will be implemented in the rest of this paper. Our aim is to give precise definitions as related to the underlying language, state the axioms and rules of the arising logic, prove soundness and completeness with respect to the intended class of domains, and reason about the intrinsic effectiveness and efficiency properties. (However, we must omit elaborate examples, due to the lack of space; in this respect, the reader is referred to the quoted literature.) The outcome we strive for is, in fact, an alternative modal description of implicit knowledge, and in the presence of a rather general operator describing increase of individual knowledge. (Thus, we are not ambitious in producing a system 'beating' others (in particular, more differentiated ones) here. But note that it is very desirable to have to hand distinct (e.g., differently fine-grained) ways of seeing a subject: this would allow one to react on varying problems flexibly; hence such a broadening of the horizon is a widespread practice in many mathematically oriented fields.)

The subsequent technical part of the paper is organized as follows. In the next section, we recapitulate the basics of multi-agent epistemic logic. The facts we need from the logic of subset spaces are then listed in Section 3. Section 4 contains the new multi-agent setting of LSS. Our main results, including some proof sketches, follow in the next two sections. Finally, we conclude with a summing up and a couple of additional remarks. – An attempt has been made to keep the paper largely self-contained. However, acquaintance of the reader with basic modal logic has to be assumed. As to that, the textbook [4] may serve as a standard reference.

2   Revisiting the Most Common Logic of Knowledge

All languages we consider in this paper are based on a denumerably infinite set $\mathsf{Prop} = \{p, q, \ldots\}$ of symbols called *proposition variables* (which should represent the basic facts about the states of the world). Let $n \in \mathbb{N}$ be given (the number of agents under discussion). Then, our modal language for knowledge contains, among other things, a one-place operator K_i representing the i-th agent's knowledge, for every $i \in \{1, \ldots, n\}$. The set KF of all *knowledge formulas* is defined by the rule

$$\alpha ::= \top \mid p \mid \neg\alpha \mid \alpha \wedge \alpha \mid \mathsf{K}_i\alpha \mid \mathsf{I}\alpha,$$

where $i \in \{1, \ldots, n\}$. The missing boolean connectives will be treated as abbreviations, as needed. The connective I is called the *implicit knowledge operator*. Moreover, the modal duals of K_i and I are denoted by L_i and J, respectively.

As was indicated right at the outset, each of the operators K_i comes along with a binary relation R_i on the set X of all states of the world. The kind of knowledge we would like to model should certainly be mirrored in the characteristics of these relations. Having multi-agent systems à la [7] in mind where 'accessibility' means 'indistinguishability of the local states of the other agents', one is led to *equivalence relations* actually. Furthermore, the *intersection* of these equivalences is the relation associated with the implicit knowledge operator. Thus, the multi-modal frames for interpreting the above formulas are tuples $F = (X, R_1, \ldots, R_n, R_\mathsf{I})$, where X is a non-empty set, $R_i \subseteq X \times X$ is an equivalence relation for every $i \in \{1, \ldots, n\}$, and $R_\mathsf{I} = \bigcap_{i=1,\ldots,n} R_i$. And a model M based on such a frame is obtained by adding a valuation to the frame, i.e., a mapping V from Prop into the powerset of X, determining those states where the respective proposition variables become valid. Satisfaction of formulas is then defined *internally,* i.e., in models at particular states. We here remind the reader of the case of a modal operator, say the one for implicit knowledge:

$$M, x \models \mathsf{I}\alpha : \Longleftrightarrow \text{ for all } y \in X : \text{if } (x, y) \in R_\mathsf{I}, \text{ then } M, x \models \alpha,$$

for all $x \in X$ and $\alpha \in \mathsf{KF}$. – The just described semantics is accompanied by a logic which is a slight extension of the multi-modal system S5$_{n+1}$. This means that we have, in particular, the well-known and much-debated knowledge and introspection axioms for each of the modalities involved, for example, those relating to the I-operator:

- $I\alpha \to \alpha$
- $I\alpha \to II\alpha$
- $J\alpha \to IJ\alpha$,

where $\alpha \in \mathsf{KF}$. The schemata

- $\mathsf{K}_i\alpha \to I\alpha$, for every $i \in \{1, \dots, n\}$,

designated $(\mathbb{I})$, constitute the extension of $\mathrm{S5}_{n+1}$ addressed a moment ago. The following is taken from [7], Theorem 3.4.1 (see also [13], Theorem 2.3.2).

Theorem 1. *The logic* $\mathrm{S5}_{n+1} + (\mathbb{I})$ *is sound and complete with respect to the class of models described above.*

While sketching a proof of this theorem, the authors of [7] point to the difficulties related to the intersection property (i.e., $R_I = \bigcap_{i=1,\dots,n} R_i$) on the way towards completeness. We, too, shall encounter this problem, in Section 5 (albeit in weakened form).

3   The Language and the Logic of Subset Spaces

In this section, we first fix the language for subset spaces, $\mathcal{L}$. After that, we link the semantics of $\mathcal{L}$ with the common relational semantics of modal logic. (This link will be utilized later in this paper.) Finally, we recall some facts on the logic of subset spaces needed subsequently. – The proceeding in this section is a bit more rigorous than that in the previous one, since $\mathcal{L}$ and LSS are assumed to be less established.

To begin with, we define the syntax of $\mathcal{L}$. Let the set SF of all *subset formulas*[1] over **Prop** be defined by the rule

$$\alpha ::= \top \mid p \mid \neg\alpha \mid \alpha \wedge \alpha \mid \mathsf{K}\alpha \mid \Box\alpha.$$

Here, the duals of K and $\Box$ are denoted by L and $\Diamond$, respectively. In view of our considerations in the introduction, K is called the *knowledge operator* and $\Box$ the *effort operator*.

Second, we examine the semantics of $\mathcal{L}$. For a start, we define the relevant domains. We let $\mathcal{P}(X)$ designate the powerset of a given set X.

Definition 1 (Semantic Domains)

1. *Let X be a non-empty set (of* states*) and $\mathcal{O} \subseteq \mathcal{P}(X)$ a set of subsets of X. Then, the pair $\mathcal{S} = (X, \mathcal{O})$ is called a* subset frame.

---

[1] The prefix 'subset' will be omitted provided there is no risk of confusion.

2. Let $\mathcal{S} = (X, \mathcal{O})$ be a subset frame. Then the set $\mathcal{N}_{\mathcal{S}} := \{(x, U) \mid x \in U \text{ and } U \in \mathcal{O}\}$ is called the set of neighborhood situations of $\mathcal{S}$.

3. Let $\mathcal{S} = (X, \mathcal{O})$ be a subset frame. An $\mathcal{S}$-valuation is a mapping $V : \mathsf{Prop} \to \mathcal{P}(X)$.

4. Let $\mathcal{S} = (X, \mathcal{O})$ be a subset frame and V an $\mathcal{S}$-valuation. Then, $\mathcal{M} := (X, \mathcal{O}, V)$ is called a subset space (based on $\mathcal{S}$).

Note that neighborhood situations denominate the semantic atoms of the bi-modal language $\mathcal{L}$. The first component of such a situation indicates the actual state of the world, while the second reflects the uncertainty of the agent in question about it. Furthermore, Definition 1.3 shows that values of proposition variables depend on states only. This is in accordance with the common practice in epistemic logic; see [7] or [13] once more.

For a given subset space $\mathcal{M}$, we now define the relation of *satisfaction*, $\models_{\mathcal{M}}$, between neighborhood situations of the underlying frame and formulas from SF. Based on that, we define the notion of *validity* of formulas in subset spaces. In the following, neighborhood situations are often written without parentheses.

Definition 2 (Satisfaction and Validity). *Let $\mathcal{S} = (X, \mathcal{O})$ be a subset frame.*

1. *Let $\mathcal{M} = (X, \mathcal{O}, V)$ be a subset space based on $\mathcal{S}$, and let $x, U \in \mathcal{N}_{\mathcal{S}}$ be a neighborhood situation. Then*

$$
\begin{array}{lll}
x, U \models_{\mathcal{M}} \top & & \text{is always true} \\
x, U \models_{\mathcal{M}} p & :\iff & x \in V(p) \\
x, U \models_{\mathcal{M}} \neg\alpha & :\iff & x, U \not\models_{\mathcal{M}} \alpha \\
x, U \models_{\mathcal{M}} \alpha \wedge \beta & :\iff & x, U \models_{\mathcal{M}} \alpha \text{ and } x, U \models_{\mathcal{M}} \beta \\
x, U \models_{\mathcal{M}} \mathsf{K}\alpha & :\iff & \forall y \in U : y, U \models_{\mathcal{M}} \alpha \\
x, U \models_{\mathcal{M}} \Box\alpha & :\iff & \forall U' \in \mathcal{O} : [x \in U' \subseteq U \Rightarrow x, U' \models_{\mathcal{M}} \alpha],
\end{array}
$$

where $p \in \mathsf{Prop}$ and $\alpha, \beta \in \mathsf{SF}$. In case $x, U \models_{\mathcal{M}} \alpha$ is true we say that α holds in $\mathcal{M}$ at the neighborhood situation x, U.

2. *Let $\mathcal{M} = (X, \mathcal{O}, V)$ be a subset space based on $\mathcal{S}$. A subset formula α is called* valid *in $\mathcal{M}$ iff it holds in $\mathcal{M}$ at every neighborhood situation of $\mathcal{S}$.*

Note that the idea of knowledge and effort described in the introduction is made precise by Item 1 of this definition. In particular, knowledge is defined as validity at all states that are indistinguishable to the agent here, too.

Subset frames and spaces can be considered from a different perspective, as is known since [5] and reviewed in the following. Let a subset frame $\mathcal{S} = (X, \mathcal{O})$ and a subset space $\mathcal{M} = (X, \mathcal{O}, V)$ based on $\mathcal{S}$ be given. Take $X_{\mathcal{S}} := \mathcal{N}_{\mathcal{S}}$ as a set of worlds, and define two accessibility relations $R_{\mathcal{S}}^{\mathsf{K}}$ and $R_{\mathcal{S}}^{\Box}$ on $X_{\mathcal{S}}$ by

$$
\begin{array}{ll}
(x, U)\, R_{\mathcal{S}}^{\mathsf{K}}\, (x', U') : \iff & U = U' \text{ and} \\
(x, U)\, R_{\mathcal{S}}^{\Box}\, (x', U') : \iff & (x = x' \text{ and } U' \subseteq U),
\end{array}
$$

for all $(x, U), (x', U') \in X_{\mathcal{S}}$. Moreover, let $V_{\mathcal{M}}(p) := \{(x, U) \in X_{\mathcal{S}} \mid x \in V(p)\}$, for every $p \in \mathsf{Prop}$. Then, bi-modal Kripke structures $\mathcal{S}_{\mathcal{S}} := (X_{\mathcal{S}}, \{R_{\mathcal{S}}^{\mathsf{K}}, R_{\mathcal{S}}^{\Box}\})$ and $\mathcal{M}_{\mathcal{M}} := (X_{\mathcal{S}}, \{R_{\mathcal{S}}^{\mathsf{K}}, R_{\mathcal{S}}^{\Box}\}, V_{\mathcal{M}})$ result in such a way that $\mathcal{M}_{\mathcal{M}}$ is equivalent to $\mathcal{M}$ in the following sense.

Proposition 1. *For all $\alpha \in$ SF and $(x, U) \in X_{\mathcal{S}}$, we have that $x, U \models_{\mathcal{M}} \alpha$ iff $M_{\mathcal{M}}, (x, U) \models \alpha$.*

Here (and later on as well), the non-indexed symbol '$\models$' denotes the usual satisfaction relation of modal logic (as it was the case in Section 2 already). – The proposition is easily proved by induction on α. We call $S_{\mathcal{S}}$ and $M_{\mathcal{M}}$ the Kripke structures *induced* by $\mathcal{S}$ and $\mathcal{M}$, respectively.[2]

We now turn to the *logic* of subset spaces, LSS. Here is the appropriate axiomatization from [5], which was proved to be sound and complete in Sect. 1.2 and, respectively, Sect. 2.2 there:

1. All instances of propositional tautologies
2. $\mathsf{K}(\alpha \to \beta) \to (\mathsf{K}\alpha \to \mathsf{K}\beta)$
3. $\mathsf{K}\alpha \to (\alpha \wedge \mathsf{KK}\alpha)$
4. $\mathsf{L}\alpha \to \mathsf{KL}\alpha$
5. $(p \to \Box p) \wedge (\Diamond p \to p)$
6. $\Box(\alpha \to \beta) \to (\Box\alpha \to \Box\beta)$
7. $\Box\alpha \to (\alpha \wedge \Box\Box\alpha)$
8. $\mathsf{K}\Box\alpha \to \Box\mathsf{K}\alpha$,

where $p \in$ Prop and $\alpha, \beta \in$ SF. – The last schema is by far the most interesting in this connection, as the interplay between knowledge and effort is captured by it. The members of this schema are called the *Cross Axioms* since [14]. Note that the schema involving only proposition variables is in accordance with the remark on Definition 1.3 above.

As the next step, let us take a brief look at the effect of the axioms from the above list within the framework of common modal logic. To this end, we consider bi-modal Kripke models $M = (X, R, R', V)$ satisfying the following four properties:

- the accessibility relation R of M belonging to the knowledge operator K is an equivalence,
- the accessibility relation R' of M belonging to the effort operator $\Box$ is reflexive and transitive,
- the composite relation $R' \circ R$ is contained in $R \circ R'$ (this is usually called the *cross property*), and
- the valuation V of M is constant along every R'-path, for all proposition variables.

Such a model M is called a *cross axiom model* (and the frame underlying M a *cross axiom frame*). Now, it can be verified without difficulty that LSS is sound with respect to the class of all cross axiom models. And it is also easy to see that every induced Kripke model is a cross axiom model (and every induced Kripke frame a cross axiom frame). Thus, the completeness of LSS for cross axiom models follows from that of LSS for subset spaces (which is Theorem 2.4 in [5]) by means of Proposition 1. This completeness result will be used below, in Section 6.

---

[2] It is an interesting question whether one can identify the induced Kripke structures amongst all bi-modal ones; see the paper [12] for an answer to this.

4   A Multi-agent Semantics Based on Subset Spaces

In this section, subset spaces for multiple agents are shifted into the focal point of interest. We first introduce the class of domains we consider relevant in this connection. The members of this class turn out to be slightly different from those 'multi-agent structures' that were taken as a basis in the paper [10]. The main difference, however, concerns the semantic atoms, into which the actual knowledge states of the agents are incorporated now. In what follows, we discuss the possible ways of interpreting knowledge or subset formulas within the new framework. We argue why we should confine ourselves to formulas from SF here, and how implicit knowledge comes into play then. Finally in this section, we prove that the logic LSS is sound with respect to the novel semantics.

For a start, we modify Definition 1 accordingly. Let $n \in \mathbb{N}$ be the number of the involved agents again.

Definition 3 (Multi-agent Subset Spaces)

1. Let X be a *non-empty set and* $\mathcal{O}_i \subseteq \mathcal{P}(X)$ *a set of subsets of* X, *for every* $i \in \{1, \ldots, n\}$. *Then, the tuple* $\mathcal{S} = (X, \mathcal{O}_1, \ldots, \mathcal{O}_n)$ *is called a* multi-agent subset frame.
2. *Let* $\mathcal{S} = (X, \mathcal{O}_1, \ldots, \mathcal{O}_n)$ *be a multi-agent subset frame. Then the set*

$$\mathcal{K}_{\mathcal{S}} := \{(x, U_1, \ldots, U_n) \mid x \in U_i \text{ and } U_i \in \mathcal{O}_i \text{ for all } i = 1, \ldots, n\}$$

    is called the set of knowledge situations *of* $\mathcal{S}$.
3. *The notion of* $\mathcal{S}$*-valuation is the same as in Definition 1.*
4. *Let* $\mathcal{S} = (X, \mathcal{O}_1, \ldots, \mathcal{O}_n)$ *be a multi-agent subset frame and* V *an* $\mathcal{S}$*-valuation. Then,* $\mathcal{M} := (X, \mathcal{O}_1, \ldots, \mathcal{O}_n, V)$ *is called a* multi-agent subset space (based on $\mathcal{S}$).

The second item of Definition 3 deserves a comment. Clearly, the *meaning* of every component of a knowledge situation remains unaltered in principle; but each individual agent is taken into account now. The *name*, however, is changed because the epistemic aspect, compared to the spatial one, comes more to the fore here.

Now, we would like to evaluate formulas in multi-agent subset spaces $\mathcal{M}$. For that purpose, let $x, U_1, \ldots, U_n$ be a knowledge situation of some multi-agent subset frame $\mathcal{S}$ (on which $\mathcal{M}$ is based). As no difficulties are raised in the propositional cases, we may proceed to the modalities directly. First, the case $\mathsf{K}_i \alpha$ is considered, where $i \in \{1, \ldots, n\}$. In order to retain the intended meaning, K_i should quantify across all the states that agent i considers possible at the world x, i.e., across U_i. But for some $y \in U_i$ it could be the case that $y, U_1, \ldots, U_n$ does not belong to $\mathcal{K}_{\mathcal{S}}$, for the simple reason that $y \notin U_j$ for some $j \in \{1, \ldots, n\}$. Thus, such a quantification is impossible in general. We conclude that we must drop formulas of the type $\mathsf{K}_i \alpha$ because of the new semantics (unless we add additional agent-specific modalities as in [10]). However, note that the knowledge of the individual agents is still represented, namely by the corresponding domains and, in particular, the semantic atoms.

Fortunately, the just detected problem does not appear in the case of the operator I. In fact, quantification now concerns states from the intersection of all the actual knowledge states; hence every such state leads to a knowledge situation as defined above. Thus, we let

$$x, U_1, \ldots, U_n \models_{\mathcal{M}} \mathsf{I}\alpha : \iff \forall y \in \bigcap_{i=1,\ldots,n} U_i : y, U_1, \ldots, U_n \models_{\mathcal{M}} \alpha.$$

Regarding formulas from KF, we have got a restricted correspondence between the syntax and the semantics that way. But what can be said in the case of SF? – An easy inspection shows that the knowledge operator K causes as much a problem as K_i : because of the semantic defaults, quantifying is only possible over those states that are common to *all* of the agents. But *doing so* obviously means *turning* K *into* I. Consequently, we really *set*

$$\mathsf{K} = \mathsf{I}$$

henceforth. This is additionally justified by the fact that both operators share the same properties of knowledge (expressed by the S5-axioms).

The remaining case to be treated is that of the effort operator $\square$. It becomes clear on second thought that this knowledge increasing modality should represent a *system component* here and may have an effect on each of the agents thus. For this reason, we define

$$x, U_1, \ldots, U_n \models_{\mathcal{M}} \square\alpha : \iff \begin{cases} \forall U_1' \in \mathcal{O}_1 \cdots \forall U_n' \in \mathcal{O}_n : \left[x \in U_i' \subseteq U_i \text{ for} \right. \\ \left. i = 1, \ldots, n \Rightarrow x, U_1', \ldots, U_n' \models_{\mathcal{M}} \alpha \right]. \end{cases}$$

In this way, the definition of the multi-agent semantics based on subset spaces is completed. The set SF has proved to be the relevant set of formulas, after identifying K and I.[3]

We are going to show that the logic LSS is sound for multi-agent subset spaces.

Proposition 2 (Soundness). *All formulas from* LSS *are valid in every multi-agent subset space.*

Proof. We only care about the Cross Axioms, since everything else is quite straightforward. Actually, we consider the dual schemata. Let $\mathcal{M}$ be an arbitrary multi-agent subset space and $x, U_1, \ldots, U_n$ a knowledge situation of the underlying frame. Suppose that $x, U_1, \ldots, U_n \models_{\mathcal{M}} \Diamond\mathsf{J}\alpha$. Then there are $U_1' \in \mathcal{O}_1, \ldots, U_n' \in \mathcal{O}_n$ such that $x \in U_i' \subseteq U_i$ for $i = 1, \ldots, n$ and $x, U_1', \ldots, U_n' \models_{\mathcal{M}} \mathsf{J}\alpha$. This means that there exists some $y \in \bigcap_{i=1,\ldots,n} U_i'$ for which $y, U_1', \ldots, U_n' \models_{\mathcal{M}} \alpha$. The world y is also contained in the intersection $\bigcap_{i=1,\ldots,n} U_i$. Thus, the tuple $y, U_1, \ldots, U_n$ is a knowledge situation satisfying $y, U_1, \ldots, U_n \models_{\mathcal{M}} \Diamond\alpha$. Consequently, $x, U_1, \ldots, U_n \models_{\mathcal{M}} \mathsf{J}\Diamond\alpha$. This proves the validity of $\Diamond\mathsf{J}\alpha \to \mathsf{J}\Diamond\alpha$. It follows that all Cross Axioms are valid in every multi-agent subset space.

---

[3] We retain the notation SF although we shall use I in place of K (and J in place of L, respectively) as from now.

The much more difficult question of completeness is tackled in the following section.

5   Completeness

As in the single-agent case, an infinite step-by-step construction is used for proving the completeness of LSS with respect to the new semantics, too; cf. [5], Sect. 2.2 (and [4], Sect. 4.6, for the method in general).[4] For that, it is natural to bring the canonical model of LSS into play in some way. Thus, we fix several notations concerning that model first. Let $\mathcal{C}$ be the set of all maximal LSS-consistent sets of formulas. Furthermore, let $\xrightarrow{\mathsf{I}}$ and $\xrightarrow{\square}$ be the accessibility relations induced on $\mathcal{C}$ by the modalities I and $\square$, respectively. Let $\alpha \in \mathsf{SF}$ be a non-LSS-derivable formula. Then, a multi-agent subset space falsifying α is to be built incrementally. In order to ensure that the resulting limit structure behaves as desired, several requirements on the approximations have to be met at every stage.

Suppose that $\neg\alpha \in \Gamma \in \mathcal{C}$, i.e., Γ is to be realized. We choose a denumerably infinite set of points, Y (the possible worlds of the desired model), fix an element $x_0 \in Y$, and construct inductively a sequence of quadruples $\left(X_m, (P_m^1, \ldots, P_m^n), (j_m^1, \ldots, j_m^n), (t_m^1, \ldots, t_m^n)\right)$ such that, for all $m \in \mathbb{N}$ and $i \in \{1, \ldots, n\}$,

1. X_m is a finite subset of Y containing x_0,
2. P_m^i is a finite set carrying a partial order $\leq_m^i$, with respect to which there is a least element $\bot \in X_m$,
3. $j_m^i : P_m^i \to \mathcal{P}(X_m)$ is a function such that $p \leq_m^i q \iff j_m^i(p) \supseteq j_m^i(q)$, for all $p, q \in P_m^i$,
4. $t_m^i : X_m \times P_m^i \to \mathcal{C}$ is a partial function such that, for all $x, y \in X_m$ and $p, q \in P_m^i$,
   (a) $t_m^i(x, p)$ is defined iff $x \in j_m^i(p)$; in this case it holds that
      i. if $y \in j_m^i(p)$, then $t_m^i(x, p) \xrightarrow{\mathsf{I}} t_m^i(y, p)$,
      ii. if $p \leq_m^i q$, then $t_m^i(x, p) \xrightarrow{\square} t_m^i(x, q)$,
   (b) $t_m^i(x_0, \bot) = \Gamma$.

The next four conditions say to what extent the final model is approximated by the structures $\left(X_m, (P_m^1, \ldots, P_m^n), (j_m^1, \ldots, j_m^n), (t_m^1, \ldots, t_m^n)\right)$. Actually, it will be guaranteed that, for all $m \in \mathbb{N}$ and $i \in \{1, \ldots, n\}$,

5. $X_m \subseteq X_{m+1}$,
6. P_{m+1}^i is an *end extension* of P_m^i (i.e., a superstructure of P_m^i such that no element of $P_{m+1}^i \setminus P_m^i$ is strictly smaller than any element of P_m^i),
7. $j_{m+1}^i(p) \cap X_m = j_m^i(p)$ for all $p \in P_m^i$,
8. $t_{m+1}^i \mid_{X_m \times P_m^i} = t_m^i$.

---

[4] Due to the lack of space, we can only give a proof sketch here; however, some of the technical differences will be highlighted below.

Finally, the construction complies with the following requirements on existential formulas: for all $n \in \mathbb{N}$ and $i \in \{1, \ldots, n\}$,

9. if $\mathsf{J}\beta \in t_m^i(x, p)$, then there are $m < k \in \mathbb{N}$ and $y \in j_k^i(p)$ such that $\beta \in t_k^i(y, p)$,
10. if $\diamond\beta \in t_m^i(x, p)$, then there are $m < k \in \mathbb{N}$ and $p \leq_k^i q \in P_k^i$ such that $\beta \in t_k^i(x, q)$.

With that, the final model refuting α can be defined easily. Furthermore, a relevant *Truth Lemma* (see [4], 4.21) can be proved for it, from which the completeness of LSS with respect to the multi-agent semantics follows immediately. Thus, it remains to specify, for all $m \in \mathbb{N}$, the approximating structures $\left(X_m, (P_m^1, \ldots, P_m^n), (j_m^1, \ldots, j_m^n), (t_m^1, \ldots, t_m^n)\right)$ in a way that all the above requirements are met. This makes up one of the crucial parts of the proof.

Since the case $m = 0$ is rather obvious, we focus on the induction step. Here, some existential formula contained in some maximal LSS-consistent set $t_m^i(x, p)$ must be made true, where $x \in X_m$ and $p \in P_m^i$; see item 9 and item 10 above. We confine ourselves to the case of the implicit knowledge operator. So let $\mathsf{J}\beta \in t_m^i(x, p)$. We choose a new point $y \in Y$ and let $X_{m+1} := X_m \cup \{y\}$. The sets $P_m^1, \ldots, P_m^n$ remain unchanged (and the associated partial orders therefore as well), i.e., we define $P_{m+1}^i := P_m^i$ for $i = 1, \ldots, n$. However, the mappings $j_m^1, \ldots, j_m^n$ are modified as follows. We let $j_{m+1}^i(q) := j_m^i(q) \cup \{y\}$, for all $q \in P_m^i$ satisfying $q \leq_m^i p$ and all $i \in \{1, \ldots, n\}$. The latter requirement obviously guarantees that the new point is really in the intersection of the local knowledge states. Finally, the mappings $t_m^1, \ldots, t_m^n$ are adjusted. From the *Existence Lemma* of modal logic (see [4], 4.20) we know that, for every $i \in \{1, \ldots, n\}$, there is some point Γ_i of $\mathcal{C}$ such that $t_m^i(x, p) \xrightarrow{\;i\;} \Gamma_i$ and $\beta \in \Gamma_i$. Thus, we define $t_{m+1}^i(y, p) := \Gamma_i$. Moreover, the maximal consistent sets which are to be assigned to the pairs (y, q) where $q \leq_m^i p$ and $q \neq p$, are obtained by means of the cross property (which in fact holds on the canonical model); for all other pairs $(z, r) \in X_{m+1} \times P_{m+1}^i$, we let $t_{m+1}^i(z, r) := t_m^i(z, r)$. This completes the definition of $\left(X_{m+1}, (P_{m+1}^1, \ldots, P_{m+1}^n), (j_{m+1}^1, \ldots, j_{m+1}^n), (t_{m+1}^1, \ldots, t_{m+1}^n)\right)$ in the case under consideration.

We must now check that the validity of the properties stated in items 1 – 8 above is transferred from m to $m + 1$. Doing so, several items prove to be evident from the construction. In some cases, however, the particularities of the accessibility relations on $\mathcal{C}$ (like the cross property) have to be applied. Further details regarding this must be omitted here.

As to the validity of item 9 and item 10, it has to be ensured that *all* possible cases are eventually exhausted. To this end, processing must suitably be scheduled with regard to both modalities. This can be done with the aid of appropriate enumerations. The reader is referred to the paper [5] to see how this works in the single-agent case. – In the following theorem, the above achievements are summarized.

Theorem 2 (Completeness). *If the formula $\alpha \in$ SF is valid in all multi-agent subset spaces, then α is LSS-derivable.*

Proposition 2 and Theorem 2 together constitute the main result of this paper, saying that LSS is sound and complete with respect to the class of all multi-agent subset spaces.

6   Remarks on Decidability and Complexity

While the soundness and the completeness of a logic depend on the underlying semantics by definition, *decidability* is a property of the logic (as a set of formulas) by itself. Thus, this property could be established by using a different semantics, and this is actually the case with LSS.

At the end of Section 2, it was stated that LSS is sound and complete for cross axiom models. In addition, LSS satisfies the *finite model property* with respect to this class of models, as was shown in [5], Sect. 2.3. Now, it is known from modal logic that both properties together imply decidability (see [4], Sect. 6.2). Thus, we can readily adopt this fact with reference to the present context.

Theorem 3 (Decidability). LSS *is a decidable set of formulas.*

And what is true of decidability is just as true of *complexity:* being a property of the logic alone. Unfortunately, the precise complexity of LSS has not yet been determined. Quite recently, the *weak logic of subset spaces*, which results from LSS by forgetting the Cross Axioms, was proved to be PSPACE-complete; see [2]. Thus, we have a partial corollary at least. The general case, however, still awaits a solution.

7   Conclusion

In this paper, we have introduced a new description of the implicit knowledge of a group of agents on the basis of subset spaces. Actually, the usual logic of implicit knowledge and Moss and Parikh's logic of subset spaces have been synthesized. The result is a novel semantics for implicit knowledge first, where the actual knowledge states of the individual agents are represented by the semantic atoms. We have argued that, relating to this framework, the implicit knowledge operator I takes over the role of the knowledge operator K from the language $\mathcal{L}$ for (single-agent) subset spaces. Thus, subset space formulas can speak about implicit knowledge and its dynamic change when interpreting them in multi-agent subset spaces.

The second outcome of this paper is a meta-theorem on the logic accompanying this non-standard semantics of I. We have proved that the logic of subset spaces, LSS (see Section 3), is sound and complete with respect to multi-agent subset spaces, too. Moreover, this logic is even decidable, which is obtained as a consequence of earlier results.

It has been argued here and there that subset spaces provide an alternative basis for reasoning about knowledge, complementing the most common and well-established epistemic logic as proposed, e.g., in [13]. It appears to us that the present paper as well makes a contribution underpinning this thesis.

An important open problem regarding LSS was addressed near the end of Section 6. Yet a lot remains to be done beyond answering that question. In particular, one should try to bring subset spaces into line with as many theoretical or practice-oriented epistemic concepts as possible, according to the thesis which has just been mentioned. (An additional justification for such a project was already indicated above, right before the final section of the introduction.)

References

1. Aiello, M., Pratt-Hartmann, I.E., van Benthem, J.F.A.K.: Handbook of Spatial Logics. Springer, Dordrecht (2007)
2. Balbiani, P., van Ditmarsch, H., Kudinov, A.: Subset space logic with arbitrary announcements. In: Lodaya, K. (ed.) ICLA 2013. LNCS, vol. 7750, pp. 233–244. Springer, Heidelberg (2013)
3. van Benthem, J.: Logical Dynamics of Information and Interaction. Cambridge University Press, Cambridge (2011)
4. Blackburn, P., de Rijke, M., Venema, Y.: Modal Logic. Cambridge Tracts in Theoretical Computer Science, vol. 53. Cambridge University Press, Cambridge (2001)
5. Dabrowski, A., Moss, L.S., Parikh, R.: Topological reasoning and the logic of knowledge. Annals of Pure and Applied Logic 78, 73–110 (1996)
6. van Ditmarsch, H., van der Hoek, W., Kooi, B.: Dynamic Epistemic Logic. Synthese Library, vol. 337. Springer, Dordrecht (2007)
7. Fagin, R., Halpern, J.Y., Moses, Y., Vardi, M.Y.: Reasoning about Knowledge. MIT Press, Cambridge (1995)
8. Georgatos, K.: Knowledge theoretic properties of topological spaces. In: Masuch, M., Polos, L. (eds.) Logic at Work 1992. LNCS (LNAI), vol. 808, pp. 147–159. Springer, Heidelberg (1994)
9. Georgatos, K.: Knowledge on treelike spaces. Studia Logica 59, 271–301 (1997)
10. Heinemann, B.: Topology and knowledge of multiple agents. In: Geffner, H., Prada, R., Machado Alexandre, I., David, N. (eds.) IBERAMIA 2008. LNCS (LNAI), vol. 5290, pp. 1–10. Springer, Heidelberg (2008)
11. Heinemann, B.: Logics for multi-subset spaces. Journal of Applied Non-Classical Logics 20(3), 219–240 (2010)
12. Heinemann, B.: Characterizing subset spaces as bi-topological structures. In: McMillan, K., Middeldorp, A., Voronkov, A. (eds.) LPAR 2013. LNCS, vol. 8312, pp. 373–388. Springer, Heidelberg (2013)
13. Meyer, J.J.C., van der Hoek, W.: Epistemic Logic for AI and Computer Science. Cambridge Tracts in Theoretical Computer Science, vol. 41. Cambridge University Press, Cambridge (1995)
14. Moss, L.S., Parikh, R.: Topological reasoning and the logic of knowledge. In: Moses, Y. (ed.) Theoretical Aspects of Reasoning about Knowledge (TARK 1992), pp. 95–105. Morgan Kaufmann, Los Altos (1992)
15. Wáng, Y.N., Ågotnes, T.: Subset space public announcement logic. In: Lodaya, K. (ed.) ICLA 2013. LNCS, vol. 7750, pp. 245–257. Springer, Heidelberg (2013)

Knowledge Preemption and Defeasible Rules

Éric Grégoire

CRIL
Université d'Artois, CNRS
rue Jean Souvraz SP18 F-62307 Lens France
gregoire@cril.univ-artois.fr

Abstract. Preemption is a reasoning mechanism that makes an incoming logically weaker piece of information prevail over pre-existing stronger knowledge. In this paper, recent results about preemption are extended to cover a family of knowledge representation formalisms that accommodate defeasible rules through reasoning on minimal models and abnormality propositions to represent exceptions. Interestingly, despite the increase of expressiveness and computational complexity of inference in this extended setting, reasonable working conditions allow the treatment of preemption in standard logic to be directly imported as such at no additional computing cost.

Keywords: Artificial Intelligence, Knowledge Representation and Reasoning, Preemption, Non-monotonic logics, Defeasible Rules.

1 Introduction

This paper is concerned with the issue of designing a system that must handle an incoming piece δ of symbolic information[1] that must prevail over all the logically stronger information within some preexisting source of knowledge Δ.

For example, assume that Δ contains some beliefs about the location of the KSEM conference for next year. The system believes that KSEM will take place either in the USA or in Germany. Namely $\Delta \supseteq \{USA\ or\ Germany\}$. Now, an incoming piece of information δ comes up, asserting that an additional equally credible hypothesis is that KSEM could also take place in Sweden, namely $\delta = USA\ or\ Germany\ or\ Sweden$. Clearly, when δ is intended to prevail over the initial belief, the expected fused information should not be $\Delta \cup \{\delta\}$ since in such a case the system would keep its initial belief USA or $Germany$ in addition to the new one. Actually, δ is a logical consequence of Δ since whenever α or β is true, α or β or γ is also true for any γ. In other words, a deductive system was already capable of inferring δ from Δ. On the contrary, preemption is intended to make sure that no stronger knowledge like USA or $Germany$ could be derived anymore when δ is inserted in Δ. δ is thus intended to prevail over the pre-existing strictly stronger knowledge that was already allowing δ to be deduced. Accordingly, handling δ in the expected intended way might conduct some preexisting information to be retracted from Δ.

---

[1] In this paper, words *information*, *belief* and *knowledge* are used interchangeably.

R. Buchmann et al. (Eds.): KSEM 2014, LNAI 8793, pp. 13–24, 2014.

As another example, assume that Δ contains the information *If the paper is accepted then the paper will be presented orally*. Now, suppose $\delta = $ *If the paper is accepted and it is not a poster then the paper will be presented orally* is asserted and is intended to prevail. Clearly, it is not sufficient to insert δ within Δ since the first rule would still remain and, provided that the paper is accepted, the resulting information would still allow one to conclude that the paper will be presented orally (even if Δ contains the information that the paper is a poster).

As illustrated through these basic examples, making δ prevail over the preexisting contents of Δ requires the information in Δ that is logically strictly subsuming δ to be removed or weakened. Obviously, this needs all the deductive interactions within $\Delta \cup \{\delta\}$ to be taken into account.

Recently, a family of approaches to logic-based preemption have been introduced in the context of standard logic [1,2]. In this paper, we investigate how preemption can cope with a class of more expressive formalisms that allow forms of defeasible rules through non-monotonic reasoning based on minimal models and abnormality propositions to represent exceptions. For example, in such an extended framework, δ can be *If the paper is accepted and if it is consistent to assume that the paper is not a poster, which is an exception to the rule, then the paper will be presented in an oral session*. The second condition is based on a consistency assumption and is defeasible: here, it relies on the inability to derive that the paper is a poster. If some novel incoming information asserts that the paper is a poster then the rule will no longer allow us to derive that the paper will be presented orally. Such defeasible inferences pertain to non-monotonic logic and not to standard logic, which requires any deduction to remain correct whatever additional premises could be. In the paper, the focus is on formalisms based on abnormality propositions *à la* McCarthy [3] to represent defeasible exceptions to rules and reasoning on minimal models. These formalisms are also equivalent to a specific use of ground ASP (Answer Set Programming) [4] allowing for disjunction and standard negation. Interestingly, despite the increase of expressiveness and computational complexity of inference in this non-monotonic setting, we show that reasonable working conditions allow the treatment of preemption in standard logic to be directly imported without any increase of computational complexity.

2   Logical Preliminaries

The knowledge representation formalism that we will use is an extension of standard clausal Boolean logic. First, let us recall basic concepts of Boolean logic. Let $\mathcal{L}$ be a language of formulas over a finite alphabet $\mathcal{P}$ of Boolean variables, also called *atoms*. Atoms are denoted by $a, b, c, \ldots$ The $\wedge, \vee, \neg, \rightarrow$ and $\Leftrightarrow$ symbols represent the standard conjunctive, disjunctive, negation, material implication and equivalence connectives, respectively. A *literal* is an atom or a negated atom. Formulas are built in the usual way from atoms, connectives and parentheses; they are denoted by $\alpha, \beta, \gamma, \ldots$. Sets of formulas are denoted by $\Delta, \Gamma, \ldots$ An *interpretation* I is a truth assignment function that assigns values from $\{true, false\}$ to every Boolean variable, and thus, following usual compositional rules, to all formulas of $\mathcal{L}$. A formula δ is *consistent* (also called *satisfiable*) when there exists at least one interpretation that satisfies δ, i.e., that makes

δ become *true*: such an interpretation is called a *model* of δ. By convenience, an interpretation and a model are represented by the set of atoms that they satisfy. $\models$ denotes deduction, i.e., $\Delta \models \delta$ denotes that δ is a logical consequence of Δ, namely that δ is satisfied in all models of Δ. Two formulas α and β are logically equivalent, written $\alpha \equiv \beta$, iff $\alpha \models \beta$ and $\beta \models \alpha$. $\models \alpha$ means that α is tautologous (i.e., *true* in all interpretations) and $\models \neg\alpha$ that α is a contradiction. $\top$ stands for a tautology and $\bot$ stands for a contradiction. Note that a set of formulas Δ is consistent iff $\Delta \not\models \bot$. Without loss of generality, formulas can be represented in Conjunctive Normal Form (CNF). A CNF is a conjunction of clauses, where a *clause* is a disjunction of literals. A DNF is a disjunction of *terms*, where a term is a conjunction of literals. We always assume that any clause contains at most one occurrence of a given literal. The empty clause denotes $\bot$. For convenience purpose, a clause can be identified with the set of its literals. A *rule* is a formula of the form $\alpha \Rightarrow \beta$, where α is a term and β is a clause; α and β are called the *body* and the *head* of the rule, respectively. A clause can always be rewritten as a rule, and conversely. Deduction in clausal Boolean logic is co-NP-complete. Indeed, $\Delta \models \alpha$ iff $\Delta \cup \{\neg\alpha\}$ is unsatisfiable and checking whether a set of clauses is satisfiable is NP-complete. A set of formulas Δ is a theory iff it is a deductively closed set of formulas: $\Delta = \{\alpha \mid \Delta \models \alpha\}$. The deductive closure of a set Γ of formulas is denoted by $Cn(\Gamma)$, so that a theory Δ is such that $\Delta = Cn(\Delta)$.

3   Implicants and Implicates as Basic Tools for Preemption

Preemption is deep-rooted in *subsumption*, *strict implicants*, *prime implicants* and *prime implicates*, which are well-known concepts in Boolean logic. From now on, we assume that Δ is a satisfiable non-tautologous CNF and that α, β and δ are satisfiable non-tautologous clauses. Moreover, we assume that $\Delta \cup \{\delta\}$ is satisfiable.

Definition 1. α *is a* strict implicant *of* β *iff* $\alpha \models \beta$ *but* $\beta \not\models \alpha$.

Definition 2. Δ strictly subsumes β *iff* $\Delta \models \alpha$ *for some strict implicant* α *of* β.

By abuse of words, we will use "subsume" in place of "strictly subsume".

Definition 3. α *is a* prime implicant *of* β *iff* α *is a strict implicant of* β *and there does not exist any strict implicant* δ *of* β *such that* α *is a strict implicant of* δ.

Interestingly, when α and β are under their set-theoretical representation, α is an (a strict) implicant of β iff α is a (strict) subset of β. Moreover, when β is made of n literals, the prime implicants of β are the sub-clauses of β made of $n - 1$ literals; β is not subsumed by Δ iff none of the prime implicants of β can be deduced from the same Δ. Strict implicants and subsumption are the basic concepts founding *preemption operators* defined in [1,2], which can handle the KSEM conference location and the accepted papers examples, and, more generally, situations where an additional incoming logically weaker (i.e., subsumed by Δ) piece of information β must belong to the resulting set of clauses Δ' but cannot be subsumed by Δ'. Such a clause β is a *prime implicate* of Δ'.

Definition 4. *A prime implicate of Δ is any clause δ such that*

(1) $\Delta \models \delta$, *and*
(2) $\delta' \equiv \delta$ *for every clause δ' such that $\Delta \models \delta'$ and $\delta' \models \delta$.*

Notation. Δ_{PI} denotes the set of all prime implicates of Δ.

Since we assume that Δ is a consistent set of non-tautological clauses, δ is a prime implicate of Δ iff δ is a minimal (w.r.t. $\subseteq$) non-tautological clause amongst the set formed of the clauses β such that $\Delta \models \beta$. In our first motivating example, under the set-theoretical representation of clauses, $\alpha = \{USA, Germany\}$ is a prime implicant of $\beta = \{USA, Germany, Sweden\}$. Thus, $\Delta = Cn(\{\alpha\})$ subsumes β. To make β prevail, preemption delivers a set of clauses Δ' such that Δ' entails β but does not subsume it. Especially, Δ' cannot contain α and β is a prime implicate of Δ'. The other motivating example about an accepted paper to be presented orally or not, is of the same vein. It deals with an additional incoming rule $\delta = (\alpha \wedge \beta) \Rightarrow \gamma$ intended to prevail over any piece of stronger knowledge, especially the rule $\delta = \alpha \Rightarrow \gamma$. To reach this, the resulting knowledge must contain (the clausal form of) $(\alpha \wedge \beta) \Rightarrow \gamma$ as a prime implicate. Consequently, no strict sub-clause of the clause representing the rule can also be a prime implicate of the resulting knowledge.

Definition 5. *δ prevails in Δ iff*

1. $\Delta \models \delta$, *and*
2. $\nexists \delta'$ *such that $\delta' \subset \delta$ and $\Delta \models \delta'$.*

Equivalently, we have that δ prevails in Δ iff $\delta \in \Delta_{PI}$. Prime implicates have already been investigated in belief revision mainly because they provide a compact and syntax-independent yet complete representation of a belief base (see e.g., [5,6]) and because interesting computational tasks (like satisfiability checking and entailment) are tractable in this framework [7]. In the worst case, computing the set of prime implicates of Δ containing a clause β is however not in polynomial total time unless P=NP (it is in polynomial total time when for example the clause is positive and Δ is Horn) [8]. Computational techniques and experimentations about preemption in standard logic have been presented in [1].

Definition 6. *Enforcing the preemption of δ w.r.t. Δ consists in finding one subset Δ' of $Cn(\Delta \cup \{\delta\})$ such that δ prevails in Δ'.*

In [2], this definition is refined by taking into account additional possibly desired properties for Δ'.

4   A Simple Non-monotonic Framework

The language remains $\mathcal{L}$ but a specific set of atoms $\mathcal{A} = \{Ab_1, \ldots, Ab_n\}$ is identified and used as so-called abnormality propositions [3] to represent defeasible conditions and exceptions to rules. For example, $(a \wedge \neg Ab_1) \rightarrow b$ is intended to represent the rule

If a is true and if it is consistent to believe that not Ab_1 is true, then b is true. $Ab_1 \rightarrow c$
is intended to represent the rule *If Ab_1 is true then c is true.* Such an intended meaning
of these formulas requires $\neg Ab_1$ to be given a kind of negation as failure interpretation.
When no specific priorities need be applied amongst the various Ab_i propositions, a
basic inference relationship in this logic, noted $|\sim$, can defined as follows.

Interpretations and models are identical to the corresponding concepts in Boolean
logic. Let us recall that we represent them by the set of atoms that they assign *true*.

Definition 7. *Let M be an interpretation of $\mathcal{L}$. M is a minimal model of Δ iff*

1. *M is a model of Δ, and*
2. *$\nexists M'$ a model of Δ such that $(M' \cap \mathcal{A}) \subset (M \cap \mathcal{A})$.*

Accordingly, minimal models of Δ are models of Δ such that there is no model of
Δ that assigns a strict subset of abnormality propositions to *true*. A minimal model of
Δ thus minimizes exceptions to rules while satisfying Δ.

Definition 8. $\Delta |\sim \delta$ *iff δ is true in all minimal models of Δ.*

Let us illustrate the extension to standard logic that this framework provides. Assume
$\Delta = \{(a \wedge \neg Ab_1) \rightarrow b, a\}$. In standard logic, we have that $\Delta \not\models b$. Indeed, some (but
not all) models of Δ contain Ab_1, preventing $\neg Ab_1$ to be derivable in standard logic. On
the contrary, all minimal models of Δ falsify Ab_1. This translates the idea that $\neg Ab_1$
can be assumed *true* by default. Accordingly, in the non-monotonic setting, we have
that $\Delta |\sim b$.

While we are still using the language of Boolean logic, an increase of expressiveness
is provided by the non-monotonic setting since this one allows a kind of negation as
failure to be used. The price to pay is the dramatic increase of computational complexity of inference. Indeed, finding a minimal model is $P^{NP[O(log\ n)]}$-hard [9] and thus
belongs to the class of decision problems that are solved by polynomial-time bounded
deterministic Turing machines making at most a logarithmic number of calls to an or-
acle in NP. Checking whether a model is minimal is co-NP-complete [10]. Checking if
δ follows from Δ according to $|\sim$ is Π_2^p-complete [11]. Recent results about the worst-
case complexity of finding minimal models of fragments of clausal Boolean logics can
be found in [12]. Clearly, minimal models as we define them are so-called *stable models* [13,14] provided that all the occurrences of negated abnormality propositions are
treated as negation by default and no other occurrence of negation by default occurs in
$\Delta \cup \{\delta\}$. The above representation formalism could be easily encoded as ASP (Answer
Set Programming) programs [4] when they allow for disjunction and standard negation.
Note however that we make a specific use of negation by default as we only make this
connective apply on abnormality atoms. We will justify this restriction in the context of
preemption in a further section and take advantage of it. Indeed, the question that we
will address is when could preemption be achieved in this non-monotonic setting while,
at the same time, a dramatic increase of computational complexity is avoided.

5 When Does Preemption Make Sense?

In this paper, we do not address the issue of determining whether or not preemption must occur when an additional clause δ comes in. We assume that some extra-

logical contextual information is available together with δ, allowing the system to decide whether δ must simply be inserted in Δ, or whether $\Delta \cup \{\delta\}$ must be transformed into some Δ' so that δ prevails in Δ'.

However, it should be noted that preemption is a concept that can require dramatic removal effects inside Δ. Preemption must thus be used carefully; we recommend to use it in specific situations, only. Let us explain this here, since this restriction will be also a cornerstone allowing preemption to handle these situations within the above non-monotonic setting at no significant additional computational cost.

In the Boolean logic framework, when a clause δ must prevail in some set Δ' of clauses, any strict sub-clause of δ mut be non-derivable from Δ'. This entails that any literal in δ must be non-derivable from Δ' unless this literal is δ itself. This is a very strong requirement whose consequences are best illustrated when clauses are intended to represent rules. Assume for example that the clause $\delta = \neg a \vee \neg b \vee \neg c \vee d$ representing the rule $(a \wedge b \wedge c) \to d$ needs to prevail. This entails that no literal in $\neg a \vee \neg b \vee \neg c \vee d$ can be a consequence of Δ'. Indeed, for example if $\Delta' \models d$ then we also have e.g. $\Delta' \models (a \wedge b) \to d$, which contradicts the goal of having δ not subsumed. More generally, whenever $\Delta' \models x$, we also have, among other things, $\Delta' \models (\neg x \wedge anything) \to what\text{-}you\text{-}want$ and $\Delta' \models x \vee anything$ and $\Delta' \models what\text{-}you\text{-}want \to x$. Hence, intuitively, preemption makes sense mainly when Δ is made of generic information (e.g., rules) but does not yet allow elementary facts (literals) to be derived when these facts subsume δ. This coincides for example to applications like knowledge-based diagnosis systems before they are being instantiated and run together with facts about a specific device to be diagnosed: preemption should concern δ when δ is for example a more precise rule that must prevail over less precise (shorter) ones. Such a preemption task is generally performed at the conception stage or upgrade of the diagnosis system, not when it is run with specific data about a device to be diagnosed. If preemption is to be performed in some other situations then all the logical consequences described above might occur and should thus be accepted by the user. Accordingly, in the sequel, we first consider situations corresponding to our suggested use of preemption.

Definition 9. *A normal context for making δ prevail by transforming $\Delta \cup \{\delta\}$ is when for every literal x we have that $\Delta \cup \{\delta\} \not\models x$ or $x = \delta$.*

Clearly, such an assumption can be relaxed so that it only precludes the deduction of literals occurring in δ when δ is not a single literal.

Definition 10. *A normal pointwise context for making δ prevail by transforming $\Delta \cup \{\delta\}$ is when for every literal $x \in \delta$ we have that $\Delta \cup \{\delta\} \not\models x$ or $x = \delta$.*

As a case study, we will only consider Definition 9, based on the assumption that Δ should be ready to accommodate *any* non-conflicting and non-tautological clause δ that must prevail. As we will extend some subsequent definitions from the standard logic setting to cover the aforedefined non-monotonic framework, it is convenient to assume that a set of clauses, noted Σ, contains the implicitly assumed information. In standard logic, $\Sigma = \emptyset$. More generally, Σ is thus a set of additional premises to be taken into account when inference is performed. Accordingly, the definition of normal context for preemption is adapted as follows.

Definition 11 (NCP). *A normal context for making δ prevail under Σ by transforming $\Delta \cup \{\delta\}$, in short NCP for Normal Context for Preemption, is when for every literal x we have that $\Delta \cup \{\delta\} \cup \Sigma \not\models x$ or $x = \delta$.*

By monotony of $\models$, we have that

Proposition 1. *When NCP holds, we have that $\Delta \not\models x$ for every literal x such that $\delta \neq x$.*

In the following, we only consider preemption techniques that can only remove clauses from Δ to make δ prevail. We do not consider other techniques transforming $Cn(\Delta \cup \{\delta\})$. The following definition translates this restriction.

Definition 12. *Let Δ' be a set of clauses. Δ' is a preemption-enforcement of δ w.r.t. $\langle \Delta, \models, \Sigma \rangle$ when the following conditions are satisfied.*

1. *$\delta \in \Delta'$ and $\Delta' \subseteq (\Delta \cup \{\delta\})$, and*
2. *$\nexists \ \delta'$ such that $\delta' \subset \delta$ and $\Delta' \cup \Sigma \models \delta'$.*

Note that there might exist several preemption-enforcements Δ' of δ w.r.t. $\langle \Delta, \models, \Sigma \rangle$ and that Definition 12 does not require Δ' to be as large as possible. We will come back to this possible additional requirement and discuss it in the section about experimentations. By monotony of $\models$, we have that

Proposition 2. *Let Δ' be any preemption-enforcement of δ w.r.t. $\langle \Delta, \models, \Sigma \rangle$. When NCP holds, for every literal x we have that $\Delta' \not\models x$ or $x = \delta$.*

6 Preemption in the Non-monotonic Framework

Let us now consider δ and Δ under the aforedescribed simple non-monotonic framework: the inference relationship is now $\mid\sim$ and the subset of propositional atoms $\{Ab_1, \ldots, Ab_n\}$ is assigned its dedicated role. For ease of presentation, we assume that δ is not a single literal formed with some possibly negated Ab_i.

Definition 13. *Δ' is a preemption-enforcement of δ w.r.t. $\langle \Delta, \mid\sim, \Sigma \rangle$ when the following conditions are satisfied.*

1. *$\delta \in \Delta'$ and $\Delta' \subseteq (\Delta \cup \{\delta\})$, and*
2. *$\nexists \ \delta'$ such that $\delta' \subset \delta$ and $\Delta' \cup \Sigma \mid\sim \delta'$.*

We assume that the implicit information is encapsulated within the definition of $\mid\sim$ itself, which selects models that minimize the set of satisfied Ab_i abnormality atoms: accordingly, unless explicitly stated, we assume that $\Sigma = \emptyset$. Actually, Σ will be non-empty only when we attempt to model inferences based on $\mid\sim$ by making use of $\models$.

Case 1. First, let us assume that we keep NCP as such in the non-monotonic framework. This natural assumption translates the idea that $\Delta \cup \{\delta\}$ encodes generic knowledge that, by itself, does not allow any fact (different from δ) to be deduced by using $\models$. By definition of NCP, we have that:

Proposition 3. *When NCP holds, $\Delta' \not\models \neg Ab_i$ and $\Delta' \not\models Ab_i$ for any $i \in [1..n]$ and for any $\Delta' \subseteq (\Delta \cup \{\delta\})$.*

Obviously, it would not make sense to extend NCP by replacing $\models$ by $\mid\sim$ and require $\Delta \cup \{\delta\} \not\mid\sim x$ for any literal x, including literals built from abnormality propositions.

Proposition 4. *Assume NCP holds, $\delta \in \Delta'$ and $\Delta' \subseteq (\Delta \cup \{\delta\})$.*
If $\Delta' \cup \{\bigwedge_{i=1}^{n} \neg Ab_i\}$ is consistent then $\Delta' \mid\sim \bigwedge_{i=1}^{n} \neg Ab_i$ and $\Delta' \not\mid\sim Ab_j$ for any $j \in [1 \ldots n]$.

Notice that $\Delta' \cup \{\bigwedge_{i=1}^{n} \neg Ab_i\}$ is inconsistent is equivalent to $\Delta' \models \bigvee_{i=1}^{n} Ab_i$. Before we comment on the motivation for this additional condition, let us examine its logical effects. NCP requires among other things that no Ab_i can be deduced: the left-hand-side of Proposition 4 additionally requires that no disjunction made of some Ab_i abnormality propositions could be deduced from Δ'. Under the applicability conditions of Proposition 4 we never have $\Delta' \cup \{\delta\} \mid\sim Ab_i$ for any $i \in [1 \ldots n]$. To some extent, this prevents some logical side-effects about disjunction from occurring in the preemption enforcement process, namely having that $\Gamma \mid\sim Ab_i \vee x$ for any x as a consequence of simply having $\Gamma \mid\sim Ab_i$.

Accordingly, we define:

Definition 14   (NCPN). *A normal context under $\mid\sim$ for making δ prevail by transforming $\Delta \cup \{\delta\}$, in short NCPN for Normal Context for Preemption in the Non-monotonic framework, is when $\Delta \cup \{\delta\} \cup \Sigma \not\models x$ for any literal x different from δ and, at the same time, $\Delta \cup \{\delta\} \cup \{\bigwedge_{i=1}^{n} \neg Ab_i\}$ is consistent.*

Interestingly, NCPN corresponds to a very natural requirement about the use of preemption. NCP requires to handle generic knowledge only, in the sense that no fact could be derived using Boolean logic. NCPN requires that $\Delta \cup \{\delta\}$ remains uninstantiated in the sense that the generic knowledge that it contains remains compatible with a situation where no exception does occur. This naturally corresponds again for example to diagnostic systems where, before being instantiated to the data from a specific device to be diagnosed, the generic knowledge entails that no fault is present or assumed by default (faults being represented by abnormality propositions being *true*).

Now, the main result is as follows. Under NCPN, preemption-enforcements in standard logic, with $\Sigma = \bigwedge_{i=1}^{n} \neg Ab_i$, are preemption-enforcements in the non-monotonic framework as the following result ensures it.

Proposition 5. *Assume NCPN holds.*
If Δ' is a preemption-enforcement of δ w.r.t. $\langle \Delta, \models, \bigwedge_{i=1}^{n} \neg Ab_i \rangle$ then Δ' is a preemption-enforcement of δ w.r.t. $\langle \Delta, \mid\sim, \emptyset \rangle$.

This result is of nice practical consequences. Despite a dramatic increase of computational complexity when moving from $\models$ to $\mid\sim$, under very natural assumptions, it remains possible to compute preemption-enforcements, using $\models$, only despite the fact that the preemption-enforcement process requires some $\mid\sim$ and $\not\mid\sim$ conditions to be checked.

Case 2. Assume that NCP does not hold. Actually, NCP is not a necessary condition for allowing preemption enforcement with respect to $|\sim$ to be completed within standard logic. Proposition 5 still applies when we only have that $\Delta \cup \{\delta\} \cup \{\bigwedge_{i=1}^{n} \neg Ab_i\}$ is consistent.

Proposition 6. *Assume $\Delta \cup \{\delta\} \cup \{\bigwedge_{i=1}^{n} \neg Ab_i\}$ is consistent.*
If Δ' is a preemption-enforcement of δ w.r.t. $\langle \Delta, \models, \bigwedge_{i=1}^{n} \neg Ab_i \rangle$ then Δ' is a preemption-enforcement of δ w.r.t. $\langle \Delta, |\sim, \emptyset \rangle$.

The scope of Proposition 6 concerns situations with absence of $\models$ provable (disjunction of) exceptions. Now, when $\Delta \cup \{\delta\} \models Ab_i$, it should be noted that we are faced to situation involving an exception. Again, if δ contains an occurrence of Ab_i then any preemption-enforcement will have to yield a set of clauses that blocks the deduction of Ab_i, since Ab_i subsumes δ. This is some kind of impossibility (or at a least, paradoxal) result: in order to make a piece of information about an exception Ab_i prevail, preemption enforcement would require to move from a situation where this exception is established (i.e., can be deduced) to a situation where this exception can no longer be deduced and cannot thus be established.

When neither NCP is assumed nor $\Delta \cup \{\delta\} \cup \{\bigwedge_{i=1}^{n} \neg Ab_i\}$ is consistent, this does not mean that preemption-enforcements with respect to $|\sim$ can never be reduced to preemption-enforcements with respect to $\models$. A simple case occurs when $\neg\delta'$ can be deduced from Δ'. This condition is sufficient to prevent δ' from being inferable from Δ' by means of $|\sim$.

Proposition 7. *Let $\delta' \subset \delta$. If $\Delta' \models \neg\delta'$ then $\Delta' \not\mid\sim \delta'$.*

7   Experimental Results

The most direct way to transform $\Delta \cup \{\delta\}$ into some Δ' such that δ prevails in Δ' according to standard logic is by removing a sufficient number of clauses from Δ so that no strict implicant of δ could be deduced anymore. A simple way to achieve this is by refutation: whenever a prime implicant δ' of δ can be deduced from the current Δ', $\Delta' \cup \{\neg\delta'\}$ is inconsistent; restoring consistency (while keeping δ in the resulting set) will block the inference of δ'. Accordingly, we have presented in [1] a computational method that extracts a sufficient number of MUSes (Minimal Unsatisfiable Subsets) so that removing one clause per MUS suffices to restore consistency. Obviously enough, such a technique does not ensure that a globally minimal number of clauses is removed. In this paper we present an alternative technique also based on the same refutation approach. However, instead of finding out MUSes, we extract maximal consistent subsets (MSS) and their complements, called co-MSS, which proves more efficient. For each prime implicant δ', removing the whole co-MSS ensures that no strict subset can be removed instead, while blocking the deduction of δ'. However, similarly to the MUS-based technique, there is no guarantee that a globally minimal number of clauses is removed in this way: this number might depend both on the extracted MSS and the order according to which prime implicants are considered. However, dropping the global criterion of the minimality of the number of removed clauses allows for efficient

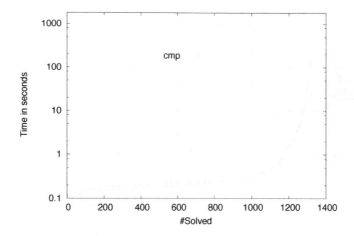

Fig. 1. CMP on various benchmarks

techniques in most cases, while keeping very close to the optimal result in terms of the number of removed clauses, most often.

We have conducted extensive experimentations of this approach and made use of our recent tool to partition a set of clauses into one MSS and its corresponding co-MSS. This partitioning algorithm is called CMP and appears to be the currently most efficient tool for extracting one MSS or one co-MSS [15]. It is implemented in C++ and make use of MINISAT [16] as the CDCL SAT-solver. CMP and all experimentation data are available from www.cril.univ-artois.fr/documents/cmp/. All experimentations have been conducted on Intel Xeon E5-2643 (3.30GHz) processors with 7.6Gb RAM on Linux CentOS. Time limit was set to 30 minutes for each single partitioning test. Two series of benchmarks have been considered. The first one was made of the 1343 benchmarks used and referred to in [17]: they are small-sized industrial-based instances from SAT competitions www.satcompetition.org and structured instances from the MAX-SAT evaluations maxsat.ia.udl.cat:81. We enriched this experimentation setting by also considering a second series of benchmarks, made of *all* the 295 instances used for the 2011 MUS competition organized in parallel with the SAT one. In Figure 1, the y-values represent the CPU time by CMP to partition x instances. CMP allows a Co-MSS to be extracted within 30 minutes for 1327 of the 1343 SAT/MAX-SAT benchmarks (293 of 295 MUS benchmarks), clearly showing the practical feasibility of the approach. Note that the number of prime implicants of δ is given by the number of literals in δ, which corresponds to the number of times CMP must be run.

8   Related Work

Various subfields of Artificial Intelligence have long been focusing on how to handle an incoming piece of information δ within Δ, especially when Δ and δ are logically conflicting (see non-monotonic logics [18], belief revision [19] and knowledge fusion [20], mainly). Indeed, when Δ and δ are mutually contradictory, a system that reasons in a deductive way from $\Delta \cup \{\delta\}$ will be able to infer any conclusion and its contrary: hence, much research efforts have concentrated on ways Δ and δ should interact to maintain logical consistency in the aggregated knowledge and prevent deductive reasoning from collapsing. When δ does not logically contradict with Δ, δ is often expected to be added inside Δ and the resulting knowledge is the (deductive closure of the) set-theoretical union of Δ and $\{\delta\}$. The issue that we address in this paper is different: δ is not logically conflicting with Δ and is not simply intended to be inserted as such in Δ. On the contrary, δ is intended to logically weaken Δ. An architecture presenting all those facets in standard Boolean logic is presented in [21]. In [22], a large-scope generic non-monotonic framework was used for preemption. It was not based on minimal-model reasoning but was built from other proof-theoretic concepts and close to variants of default logic [23]. Such a framework is very different from the one in this study and investigating conditions for importing standard logic preemption remains to be investigated.

9   Conclusions and Perspectives

In this paper, we have investigated a form of preemption paradigm for a simple generic non-monotonic framework. Interestingly, natural conditions for preemption to take place correspond to situations where preemption can still be enforced by merely using standard logic mechanisms, only. This allows us to avoid the dramatic increase of computational complexity accompanying the move from deduction to non-monotonic inference. The results presented in this paper could be extended in several directions. First, they apply to ground ASP programs with a controlled used of default negation. It is also easy to see that all results remain intact when priorities are enforced between abnormality propositions. An interesting path for further research concerns other possible syntactical constraints in ASP programs that would guarantee that preemption can be achieved at low computational cost. The results presented in this paper could also be further refined when a prime-implicates compilation of knowledge [7] is enforced on Δ and its subsequent transformations. For example when the size of δ is strictly lower than any prime implicate of Δ, its insertion within Δ makes it directly prevail. Also, making sure that no strict implicant of δ does not remain in Δ' is easy since it suffices to examine the presence of such implicates within the compiled base.

References

1. Besnard, P., Grégoire, É., Ramon, S.: Enforcing logically weaker knowledge in classical logic. In: Xiong, H., Lee, W.B. (eds.) KSEM 2011. LNCS, vol. 7091, pp. 44–55. Springer, Heidelberg (2011)
2. Besnard, P., Grégoire, É., Ramon, S.: Preemption operators. In: Proceedings of the 20th European Conference on Artificial Intelligence (ECAI 2012), pp. 893–894 (2012)
3. McCarthy, J.: Applications of circumscription to formalizing common-sense knowledge. Artificial Intelligence 28(1), 89–116 (1986)

4. Gebser, M., Kaminski, R., Kaufmann, B., Schaub, T.: Answer Set Solving in Practice. Synthesis Lectures on Artificial Intelligence and Machine Learning. Morgan and Claypool Publishers (2012)
5. Zhuang, Z.Q., Pagnucco, M., Meyer, T.: Implementing iterated belief change via prime implicates. In: Orgun, M.A., Thornton, J. (eds.) AI 2007. LNCS (LNAI), vol. 4830, pp. 507–518. Springer, Heidelberg (2007)
6. Bienvenu, M., Herzig, A., Qi, G.: Prime implicate-based belief revision operators. In: 20th European Conference on Artificial Intelligence (ECAI 2012), pp. 741–742 (2008)
7. Darwiche, A., Marquis, P.: A knowledge compilation map. Journal of Artificial Intelligence Research (JAIR) 17, 229–264 (2002)
8. Eiter, T., Makino, K.: Generating all abductive explanations for queries on propositional horn theories. In: Baaz, M., Makowsky, J.A. (eds.) CSL 2003. LNCS, vol. 2803, pp. 197–211. Springer, Heidelberg (2003)
9. Cadoli, M., Schaerf, M.: A survey of complexity results for nonmonotonic logics. Journal of Logic Programming 17(2/3&4), 127–160 (1993)
10. Cadoli, M.: The complexity of model checking for circumscriptive formulae. Information Processing Letters 44(3), 113–118 (1992)
11. Eiter, T., Gottlob, G.: On the complexity of propositional knowledge base revision, updates, and counterfactuals. Artificial Intelligence 57(2-3), 227–270 (1992)
12. Angiulli, F., Ben-Eliyahu-Zohary, R., Fassetti, F., Palopoli, L.: On the tractability of minimal model computation for some CNF theories. Artificial Intelligence 210, 56–77 (2014)
13. Lifschitz, V.: Answer set planning (abstract). In: Gelfond, M., Leone, N., Pfeifer, G. (eds.) LPNMR 1999. LNCS (LNAI), vol. 1730, pp. 373–374. Springer, Heidelberg (1999)
14. Lifschitz, V.: Answer set programming and plan generation. Artificial Intelligence 138(1-2), 39–54 (2002)
15. Grégoire, É., Lagniez, J.M., Mazure, B.: An experimentally efficient method for (MSS,CoMSS) partitioning. In: Proceedings of the 28th Conference on Artificial Intelligence (AAAI 2014). AAAI Press (2014)
16. Eén, N., Sörensson, N.: An extensible SAT-solver. In: Giunchiglia, E., Tacchella, A. (eds.) SAT 2003. LNCS, vol. 2919, pp. 502–518. Springer, Heidelberg (2004)
17. Marques-Silva, J., Heras, F., Janota, M., Previti, A., Belov, A.: On computing minimal correction subsets. In: Proceedings of the 23rd International Joint Conference on Artificial Intelligence, IJCAI 2013 (2013)
18. Ginsberg, M.: Readings in nonmonotonic reasoning. M. Kaufmann Publishers (1987)
19. Fermé, E.L., Hansson, S.O.: AGM 25 years - twenty-five years of research in belief change. Journal of Philosophical Logic 40(2), 295–331 (2011)
20. Grégoire, É., Konieczny, S.: Logic-based approaches to information fusion. Information Fusion 7(1), 4–18 (2006)
21. Besnard, P., Grégoire, É.: Handling incoming beliefs. In: Wang, M. (ed.) KSEM 2013. LNCS, vol. 8041, pp. 206–217. Springer, Heidelberg (2013)
22. Besnard, P., Grégoire, É., Ramon, S.: Overriding subsuming rules. International Journal of Approximate Reasoning 54(4), 452–466 (2013)
23. Reiter, R.: A logic for default reasoning. Artificial Intelligence 13(1-2), 81–132 (1980)

A Method for Merging Cultural Logic Systems

Xiaoxin Jing[1,2], Shier Ju[1,*], and Xudong Luo[1,*]

[1] Institute of Logic and Cognition, Sun Yat-sen University, Guangzhou, China
{hssjse,luoxd3}@mail.sysu.edu.cn
[2] College of Political Science and Law, Captital Normal University, Beijing, China
jxxaigq@gmail.com

Abstract. This paper presents a merging approach to solve the logical conflicts in cross-cultural communication. We define each agent's logic system as a tuple of the language and a binary relation over the language, which can well reflect different inference patterns in different cultures. Then based on the distance measure, merging different logic systems of this kind is a two-step process: (i) all the logic systems are expanded to be based on the same language; and (ii) the distance-based method is used to select a logic system, which is the nearest to the expanded logic systems. In this work, we define a kind of the Hamming distance to measure the distance between any two logic systems. Then based on the distance measure, we define our merging operator and we prove that the operator satisfies the properties of agreement, strong unanimity, groundedness and anonymity in social choice theory.

Keywords: merging, logic, cross-culture, agent.

1 Introduction

Merging is a process for agents to combine conflicting information coming from several sources with equal reliability to build a coherent description of the world [1]. A lot of formal theories about merging have been developed [2–6], most of which focus on the problem of belief merging and assume that belief merging scenarios take place in a same culture. However, if we take the view of cultural pluralism and consider the information processing in cross-cultural communications, these merging theories need to be extended to cross-cultural environments. For example, there are two friends, Chinese Xiaoming and American Jim, meet in the street. Then Xiaoming asks Jim: "Where are you going?" For Xiaoming, this is just a kind of daily greeting in China; but Jim may feel uncomfortable and view it as an invasion of his privacy. Therefore, in a cross-cultural communication, not only the language barrier prevents people from different cultures to communicate smoothly, but also the thinking patterns (*i.e.*, the reasoning systems) vary a lot. Often, people from different cultures reach different, even opposite, conclusions even from a same proposition.

To this end, this paper will propose a merging theory in the context of cross-cultural communication. Since different cultures have different rational logics

---

* Corresponding authors.

R. Buchmann et al. (Eds.): KSEM 2014, LNAI 8793, pp. 25–36, 2014.

[7] (*i.e.*, in different cultures there are different opinions for a same thing), we intend to apply the distance-based merging operator into logic systems. This brings a number of new challenges on this topic. That is, we need not only to merge the languages but also to reach coherent binary relations over the languages. Actually, such a binary relation represents an inference relation in a logic system, which reflects the opinion on a thing in a specific culture, *e.g.*, the question of "where are you going?" is a "daily greeting" or an "invasion of privacy".

The rest of this paper is organised as follows. Section 2 defines the logic systems of different cultures. Section 3 discusses how to merge the logic systems by using the distance-based approach and illustrates our approach by a real-life example. Section 4 reveals some properties of our merging operator. Section 5 discusses the related work to show clearly how our work advances the state-of-art in the research area. Finally, Section 7 summarises our work with future work.

2 Cultural Logic Systems

This section will define our logic system that can reflect the inference patterns of agents from different cultures.

We view a cultural logic system simply as a set of binary relations, which can reflect well different inference relations in different cultures abstractly. Specifically, we will define a cultural logic system based on a formal language $\mathcal{L}$ (a non-empty finite set of sentences) and a binary relation $R^{\vdash}$ between $\mathcal{P}(\mathcal{L})$ and $\mathcal{L}$ over the languages. Here $\mathcal{P}(\mathcal{L})$ is denoted as the power set of $\mathcal{L}$. Formally, we have:

Definition 1 (Cultural logic system). *Given a formal language $\mathcal{L}$, a cultural logic system over $\mathcal{L}$ is a pair $L = \langle \mathcal{L}, R^{\vdash}_L \rangle$, where $R^{\vdash}_L \in \mathcal{P}(\mathcal{L}) \times (\mathcal{L})$. We denote a profile of n cultural logic systems as $P = (L_1, \cdots, L_n)$, where each $L_i = \langle \mathcal{L}_i, R^{\vdash}_{L_i} \rangle$ is the cultural logic system for agent i.*

In the rest of our paper, sometimes we will use $L = \langle \mathcal{L}, R^{\vdash}_L, R^{\nvdash}_L \rangle$ to denote the complete version of a cultural logic system, where $R^{\nvdash}_L = \mathcal{P}(\mathcal{L}) \times \mathcal{L} \backslash R^{\vdash}_L$. We will not distinguish these two notations if there are no confusions arisen.

Definition 2 (Argument). *Given a formal logic language $\mathcal{L}$, an argument in $\mathcal{L}$ is a tuple of $(\Sigma, \phi) \in \mathcal{P}(\mathcal{L}) \times \mathcal{L}$, where Σ is a set of premises and ϕ is a conclusion. An argument (Σ, ϕ) is said to be valid in a cultural logic system L iff $(\Sigma, \phi) \in R^{\vdash}_L$. We denote $Ar(L)$ as all the valid arguments in L. If an argument (Σ, ϕ) is not valid in L, then $(\Sigma, \phi) \notin R^{\vdash}_L$.*

For example, ("where are you going?", "daily greeting") and ("where are you going?", "invasion of privacy") are two arguments in two cultural logic systems. In this way, a cultural logic system actually is a set of all the valid arguments in the culture. Notice that in this paper, we assume a cultural logic system is consistent, *i.e.*, for any L over $\mathcal{L}$, $\nexists(\Sigma, \phi) \in \mathcal{P}(\mathcal{L}) \times \mathcal{L}$ such that $(\Sigma, \phi) \in R^{\vdash}_L$ and $(\Sigma, \neg\phi) \in R^{\vdash}_L$. So, the above two conflictive arguments cannot be in a same cultural system.

Given a profile of multiple cultural logic systems, our purpose in this work is to characterise a logic system that minimises its total distance to each cultural logic system. Since a cultural logic system is a set of all the valid arguments in the culture, our aim is to reach a set of arguments that are accepted by all the agents with different cultural background. Since the logic systems are based on different languages, firstly we give the following definitions to expand the logic systems such that their languages can have a same basis:

Definition 3 (Expansion of a cultural logic system). *Given a profile* $P = (L_1, \cdots, L_n)$ *of* n *cultural logic systems in which each* $L_i = \langle \mathcal{L}_i, R_{L_i}^\vdash \rangle$ *and a cultural logic system* $L = \langle \mathcal{L}, R_L^\vdash \rangle$, *an expansion of* L *over* P *is defined as a cultural logic system* $exp(L, P) = \langle \mathcal{L}^e, R_{L^e}^\vdash, R_{L^e}^{\not\vdash} \rangle$, *where:*

(i) $\mathcal{L}^e = \mathcal{L} \cup (\cup_{i=1}^n \mathcal{L}_i)$;
(ii) $R_{L^e}^\vdash \supseteq R_{L_i}^\vdash$; *and*
(iii) $R_{L^e}^{\not\vdash} = P(\mathcal{L}^e) \times \mathcal{L}^e \backslash R_{L^e}^\vdash$.

Definition 4 (Consensual expansion). *For a profile* $P = (L_1, \cdots, L_n)$, *where each* $L_i = \langle \mathcal{L}_i, R_{L_i}^\vdash \rangle$, *let* $c(P) = (\cup_{i=1}^n R_{L_i}^\vdash) \cap (\cup_{i=1}^n R_{L_i}^{\not\vdash})$ *be the set of conflicting arguments among* $R_{L_1}^\vdash, \cdots, R_{L_n}^\vdash$ *and* $L = \langle \mathcal{L}, R_L^\vdash \rangle$ *be a logic system. Then the consensual expansion of* L *over* P *is defined as* $ce(L, P) = \langle \mathcal{L}^{ce}, R_{ce}^\vdash, R_{ce}^{\not\vdash} \rangle$, *where:*

(i) $\mathcal{L}^{ce} = \mathcal{L} \cup (\cup_{i=1}^n \mathcal{L}_i)$;
(ii) $R_{ce}^\vdash = R_L^\vdash \cup (\cup_{i=1}^n R_{L_i}^\vdash \backslash c(P))$; *and*
(iii) $R_{ce}^{\not\vdash} = P(\mathcal{L}^{ce}) \times \mathcal{L}^{ce} \backslash R_{ce}^\vdash$.

Intuitively, given a profile P of multiple cultural logic systems, if the arguments that are valid in one cultural logic system are not questioned by any other agent with a different cultural system, the arguments should be added into the consensual expansion for all agents involved in the profile. Formally, we have:

Definition 5 (Non-clash part). *Let* $P = (L_1, \cdots, L_n)$ *be a profile of cultural logic systems, where each* $L_i = \langle \mathcal{L}_i, R_{L_i}^\vdash, R_{L_i}^{\not\vdash} \rangle$ *is a logic system. Then the non-clash part of* P, *denoted as* $nc(P)$, *is given by:*

$$nc(P) = \langle \cup_{i=1}^n \mathcal{L}_i, \cup_{i=1}^n R_{L_i}^\vdash \backslash \cup_{i=1}^n R_{L_i}^{\not\vdash}, \cup_{i=1}^n R_{L_i}^{\not\vdash} \backslash \cup_{i=1}^n R_{L_i}^\vdash \rangle. \tag{1}$$

Definition 6 (Common part). *Let* $P = (L_1, \cdots, L_n)$ *be a profile of logic systems, where each* $L_i = \langle \mathcal{L}_i, R_{L_i}^\vdash, R_{L_i}^{\not\vdash} \rangle$ *is a logic system. Then the common part of* P, *denoted as* $cp(P)$, *is given by:*

$$cp(P) = \langle \cap_{i=1}^n \mathcal{L}_i, \cap_{i=1}^n R_{L_i}^\vdash, \cap_{i=1}^n R_{L_i}^{\not\vdash} \rangle. \tag{2}$$

We reveal a property for consensual expansion as follows:

Theorem 1. *Let* $P = (L_1, \cdots, L_n)$ *be a profile of cultural logic systems, and* L_i^{ce} *be the consensual expansion of* L_i *over* P. *Then* $\forall i \in \{1, \cdots, n\}$,

(i) $\mathcal{L}_{nc(P)} = \mathcal{L}_i^{ce}$; *and*
(ii) $R_{inc(P)}^\vdash \subseteq R_{ice}^\vdash$.

Proof. We check one by one. (i) By Definition 5, $nc(P) = \langle \cup_{i=1}^n \mathcal{L}_i, \cup_{i=1}^n R_{L_i}^\vdash \setminus \cup_{i=1}^n R_{L_i}^\nvdash, \cup_{i=1}^n R_{L_i}^\nvdash \setminus \cup_{i=1}^n R_{L_i}^\vdash \rangle$. Now let $ce(L_i, P) = \langle \mathcal{L}_i^{ce}, R_{i_{ce}}^\vdash, R_{i_{ce}}^\nvdash \rangle$ and $c(P)$ as $(\cup_{i=1}^n R_{L_i}^\vdash) \cap (\cup_{i=1}^n R_{L_i}^\nvdash)$. By Definition 5, $\mathcal{L}_{nc(P)} = \cup_{i=1}^n \mathcal{L}_i$; and by Definition 4, $\mathcal{L}_i^{ce} = \mathcal{L}_i \cup (\cup_{i=1}^n \mathcal{L}_i) = \cup_{i=1}^n \mathcal{L}_i$. Thus, we have $\mathcal{L}_{nc(P)} = \mathcal{L}_i^{ce}$. (ii) $\forall (\Sigma, \phi) \in \mathcal{P}(\cup_{i=1}^n \mathcal{L}_i) \times (\cup_{i=1}^n \mathcal{L}_i)$, if $(\Sigma, \phi) \in R_{i_{nc(P)}}^\vdash = \cup_{i=1}^n R_{L_i}^\vdash \setminus \cup_{i=1}^n R_{L_i}^\nvdash$. So, we have $(\Sigma, \phi) \notin c(P) = (\cup_{i=1}^n R_{L_i}^\vdash) \cap (\cup_{i=1}^n R_{L_i}^\nvdash)$. Thus, $(\Sigma, \phi) \in \cup_{i=1}^n R_{L_i}^\vdash \setminus c(P)$ and $(\Sigma, \phi) \in R_{L_i}^\vdash \cup (\cup_{i=1}^n R_{L_i}^\vdash \setminus c(P))$. That is, $(\Sigma, \phi) \in R_{i_{ce}}^\vdash$.   □

The above theorem presents an important property of the consensual expansion. By this property, when we take the consensual expansion over a profile, the valid arguments in the non-clash part of the profile is always preserved. Moreover, the consensual expansion is the most cautious expansion that one can define since an argument is added into the merged logic system only when all the other agents agree on it.

3   Cultural Logic System Merging

In this section, we employ the idea behind the distance-based approach of belief merging [2, 3] to merge cultural logic systems.

Intuitively, we aim to select a logic system that is the closest to the given profile of several cultural logic systems. Formally, we have:

Definition 7 (Logic system merging). *Given a profile* $P = (L_1, \cdots, L_n)$ *where each* $L_i = \langle \mathcal{L}_i, R_{L_i}^\vdash \rangle$ *is a cultural logic system, let* d *be a kind of distance between two cultural logic systems. And suppose that* L_i' *is the expansion of* L_i *over* P. *Then the merging of* P *is defined as:*

$$\triangle_d^\otimes(L_1, \cdots, L_n) = \{L \mid L \text{ over } \cup_{i=1}^n \mathcal{L}_i \text{ such that } L = \arg \min \otimes_{i=1}^n d(L, L_i')\}. \tag{3}$$

where an aggregation function $\otimes$ *is a mapping from* $(\mathbb{R}^+)^n$ *to* $\mathbb{R}^+$ *satisfying:*

(i) monotonicity: if $x_i \geq x_i'$, *then* $\otimes(x_1, \cdots, x_i, \cdots, x_n) \geq \otimes(x_1, \cdots, x_i', \cdots, x_n)$;
(ii) minimality: $\otimes(x_1, \cdots, x_n) = 0$ *iff for all* $i \in \{1, \cdots, n\}$, $x_i = 0$; *and*
(iii) identity: $\otimes(x) = x$.

Clearly, for any $(\Sigma, \phi) \in \mathcal{P}(\cup_{i=1}^n \mathcal{L}_i) \times \cup_{i=1}^n \mathcal{L}_i$, if (Σ, ϕ) is an argument in the logic system after merging, then $(\Sigma, \phi) \in R_\triangle^\vdash$. In the literature, aggregation function $\otimes$ can be in many forms [2, 8], but in this work we just use the ordinary addition operator, on real number set, as the aggregation function, because it is the most intuitive one.

So, to merge a profile of cultural logic systems is a two-step process. Firstly, each cultural logic system L_i over the language of $\mathcal{L}_i$ is expanded such that all the expansions of the logic systems are over the same language $\mathcal{L}' = \cup_{i=1}^n \mathcal{L}_i$. Next, the logic systems over $\mathcal{L}'$ are selected with the distance-based method as the result of merging. Here we assume that the expansion function of each agent's logic system over the profile is the consensual one. To do this, we will apply the idea behind the well-known Hamming distance [1, 2]. Formally, we have:

Definition 8 (Hamming distance between cultural logic systems). *Given two cultural logic systems L_1 and L_2 over the same language $\mathcal{L}$, their Hamming distance is the number of arguments on which the two logic systems differ, i.e.,*

$$d(L_1, L_2) = |\{(\Sigma, \phi) \in P(\mathcal{L}) \times \mathcal{L} \,|\, (\Sigma, \phi) \in R_{L_1}^{\vdash} \text{ but } (\Sigma, \phi) \in R_{L_2}^{\nvdash} \text{ or vice-versa}\}| \,. \tag{4}$$

For example, given two logic systems: $L_1 = \langle \mathcal{L}, R_{L_1}^{\vdash}, R_{L_1}^{\nvdash} \rangle$ where $R_{L_1}^{\vdash} = \{(\Sigma_1, \phi_1), (\Sigma_2, \phi_2)\}$, $R_{L_1}^{\nvdash} = \{(\Sigma_3, \phi_3), (\Sigma_4, \phi_4)\}$, and $L_2 = \langle \mathcal{L}, R_{L_2}^{\vdash}, R_{L_2}^{\nvdash} \rangle$ where $R_{L_2}^{\vdash} = \{(\Sigma_1, \phi_1), (\Sigma_3, \phi_3)\}$, $R_{L_2}^{\nvdash} = \{(\Sigma_2, \phi_2), (\Sigma_4, \phi_4)\}$, then their Hamming distance is $d(L_1, L_2) = 2$. This is because (Σ_2, ϕ_2) is valid in L_1 but not in L_2, while (Σ_3, ϕ_3) is valid in L_2 but not in L_1.

Generally speaking, given two cultural logic systems L_1 and L_2 over the same language of $\mathcal{L}$, since language $\mathcal{L}$ is finite, suppose the number of the sentences in $\mathcal{L}$ is m, then in total $m(C_m^1 + C_m^2 + \cdots + C_m^m)$ arguments can be generated by language $\mathcal{L}$. Some arguments are valid in a cultural logic system, while the others are not. Therefore, given any two cultural logic systems, we can always count the number of the conflicting arguments between them (*i.e.*, valid in one logic but not in the other).

We us the idea behind Hamming distance to define the distance between two cultural logic systems. Then we need to justify what we did is proper. That is, we need to prove that our definition of the distance satisfies the basic axioms of a distance measure:

Theorem 2. *Given any cultural logic systems L_i, L_j and L_k over language $\mathcal{L}$ ($i,j,k \in \{1, \cdots, n\}$), their Hamming distance satisfies:*

(i) non-negativity: $d(L_i, L_j) \geq 0$;
(ii) identity of indiscernible: $d(L_i, L_j) = 0$ if and only if $L_i = L_j$;
(iii) symmetry: $d(L_i, L_j) = d(L_j, L_i)$; and
(iv) triangular inequality: $d(L_i, L_k) \leq d(L_i, L_j) + d(L_j, L_k)$.

Proof. Let $L_i = \langle \mathcal{L}, R_{L_i}^{\vdash} \rangle$, $L_j = \langle \mathcal{L}, R_{L_j}^{\vdash} \rangle$ and $L_k = \langle \mathcal{L}, R_{L_k}^{\vdash} \rangle$. We check theses properties one by one. (i) Non-negativity. By formula (2), $d(L_i, L_j) \geq 0$ is obvious. (ii) Identity of indiscernible. ($\Rightarrow$) If $L_i = L_j$, then $R_{L_i}^{\vdash} = R_{L_j}^{\vdash}$. That is, $\forall (\Sigma, \phi) \in P(\mathcal{L}) \times \mathcal{L}$, $(\Sigma, \phi) \in R_{L_i}^{\vdash}$ if and only if $(\Sigma, \phi) \in R_{L_j}^{\vdash}$. Thus, $\nexists (\Sigma, \phi) \in P(\mathcal{L}) \times \mathcal{L}$ such that $(\Sigma, \phi) \in R_{L_i}^{\vdash}$ while $(\Sigma, \phi) \in R_{L_j}^{\nvdash}$, or $(\Sigma, \phi) \in R_{L_i}^{\nvdash}$ while $(\Sigma, \phi) \in R_{L_j}^{\vdash}$. Thus, $d(L_i, L_j) = 0$. ($\Leftarrow$) If $d(L_i, L_j) = 0$, by formula (2), $\nexists (\Sigma, \phi) \in P(\mathcal{L}) \times \mathcal{L}$ such that $(\Sigma, \phi) \in R_{L_i}^{\vdash}$ while $(\Sigma, \phi) \in R_{L_j}^{\nvdash}$, or $(\Sigma, \phi) \in R_{L_i}^{\nvdash}$ while $(\Sigma, \phi) \in R_{L_j}^{\vdash}$. That is, $\forall (\Sigma, \phi) \in P(\mathcal{L}) \times \mathcal{L}$, $(\Sigma, \phi) \in R_{L_i}^{\vdash}$ if and only if $(\Sigma, \phi) \in R_{L_j}^{\vdash}$. Thus, $R_{L_i}^{\vdash} = R_{L_j}^{\vdash}$. Finally, since $\mathcal{L}_i = \mathcal{L}_j = \mathcal{L}$, $L_i = L_j$. (iii) Symmetry. By formula (2), it is obvious that $d(L_i, L_j) = d(L_j, L_i)$. (iv) Triangular inequality. Since $d(L_i, L_j)$ is the minimal number of coordinate changes of arguments (generated in $\mathcal{L}$) necessary to get from L_i to L_j, and $d(L_j, L_k)$ is the minimal number of coordinate changes of arguments necessary to get from L_j to L_k, we know that $d(L_i, L_j) + d(L_j, L_k)$ changes of arguments in

$\mathcal{L}$ will get us from L_i to L_k. Moreover, since $d(L_i, L_k)$ is the minimal number of coordinate changes of arguments necessary to get from L_i to L_k, we have $d(L_i, L_k) \leq d(L_i, L_j) + d(L_j, L_k)$. $\qquad\qquad\qquad\square$

For a better understanding of our merging operator of cultural logic systems, let us consider the following example. Suppose American Amy (A), Russian Blair (B), and Chinese Chris (C) have a short conversation about the cultural phenomena of their own countries. In Amy's opinion, the sentence "Where are you going?" (s_1) is a invasion of privacy (s_2), and she thinks "Dragon (s_3) is a representation of evil (s_4)"; while for Chris, she does not think the question has any thing to do with the privacy issue and dragon could not be the symbol of evil. At last, Blair agrees with Amy on the privacy issue but agree with Chris on the dragon. Obviously, their different opinions are due to their different languages and inference systems.

Putting all the information together, Amy's cultural logic system can be written as $L_A = \langle \mathcal{L}_A, R^\vdash_{L_A} \rangle$, where

$$\mathcal{L}_A = \{s_1, s_2, s_3, s_4\},$$
$$R^\vdash_{L_A} = \{(s_1, s_2)(s_3, s_4), (s_1, s_1), (s_2, s_2), (s_3, s_3), (s_4, s_4)\};$$

Blair's cultural logic system is: $L_B = \langle \mathcal{L}_B, R^\vdash_{L_B} \rangle$, where

$$\mathcal{L}_B = \{s_1, s_2, s_3, s_4\},$$
$$R^\vdash_{L_B} = \{(s_1, s_2), (s_1, s_1), (s_2, s_2), (s_3, s_3), (s_4, s_4)\};$$

and Chris' cultural logic system is $L_C = \langle \mathcal{L}_C, R^\vdash_{L_C} \rangle$, where

$$\mathcal{L}_C = \{s_1, s_3, s_4\},$$
$$R^\vdash_{L_C} = \{(s_1, s_1), (s_3, s_3), (s_4, s_4)\}.$$

Since the cultural logic language of Chris is different from those of Amy and Blair, by using our merging approach, we need to expand their logic systems to the same language of $\mathcal{L}'$ by the consensual expansion. Formally, we have $c(L_1, L_2, L_3) = (\cup^n_{i=1} R^\vdash_{L_i}) \cap (\cup^n_{i=1} R^\nvdash_{L_i}) = \{(s_3, s_4)\}$. Then:

- $L'_A = \langle \mathcal{L}', R'^\vdash_{L_A}, R'^\nvdash_{L_A} \rangle$, where

$$\mathcal{L}' = \mathcal{L}_A \cup \mathcal{L}_B \cup \mathcal{L}_C = \{s_1, s_2, s_3, s_4\},$$
$$R'^\vdash_{L_A} = \{(s_1, s_2)(s_3, s_4), (s_1, s_1), (s_2, s_2), (s_3, s_3), (s_4, s_4)\},$$
$$R'^\nvdash_{L_A} = (\mathcal{P}(\mathcal{L}'_A) \times \mathcal{L}'_A) \backslash R'^\vdash_{L_A}.$$

- $L'_B = \langle \mathcal{L}', R'^\vdash_{L_B}, R'^\nvdash_{L_B} \rangle$, where

$$\mathcal{L}' = \mathcal{L}_A \cup \mathcal{L}_B \cup \mathcal{L}_C = \{s_1, s_2, s_3, s_4\},$$
$$R'^\vdash_{L_B} = \{(s_1, s_2), (s_1, s_1), (s_2, s_2), (s_3, s_3), (s_4, s_4)\},$$
$$R'^\nvdash_{L_B} = (\mathcal{P}(\mathcal{L}') \times \mathcal{L}'_B) \backslash R'^\vdash_{L_B}.$$

– $L'_C = \langle \mathcal{L}', R''^{\vdash}_{L_C}, R''^{\not\vdash}_{L_C} \rangle$, where

$$\mathcal{L}' = \mathcal{L}_A \cup \mathcal{L}_B \cup \mathcal{L}_C = \{s_1, s_2, s_3, s_4\},$$

$$R''^{\vdash}_{L_C} = \{(s_1, s_2), (s_1, s_1), (s_2, s_2), (s_3, s_3), (s_4, s_4)\},$$

$$R''^{\not\vdash}_{L_C} = (\mathcal{P}(\mathcal{L}') \times \mathcal{L}'_C) \backslash R''^{\vdash}_{L_C}.$$

Table 1. Distances with Σ

	L'_A	L'_B	L'_C	Σ
L_1	6	5	5	16
L_2	5	4	4	13
L_3	4	3	3	10
L_4	3	2	2	7
L_5	2	1	1	4
L_6	1	0	0	1
. . .	. . .	. . .	. . .	. . .
$L_{2^{60}}$	59	60	60	179

In the following, we will show the result of merging A, B and C's cultural logic systems. As we have mentioned before, in total

$$4 \times (C_4^1 + C_4^2 + C_4^3 + C_4^4) = 4 \times \left(\frac{4}{1} + \frac{4 \times 3}{2 \times 1} + \frac{4 \times 3 \times 2}{3 \times 2 \times 1} + \frac{4 \times 3 \times 2 \times 1}{4 \times 3 \times 2 \times 1}\right) = 60$$

arguments can be generated in $\mathcal{L}'$, which result in 2^{60} different logic systems by taking different numbers of arguments as valid arguments. We summarise the calculations in Table 1. For each possible logic system generated over the language $\mathcal{L}'$, we give the distance between this logic system and the expanded logic systems for Amy, Blair and Chris using our Hamming distance measure. Then by using the normal addition operator as the aggregation function, we can obtain the total distance from one logic system to the profile. Since there are in total 2^{60} different logic systems could be generated, we can only display a number of examples as follows:

$L_1 = \langle \{s_1, s_2, s_3, s_4\}, \emptyset \rangle,$

$L_2 = \langle \{s_1, s_2, s_3, s_4\}, \{(s_1, s_1)\} \rangle,$

$L_3 = \langle \{s_1, s_2, s_3, s_4\}, \{(s_1, s_1), (s_2, s_2)\} \rangle,$

$L_4 = \langle \{s_1, s_2, s_3, s_4\}, \{(s_1, s_1), (s_2, s_2), (s_3, s_3)\} \rangle,$

$L_5 = \langle \{s_1, s_2, s_3, s_4\}, \{(s_1, s_1), (s_2, s_2), (s_3, s_3), (s_4, s_4)\} \rangle,$

$L_6 = \langle \{s_1, s_2, s_3, s_4\}, \{(s_1, s_1), (s_2, s_2), (s_3, s_3), (s_4, s_4), (s_1, s_2)\} \rangle,$

$L_{2^{60}} = \langle \{s_1, s_2, s_3, s_4\}, \mathcal{P}(\mathcal{L}) \times \mathcal{L} \backslash \{(s_1, s_1), (s_2, s_2), (s_3, s_3), (s_4, s_4), (s_1, s_2)\} \rangle.$

We just explain two of them above and others can be understood similarly. L_1 is the logic system over $\mathcal{L}'$ without any valid argument. L_2 is the logic system over $\mathcal{L}'$, in which (s_1, s_1) is its valid argument and other arguments that can be generated over $\mathcal{L}'$ are all invalid. Finally, from Table 1, if we take the sum as the aggregation function, then the minimum distance is 1. Thus, we have:

$$\triangle_d^{\Sigma}(L_A, L_B, L_C) = \langle \{s_1, s_2, s_3, s_4\}, \{(s_1, s_1), (s_2, s_2), (s_3, s_3), (s_4, s_4), (s_1, s_2)\} \rangle.$$

It reflects the common reference pattern after merging, which satisfies the majority of the group. That is, since two of three think the question of *"Where are you going?"* is an invasion of privacy, it is; but since two of three they do not think *dragon* symbolizes, it is not.

4   Properties of the Merging

In this section, we investigate the properties of our merging operator on cultural logic systems. More specifically, we expect that a merging operators should satisfy the following properties:

Definition 9 (Agreement). *Given a profile $P = (L_1, \cdots, L_n)$, if no disagreements appear in this profile, i.e., $\forall i, j \in \{1, \cdots, n\}$, $\forall (\Sigma, \phi) \in P(\mathcal{L}_i \cap \mathcal{L}_j) \times (\mathcal{L}_i \cap \mathcal{L}_j)$, $(\Sigma, \phi) \in R_{L_i}^{\vdash}$ iff $(\Sigma, \phi) \in R_{L_j}^{\vdash}$, then*

$$\triangle(L, \cdots, L_n) = \cup_{i=1}^n L_i = \langle \cup_{i=1}^n \mathcal{L}_i, \cup_{i=1}^n R_{L_i}^{\vdash} \rangle.$$

This property reflects if there are no conflicts among the given profile, then the merged profile is simply the union of all the logic systems in the profile.

Definition 10 (Unanimity). *Given a profile of cultural logic systems $P = (L_1, \cdots, L_n)$, a merging operator $\triangle$ is unanimous if it satisfies that $\forall (\Sigma, \phi) \in \mathcal{P}(\cup_{i=1}^n \mathcal{L}_i) \times \cup_{i=1}^n \mathcal{L}_i$, if $\forall i \in \{1, \cdots, n\}$, $(\Sigma, \phi) \in R_{L_i}^{\vdash}$ then $(\Sigma, \phi) \in R_{\triangle}^{\vdash}$.*[1]

The unanimity property reflects that if an argument is accepted by all the agents in different cultures, then it is collectively accepted. Furthermore, we can propose a strong unanimity to reflect that if an argument is not questioned by any other agent, it should be collectively accepted. Formally, we have:

Definition 11 (Strong unanimity). *Given a profile $P = (L_1, \cdots, L_n)$, let $nc(P) = \langle \mathcal{L}_{nc}, R_{nc}^{\vdash}, R_{nc}^{\not\vdash} \rangle$ be the non-clash part of the profile, a merging operator $\triangle$ is strongly unanimous iff $R_{nc}^{\vdash} \subseteq R_{\triangle}^{\vdash}$.*

Definition 12 (Groundedness). *Given a profile $P = (L_1, \cdots, L_n)$, a merging operator $\triangle$ is grounded iff $\forall (\Sigma, \phi) \in P(\cup_{i=1}^n \mathcal{L}_i) \times \cup_{i=1}^n \mathcal{L}_i$, $(\Sigma, \phi) \in R_{\triangle}^{\vdash}$ implies that there exists at least one $i \in \{1, \cdots, n\}$ such that $(\Sigma, \phi) \in R_{L_i}^{\vdash}$.*

Groundedness states that if an argument is collectively accepted, then it must be accepted in at least one of the cultures corresponding to the cultural logic systems in a profile.

---

[1] $R_{\triangle P}^{\vdash}$ is abbreviated as $R_{\triangle}^{\vdash}$, meaning the consequence relations in the resulting logic system after merging P.

Definition 13 (Anonymity). *For any two profiles $P = (L_1, \cdots, L_n)$ and $P' = (L'_1, \cdots, L'_n)$ over $\mathcal{L}$, which are permutations of each other, L is in the outcome after merging P iff it is also in the outcome after merging P'.*

This property requires that all cultures equally important in the merging.

The following theorem states that our merging operator based on the Hamming distance satisfies all the properties above, so its design is proper.

Theorem 3. *If the expansion for each logic system in a profile is the consensual one, our merging operator satisfies the properties of agreement, strong unanimity, groundedness and anonymity.*

Proof. Given a profile $P = (L_1, \cdots, L_n)$ in which each $L_i = \langle \mathcal{L}_i, R^\vdash_{L_i} \rangle$, we denote the consensual expansion for each logic system over P as $ce(L_i, P) = \langle \mathcal{L}'_i, R'^\vdash_{L_i}, R'^\nvdash_{L_i} \rangle$ and our merging operator as $\triangle^\otimes_d$. We will prove that our merging operator satisfies these properties one by one in the following:

(i) Agreement. If $\forall i, j \in \{1, \cdots, n\}$ and $\forall (\Sigma, \phi) \in P(\mathcal{L}_i \cap \mathcal{L}_j) \times (\mathcal{L}_i \cap \mathcal{L}_j)$, $(\Sigma, \phi) \in R^\vdash_{L_i}$ iff $(\Sigma, \phi) \in R^\vdash_{L_j}$, we need to prove $\triangle^\otimes_d (L_1, \cdots, L_n) = \cup^n_{i=1} L_i = \langle \cup^n_{i=1} \mathcal{L}_i, \cup^n_{i=1} R^\vdash_{L_i} \rangle$. Since $\forall i \in \{1, \cdots, n\}$, $ce(L_i, P) = \langle \mathcal{L}'_i, R'^\vdash_{L_i}, R'^\nvdash_{L_i} \rangle$, there are three cases as follows. 1) If $(\Sigma, \phi) \in R^\vdash_{L_i}$, then $(\Sigma, \phi) \in R'^\vdash_{L_i}$. 2) If $(\Sigma, \phi) \notin R^\vdash_{L_i}$ and $(\Sigma, \phi) \in P(\mathcal{L}_i) \times \mathcal{L}_i$, then $(\Sigma, \phi) \in R^\nvdash_{L_i}$, so $(\Sigma, \phi) \in R'^\nvdash_{L_i}$. And 3) if $(\Sigma, \phi) \notin R^\vdash_{L_i}$ and $(\Sigma, \phi) \notin P(\mathcal{L}_i) \times \mathcal{L}_i$, there are two cases: (a) $\exists L_j \in P$ such that $(\Sigma, \phi) \in R^\vdash_{L_j}$, and thus, because there is no disagreement in the profile, $(\Sigma, \phi) \in R'^\vdash_{L_i}$; and (b) $\nexists L_j \in P$ such that $(\Sigma, \phi) \in R^\vdash_{L_j}$, and then $(\Sigma, \phi) \in R'^\nvdash_{L_i}$. In all cases, we can find that if (Σ, ϕ) is a valid argument in at least one of these cultural logic systems in the profile, (Σ, ϕ) is also a valid argument for any L_i after consensual expansion over P. In other words, $\forall i \in \{1, \cdots, n\}$, $ce(L_i, P) = \langle \cup^n_{i=1} \mathcal{L}_i, \cup^n_{i=1} R^\vdash_{L_i} \rangle = \cup^n_{i=1} L_i$. It remains to show that $\triangle^\otimes_d (\cup^n_{i=1} L_i, \cdots, \cup^n_{i=1} L_i) = \cup^n_{i=1} L_i$. Since d is a Hamming distance given in Definition 8, by Theorem 2, it must satisfy the minimality requirement, *i.e.*, $\cup^n_{i=1} L_i$ is the unique logic system which distance to itself is 0.

(ii) Stronger unanimity. Given profile $P = (L_1, \cdots, L_n)$, let non-clash part as $nc(P) = \langle \mathcal{L}_{nc}, R^\vdash_{nc}, R^\nvdash_{nc} \rangle$ and any logic system, in the outcome after merging, be $L = \langle \cup \mathcal{L}_i, R^\vdash_\triangle, R^\nvdash_\triangle \rangle$. Thus, we need to prove that $\forall (\Sigma, \phi) \in P(\cup \mathcal{L}_i) \times \cup \mathcal{L}_i$, if $(\Sigma, \phi) \in R^\vdash_{nc}$, then $(\Sigma, \phi) \in R^\vdash_\triangle$. Suppose not, *i.e.*, $\exists (\Sigma_1, \phi_1) \in P(\cup^n_{i=1} \mathcal{L}_i) \times \cup^n_{i=1} \mathcal{L}_i$ such that $(\Sigma_1, \phi_1) \in R^\vdash_{nc}$ but $(\Sigma_1, \phi_1) \notin R^\vdash_\triangle$. By Theorem 1, we know that $R^\vdash_{nc} \subseteq R'^\vdash_{L_i}$, so $(\Sigma_1, \phi_1) \in R'^\vdash_{L_i}$. Considering another logic system $L' = \langle \cup^n_{i=1} \mathcal{L}_i, R'^\vdash_\triangle, R'^\nvdash_\triangle \rangle$ where $R'^\vdash_\triangle = R^\vdash_\triangle \cup (\Sigma_1, \phi_1)$, $\forall i \in \{1, \cdots, n\}$, we have: $d(L, ce(L_i, P)) = d(L', ce(L_i, P)) - 1$. That is because $(\Sigma_1, \phi_1) \in R'^\vdash_\triangle \cap R'^\vdash_{L_i}$ but $(\Sigma_1, \phi_1) \notin R^\vdash_\triangle \cap R'^\vdash_{L_i}$ and the other arguments valid in L and L' are the same. Moreover, since the aggregation function of $\otimes$ simply is an addition operator on real number and satisfies monotonicity, we have: $\otimes^n_{i=1} d(L, ce(L_i, P)) < \otimes^n_{i=1} d(L', ce(L_i, P))$. Now we find another logic system L', which is closer to the profile than L, so L is not in the outcome after merging. We get a contradiction. So, the supposition is false, and thus we get $R^\vdash_{nc} \subseteq R^\vdash_\triangle$.

(iii) Groundedness. Given logic system profile $P = (L_1, \cdots, L_n)$, let any logic system in the outcome after merging be $L = \langle \cup^n_{i=1} \mathcal{L}_i, R^\vdash_\triangle, R^\nvdash_\triangle \rangle$. We need to prove that $\forall (\Sigma, \phi) \in P(\cup \mathcal{L}_i) \times \cup^n_{i=1} \mathcal{L}_i, (\Sigma, \phi) \in R^\vdash_\triangle$ implies that $\exists i \in \{1, \cdots, n\}$ such that $(\Sigma, \phi) \in R^\vdash_{L_i}$. Suppose not, i.e., $\exists (\Sigma_1, \phi_1) \in P(\cup^n_{i=1} \mathcal{L}_i) \times \cup^n_{i=1} \mathcal{L}_i$ such that $(\Sigma, \phi_1) \in R^\vdash_\triangle$ while $\forall i \in \{1, \cdots, n\}, (\Sigma, \phi_1) \notin R^\vdash_{L_i}$, i.e., $(\Sigma_1, \phi_1) \in R^\nvdash_{L_i}$. By Definition 4, $R'^\vdash_{L_i} = R^\vdash_L \cup (\cup^n_{i=1} R^\vdash_{L_i} \backslash c(P))$ where $c(P) = (\cup^n_{i=1} R^\vdash_{L_i}) \cap (\cup^n_{i=1} R^\nvdash_{L_i})$. Since $\forall i \in \{1, \cdots, n\}, (\Sigma_1, \phi_1) \notin R^\vdash_{L_i}, (\Sigma_1, \phi_1) \notin \cup^n_{i=1} R^\vdash_{L_i}$, and thus $\forall i \in \{1, \cdots, n\}, (\Sigma, \phi) \notin R'^\vdash_{L_i}$, i.e., $(\Sigma, \phi) \in R''^\nvdash_{L_i}$. Consider another logic system $L' = \langle \cup^n_{i=1} \mathcal{L}_i, R'^\vdash_\triangle, R'^\nvdash_\triangle \rangle$ where $R'^\vdash_\triangle = R^\vdash_\triangle \backslash (\Sigma_1, \phi_1)$, then $\forall i \in \{1, \cdots, n\}$, we have: $(\Sigma, \phi) \in R''^\nvdash_\triangle \cap R'^\nvdash_{L_i}$ but $(\Sigma, \phi) \notin R^\nvdash_\triangle \cap R''^\nvdash_{L_i}$ and the other arguments that are valid in L and L' are the same. Therefore, we have $d(L, ce(L_i, P)) = d(L', ce(L_i, P)) - 1$. Moreover, by Definition 7, the aggregation function of $\otimes$ satisfies monotonicity, we have $\otimes^n_{i=1} d(L, ce(L_i, P)) < \otimes^n_{i=1} d(L', ce(L_i, P))$. Now we find another logic system L', which is closer to the profile than L, so L is not in the outcome after merging. We get a contradiction. So, the supposition is false. Thus, $\forall (\Sigma, \phi) \in P(\cup^n_{i=1} \mathcal{L}_i) \times \cup^n_{i=1} \mathcal{L}_i, (\Sigma, \phi) \in R^\vdash_\triangle$ implies that there exists at least one $i \in \{1, \cdots, n\}$ such that $(\Sigma, \phi) \in R^\vdash_{L_i}$.

(iv) Anonymity. ($\Rightarrow$) Given two logic system profiles $P = (L_1, \cdots, L_n)$ and $P' = (L'_1, \cdots, L'_n)$, let the consensual expansion for each logic system L_i and L'_i over P and P' as L^{ce}_i and L'^{ce}_i, respectively. Since P and P' are over the same language L and they are the permutations of each other, we have $\cup^n_{i=1} \mathcal{L}_i = \cup^n_{i=1} \mathcal{L}'_i$. Moreover, $(L^{ce}_1, \cdots, L^{ce}_n)$ is a permutation of $(L'^{ce}_1, \cdots, L'^{ce}_n)$. In other words, $\forall L^{ce}_i \in P^{ce}$, we can find $L'^{ce}_j \in P'^{ce}$ such that $L^{ce}_i = L'^{ce}_j$. Suppose L is a result after merging P, by Definition 7, we have L that minimises $\otimes^n_{i=1} d(L, L^{ce}_i)$. By replacing i in this formula with j, we have L that minimises $\otimes^n_{j=1} d(L, L'^{ce}_j)$, which says exactly that L is one of the logic systems resulted by merging the profile P'. Therefore, if L is a result after merging P then it is also a result after merging of P'. ($\Leftarrow$) The inverse proposition can be proved similarly. □

5 Related Work

Our work is closely related to the work of logic based merging in which belief merging of propositional bases are studied extensively in recent years [2–5]. For example, the work in [3] proposes a logical framework for belief base merging in the presence of integrity constraints. They identify a set of properties that a merging operator should satisfy in order to have a rational behavior. This set of properties can also be used to classify particular merging methods. In particular, a distinction between arbitration and majority operators has been identified. That is, the arbitration operator aims to minimise individual dissatisfaction and the majority operator tries to minimise global dissatisfaction. The work in [4, 5] is another example, in which they view a belief-merging problem as a game. That is, rational agents negotiate jointly consistent consensus, trying to preserve as many important original beliefs as possible. Specifically, they present a model of negotiation for belief merging, a set of rational and intuitive postulates to characterise the belief merging operators, and a representation theorem.

The all above are formal models for belief merging. However, our merging method for cultural logic systems are different from them in the following aspects. *Firstly*, we characterise the merging problem in a totally different situation from belief merging scenarios. More specifically, we intend to merge different logic systems in cross-cultural contexts but they do not. *Secondly*, the representation language of our framework is a binary relation structure other than the propositional logic in belief merging theories of their work, which is more suitable to reflect different opinions for a same thing in different cultures. *Thirdly*, we discuss several basic properties, such as agreement, strong unanimity, groundedness and anonymity, in social choice theories, while in the above work these properties are not mentioned. So, our framework of merging logic systems could benefit the research in this area as well.

Judgment aggregation is about how to aggregate individual judgments on logically interconnected propositions into a collective decision on the same propositions [9]. So, it is also an information merging process from different sources. For example, the logic aggregation theory [10] is closely related to our work. More specifically, their logic aggregation is treated as argument-wise, and some possibilities and impossibilities of aggregating logics are studied. Moreover, they prove that certain logical properties can be preserved by some desired aggregation functions, while some other logical properties cannot be preserved together under non-degenerating aggregation functions, as long as some specific conditions are satisfied. Their work aims to investigate under which conditions aggregation functions will satisfy certain logical properties. However, in our work, our purpose is to investigate how to apply the distance-based merging operator into logic systems with binary relation that can well reflect different opinions for a same thing in different cultures. Thus, we propose a particular merging framework in cross-cultural environments, which satisfies a set of logical properties and some of basic ones in social choice theory.

6   Conclusion

This paper presents a logic system merging approach to solve the cultural conflicts. More specifically, we define each agent's cultural logic system as a tuple of the language and a binary relation over the language, which reflects well different opinions on a same thing in different cultures. Then merging different such systems is a two-step process. Firstly, all the cultural logic systems are expanded to be based on the same language; and then a distance-based method is used to select a cultural logic system that is the nearest to the expanded one from the given profile of cultural logic systems. In this work, we employ the Hamming distance to measure the distance between any two cultural logical systems. Moreover, we prove that this merging operator satisfies the properties of agreement, strong unanimity, groundedness and anonymity, which are widely discussed in social choice theory. So, the design of our merging operator is proper. As a result, the conflicts in the different cultural logic systems are resolved by establishing a common logic system, which can be used as the basis for agents from different cultures to communicate with each other.

There are several things worthy doing in the future. Firstly, with our merging operator, the merged logic systems may not be unique. If a particular one among them is required, a choice function needs to be designed. Secondly, besides distance-base merging frameworks, other merging operators could be used to merge cultural logic systems. Finally, as we have seen, given a certain language, because the number of arguments and cultural logic systems that could be generated over the language grow exponentially, the computation complexity of the distance calculation is high. Therefore, a new method that can significantly reduce the computation complexity is required.

Acknowledgements. Firstly, we appreciate Xuefeng Wen for his advices on early version of this paper. Second, we should acknowledge the financial support from Major Projects of the Ministry of Education China (No. 10JZD0006), MOE Project of Key Research Institute of Humanities and Social Sciences at Universities (No. 13JJD720017) China.

References

1. Konieczny, S., Pino-Pérez, R.: On the logic of merging. In: Proceedings of the Sixth International Conference on Principles of Knowledge Representation and Reasoning, pp. 488–498 (1998)
2. Konieczny, S., Pérez, R.P.: Merging information under constraints: A logical framework. Journal of Logic and Computation 12(5), 773–808 (2002)
3. Konieczny, S., Pérez, R.P.: Logic based merging. Journal of Philosophical Logic 40(2), 239–270 (2011)
4. Tran, T.H., Nguyen, N.T., Vo, Q.B.: Axiomatic characterization of belief merging by negotiation. Multimedia Tools and Applications 65(1), 133–159 (2013)
5. Tran, T.H., Vo, Q.B.: An axiomatic model for merging stratified belief bases by negotiation. In: Nguyen, N.-T., Hoang, K., Jędrzejowicz, P. (eds.) ICCCI 2012, Part I. LNCS, vol. 7653, pp. 174–184. Springer, Heidelberg (2012)
6. Tran, T.H., Vo, Q.B., Kowalczyk, R.: Merging belief bases by negotiation. In: König, A., Dengel, A., Hinkelmann, K., Kise, K., Howlett, R.J., Jain, L.C. (eds.) KES 2011, Part I. LNCS, vol. 6881, pp. 200–209. Springer, Heidelberg (2011)
7. Ju, S.: The cultural relativity of logic: From the viewpoint of ethnography. Social Sciences in China 31(4), 73–89 (2010)
8. Luo, X., Jennings, N.R.: A spectrum of compromise aggregation operators for multi-attribute decision making. Artificial Intelligence 171(2), 161–184 (2007)
9. List, C.: The theory of judgment aggregation: An introductory review. Synthese 187(1), 179–207 (2012)
10. Wen, X., Liu, H.: Logic aggregation. In: Grossi, D., Roy, O., Huang, H. (eds.) LORI 2013. LNCS, vol. 8196, pp. 282–295. Springer, Heidelberg (2013)

Automatic Building of Socio-semantic Networks for Requirements Analysis
Model and Business Application

Christophe Thovex and Francky Trichet

Laboratory of Computer Sciences Nantes Atlantique (LINA, UMR CNRS 6241)
University of Nantes
2 rue de la Houssinière, BP 92208
44322 Nantes, France
{christophe.thovex,francky.trichet}@univ-nantes.fr

Abstract. Experts teams using large volumes of official documents need to see and understand at a glance how texts regarding a topic are redundant and depend on each other. In line with the strategic line of a consultants company, we present a decision support system for the visual analysis of requirements and regulation texts, based on a new model of semantic social networks analysis. We present our model and a business application. In this work, standard metrics of semantic and linguistic statistics are combined with measures of social networks analysis, in order to display socio-semantic networks of regulation texts supporting experts' decision.

Keywords: semantic networks building, social networks analysis.

1   Context and Purpose

Industries and governmental authorities have to deal with regulations and to meet requirements for sustainable environment, health and so on, in their processes and activities. Regulation rules and requirements are defined in official texts which represent large knowledge bases, covering numerous technical domains and countries. Expert teams and specialized companies are called upon to study cases for companies, industries and public institutions, because of the specific skills required for this task, and regarding the hardness of retrieving and crossing official texts for each case study. As this work consists in the study of complex interactions within economic and social systems, methods of social networks analysis might provide new tools for understanding these interactions.

Expert teams in consultants companies use information retrieval tools, generally based on full-text indexation and research services. Ranking methods coming out from linguistic statistics, such as JACCARD's indice and TFIDF scores, help them in the selection of official sources to be included in case studies, but are not sufficient for producing overviews of the lexical/semantic dependencies and similarities in-between sources [1] [2] [3] [4]. As official texts are tied to same and/or

R. Buchmann et al. (Eds.): KSEM 2014, LNAI 8793, pp. 37–48, 2014.

different topics, they share terms denoting knowledge related to topics. Thus, frequent co-occurrences of terms in a documents subset enable to connect texts through terms as documents networks, in which ties between documents represent lexical and semantic relationships weighted thanks to linguistic statistics. As a result, automatic tools for visualizing such document networks, showing at a glance which texts and topics are important and central or not, might be developed and delivered to consultants companies as an innovative decision support system. It is the purpose of our work in line with a research and development project funded by TENNAXIA, French specialized and leading company in the domain of Safety, Health and Hygiene (http://www.tennaxia.com).

An ontology was already built and used in previous works of the TENNAXIA company [5]. However it consumes the experts' workload for maintenance, who expect more productive methods than ontology-based ones such as defined in [5] [6] [7]. The originality of our work is to enable the automatic building of socio-semantic networks without ontology or time-consuming back-end resources. Our approach combines semantic metrics based on linguistic statistics, with measures found in Social Networks Analysis (SNA). So, we define and experiment a model and a software application for the automatic building and analysis of *socio-semantic* networks.

In the following section (2), we define our model of socio-semantic networks, embedding linguistic statistics, official texts, terms and SNA for requirements analysis. In section 3, we present an experimentation of our model as business application and software developed for specialized consultants company. Section 4 sums up our work and its perspectives.

2   Socio-semantic Networks for Requirements Analysis

The purpose of our work is to build graphical representations of official texts and regulations (requirements) including relevant terms denoting topics, and to provide an innovative way borrowing methods from SNA for the analysis of these representations, in order to facilitate the experts' work. We define a novel approach for weighting knowledge in socio-semantic networks, which enables to quickly build and to visualize a new type of knowledge graph on-demand named *keywords network*.

2.1   Linguistic Statistics and Semantic Weights

A corpus C is a set T of texts $t_{(i)}$ containing words $w_{(j)}$. In order to fulfill our goal, we define a simple heterogeneous network structure based on the recursive application of the relation in-between texts and words.

Sine many years, text analysis and mining define a set of statistical models which provide a gateway between syntactic and semantic levels in text analysis. The JACCARD index refinement defined in [1], improves the JACCARD's measure of semantic similarity in-between terms and texts [8].

The TF metric is defined in [4] as the number of occurrences of a term divided by the number of terms, for a text t. TF quantifies the relative predominance of terms denoting knowledge in texts.

The rarity of a term denoting knowledge is usually given by the linguistic metric of Inverse Document Frequency IDF. It is defined in [2] as the log of the number of documents divided by the number of documents containing a given term, in a corpus[1]. TF and IDF are frequently improved as refinements of so-called $TF.IDF$ measures, such as in [3].

For practical reasons, we choose to take advantage of refinements of JAC-CARD index and $TFIDF$ integrated to the native code of a relational database management system[2], so as to calculate semantic weights on the arcs of our socio-semantic networks. These refinements are based on [3] and [9]. They are formalized in equations (1) and (2), based on the software documentation and on their use in our model. $Sim(i,j)$ defines a semantic weight of similarity in-between nodes of socio-semantic networks.

$$Sim(i,j) = \frac{\overrightarrow{A_{i,j}} * 1000 * M}{\overrightarrow{A_{i,j}} * 1000 * M^2 + \overrightarrow{B_{i,j}}^2 - \overrightarrow{A_{i,j}} * 1000 * M} \tag{1}$$

In (1), $\overrightarrow{A_{i,j}}$ and $\overrightarrow{B_{i,j}}$ are binary vectors respectively representing the intersection and the union of the sets of terms corresponding to the pair of texts (i,j). M is a simplification of $min(1000, 16 * TF * IDF)$, counterbalancing the processing of ASCII words as binary vectors, for optimization in native code translation. $Preg(w,t)$ defines a semantic weight of predominance of a word in a text, represented by term nodes and text nodes in socio-semantic networks.

$$Preg(w,t) = log10\left(\frac{Idoc + 0,5}{Tdoc + 0,5}\right) * \frac{2,2 * TF(w,t)}{1,2 * \left(\frac{0,25*Dterms}{ADterms} + TF(w,t)\right) * \frac{9*IDF(w)}{8+IDF(w)}} \tag{2}$$

In (2), $Idoc$ is the number of indexed texts and $Tdoc$ is the number of indexed texts comprising the word w. $Dterms$ is the number of terms of an indexed text, and $ADterms$ the average number of terms in all indexed texts.

2.2   Keywords Networks and Semantic Weights Distribution

We define the equations (1) and/or (2) in order to endow the arcs of socio-semantic networks with semantic weighs. Then, we define the way nodes collect semantic weights thanks to the arcs which they are connected to. We have defined three different structures of socio-semantic networks for the consultants' uses : *global* network, *keywords* networks and *text* networks. Global networks represent the whole data available in the studied database. Keywords networks are

---

[1] We do not consider noise words and stop words as terms denoting knowledge.
[2] SQL Server 2012.

partial representations based on keywords seized by consultants. Text networks are partial representations based on typical texts chosen by consultants (a text for a network). In order to present the main aspects of our work and to respect the space allowance for publication, we merely focus on keywords networks.

Keywords are not properly represented in keywords networks, but appear as related words with the texts related to keywords. A keywords network is oriented and composed of nodes representing (1) key-texts, (2) similar texts and (3) predominant terms. The arcs represent either semantic similarity in-between texts or semantic predominance of words in texts. In Figure 1, dotted arcs are weighted with equation (2) and represent predominance in-between terms and texts. The other arc represents similarity in-between texts, which is weighted thanks to equation (1). Key-texts might also appear as similar texts when their similarity value is high enough, regarding another key-text. User-defined threshold for similarity is made available so as to control the density of the graph.

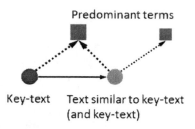

Fig. 1. Basic structure of keywords network

Figure 1 represents the basic pattern of keywords network. According to the picture, in multiple-keywords networks, key-texts are connected with one or more keywords and all texts are connected to one or more (shared) terms. Therefore, keywords networks are always represented by connate graphs. We define a method based on the weight of arcs which calculates semantic weights on nodes, in order to produce relevant indications for the consultants teams.

In a keywords network, some key-texts are also similar to some other key-texts. They are *key and similar* texts. Similar texts which are not key-texts are simple *similar* texts. As a result, we define a method for figuring out the semantic weight of all similar texts according to their role in keywords networks - *i.e.*, *key and similar* or *key* text. Weight of nodes depends on their type.

Let K be for key-text (resource), KS for key and similar text, S for similar texts, and T for term. The weight of a key-text is given by (3), and defined as the sum of predominance values of the n arcs connecting K to terms $\{T_k, T_{k+1}, \ldots T_n\}$.

$$Weight(K) = \sum_{k=1}^{n} Preg_{(K,T_k)} \qquad (3)$$

In order to avoid quantum states in weight values of key and similar text nodes, we define $Weight(KS) = Weight(K)$, otherwise key and similar text nodes should receive distinct $Weight(K) \wedge Weight(KS)$ as a unique value.

In keywords networks, all texts are tied to terms by similarity relation indicating how terms are similar to texts. Thus, all text nodes send ongoing arcs to similar term nodes. Similar texts are tied to key-texts and/or key and similar text by similarity relation. Therefore, similar texts always receive incoming arcs from other texts, and only send outgoing arcs to terms. As a result, we define the weight of a similar text as an average value based on key-text predominance, and on similarity with key-texts and/or key and similar texts. Then, we define the weight of a term as an average value based on key-texts predominance, on similar texts weight and on similarity with all texts. These definitions are presented in equations (4) and (5).

$$Weight(S) = \sum_{i=1}^{|(K,S)|} \frac{Weight(K_i) * Sim(K_i, S)}{|(K,S)|} \qquad (4)$$

In (4), $|(K,S)|$ is the degree of incoming arcs of S - $i.e.$, the number of key-texts and/or key and similar texts similar to S.

$$Weight(T) = \sum_{i=1}^{|(K,T)|+|(S,T)|} \frac{Weight(K_i)|Weight(S_i) * Sim(K_i|S_i, T)}{|(K,T)| + |(S,T)|} \qquad (5)$$

In (5), $|(K,T)| + |(S,T)|$ is the degree of incoming arcs of T - $i.e.$, the number of texts similar to T.

Thanks to this model, we are now in measure (1) to build keywords networks based on a regular graph structure, and (2) to endow nodes and arcs with relevant semantic values, within keywords networks. These values are immediately workable as indications for supporting experts' decision. Furthermore, they enable SNA measures taking weights into account, such as flow betweenness and others [10] [11] [12], to turn into semantic SNA measures when applied to keywords networks [13].

3   Requirements Analysis Using Socio-semantic Networks

Based on our model, we have developed a software application for building keywords networks - $i.e.$, socio-semantic networks on-demand for experts/consultants teams. It is currently delivered to a consultants team and should be deployed within the company after a phase of experimentation and improvement.

3.1   Overview of the Software Architecture

Our client server architecture is based on a relational database server. It embeds full-text indexing and research services, with integrated semantic statistics, management of *noise/empty* words and of thesauri[3]. Grammatical stemmers and word-breakers are part of the database management system and cover the main occidental languages. The database management systems with advanced services is the back-end part of our software architecture.

On the front-end part of our software architecture, we have developed a rich client application presented as a graphic user interface for back-end administration and front-end uses. It runs on local networks and through intranet and/or extranet connections. Graphic User Interface (GUI) is illustrated in figure 2. In the business application deployed within the experts' offices, the back-end and front-end blocks represented in figure 2 are split into distinct client applications, for administrators and experts. Applications implement the process presented in sections 3.2 and 3.3.

3.2   Full-Text Indexing and Semantic Statistics

Hierarchical correspondences usually quoted within official texts define a hierarchical organization based on versions. Experts just consider the latest versions, so in our model, we first exclude the obsolete texts before to index the studied corpus. Obsolete texts are regularly listed in the enterprise database according to rules defined by the experts, so they are easy to exclude of the indexing process.

We start the indexing process with a set of operations aiming at reducing the noise within the full-text index. The studied corpus comprises about 200 000 texts (4.5 GB). Some texts are translated in European languages, but most of the texts are French. We use three textual resources in order to populate automatically the French noise words list which allows the indexing service to avoid useless words. The first one is *Morphalou*, lexicon of French language provided by the *Centre National de Ressources Textuelles et Linguistiques*[4] (CNRTL). It defines about 500 000 terms and their metadata such as genre, plural/singular, grammatical category, etc. The second source is a list of expert terms to be included, and the third one is a list of expert terms to be excluded. Both the lists are managed by the experts.

Firstly, we run a whole indexing of the studied corpus. Then we populate the noise words list by comparing the entries of the initial index with our three textual resources of reference. According to the experts' wishes, all verbs, articles, adverbs and adjectives are considered as noise words in the studied corpus. Strongly recurrent nouns found in titles such as *reglementation* (*i.e.*, regulation) are also considered as noise words. Once the noise words list is complete, we run another indexation task taking it into account, so as to reduce and clean the research index. It allows the software to quickly build keywords networks on-demand, avoiding noise words.

---

[3] *Noise words* are words considered insignificant.
[4] http://www.cnrtl.fr/

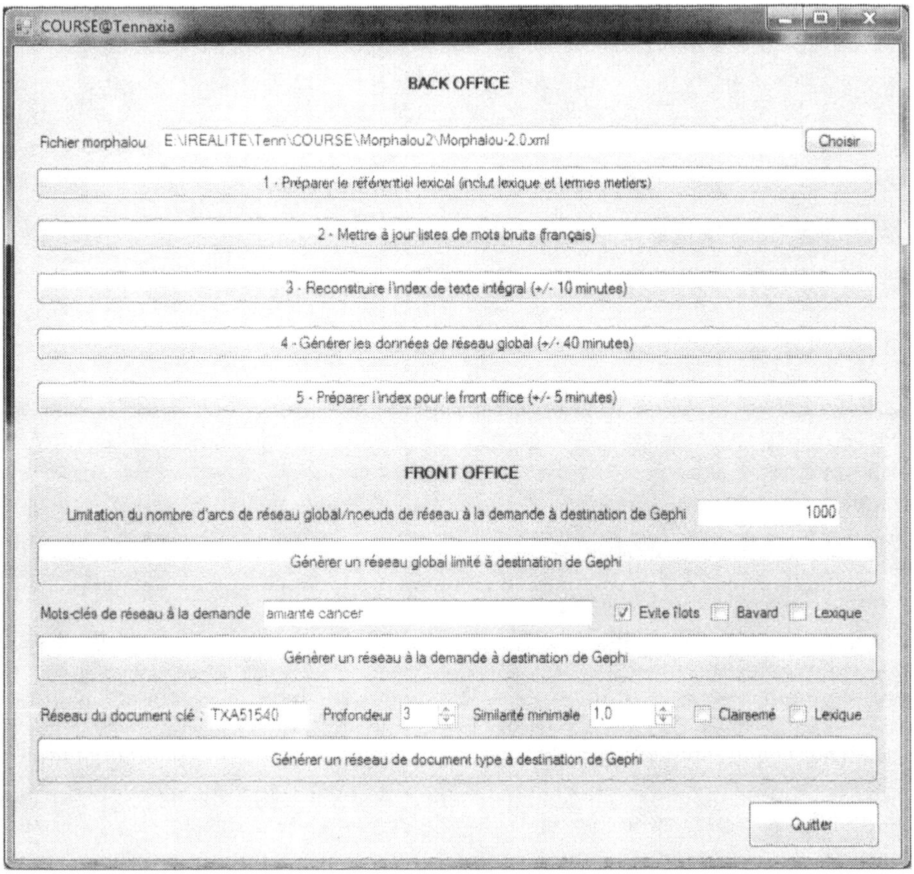

Fig. 2. Rich client application for socio-semantic networks building

A part of semantic statistics is processed with the indexing task, another part is processed on-the-fly during research tasks. Statistic semantics are calculated for each keywords research, as defined in equations (1) and (2) and avoiding possible bias resulting from noise words found within the studied corpus.

3.3   Keywords Networks Building

Experts and consultants teams can quickly seize keywords with several user-defined parameters such as similarity threshold, and get the corresponding socio-semantic network. During the networks building, inflections of terms are systematically replaced by their substantive/infinitive form, if available, thanks to the French lexicon. It avoids semantic redundancies within the networks and improves their readability.

Each keywords network is first represented as a relation which defines arcs, nodes and their respective values in the database, based on user-defined keywords. Then the software exports files from the relation, taking into account user-defined filters. Filters enable to control the diameter (*i.e.*, the length of the largest shortest path), and to make appear "islands" within sparse networks[5].

Currently, the software produces text files (a file for nodes and a file for arcs) in a format easy to load into graph visualization software. We use Gephi[6], which provides a wide choice of plugins for graph analysis and visualization, and is currently supported by an active developers community.

3.4   Experimentation

We have experimented our model and its software application in real context, first as a proof of concept, then as beta version software. A four experts team of the partner company is involved in the project since its beginning, providing technical feedback and empirical evaluations. We have measured execution times and compared several results sets with the studied corpus in order to evaluate our results[7], but the experts' approbation remains the main indicator of relevance in the evaluation of our work.

Our work does not require ontology to produce up-to-date and dynamically built socio-semantic networks. As it saves the workload consumed by experts in ontology maintenance and it provides fast response times, the originality of our work makes it difficult to compare with the few similar works found [6] [7].

Figures 3 and 4 represent a keywords network based on the French keywords "amiante cancer" (*tr.*, asbestos cancer). These keywords denote one of the knowledge fields studied by the experts. The illustrated network is filtrated in order to represent the main texts and relevant terms, and to hide minor facts related to the studied knowledge field.

In figure 3, the size and the color of nodes and labels depend on their respective value of betweenness centrality (large and red labels for high values, small and blue labels for low values). In this example, we use the definition found in [14] and implemented in Gephi. Although this centrality is not explicitly flow-based, it implicitly takes flow values into account because it is based on weights for computing shortest paths. As a result, it makes appear the most unavoidable words characterizing the texts related to asbestos and cancer, -*i.e. substance, exposition* and *affection*. This provides a relevant overview of the content at a glance. Experts and consultants teams can use it as a red wire for case studies. They will get a closer insight into other parts of the graph to find details and minor connotations - *e.g., fibres, oxyde, inhalation, alcalino-terreux* (*tr.* fibers,

---

[5] Islands are connected sub-graphs which are disconnected of the rest of a graph.

[6] Open source - http://www.gephi.org

[7] We found 8% more results with full-text researches than with usual pattern-matching, due to grammatical inflections, and response times are 5 to 70 times faster on full-text indexed content than on standard databases indexes.

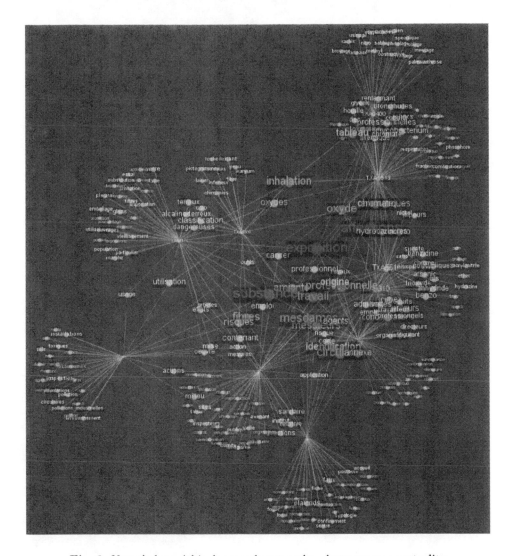

Fig. 3. Knowledge within keywords networks - betweenness centrality

oxide, inhalation, alkaline-earth). All these predominant terms can help in reformulating the research and building complementary keywords networks, useful for the experts' work.

Keywords corresponding with a subset of texts are not necessarily the most representative words of these texts. According to the experts, another relevant aspect of this experimentation is that keywords only appear as major terms within the results if they are also major terms within the retrieved texts. In figures 3 and 4, the nodes *"amiante"* and *"cancer"* (at the center of the graph) do not appear as significant terms for the corresponding texts.

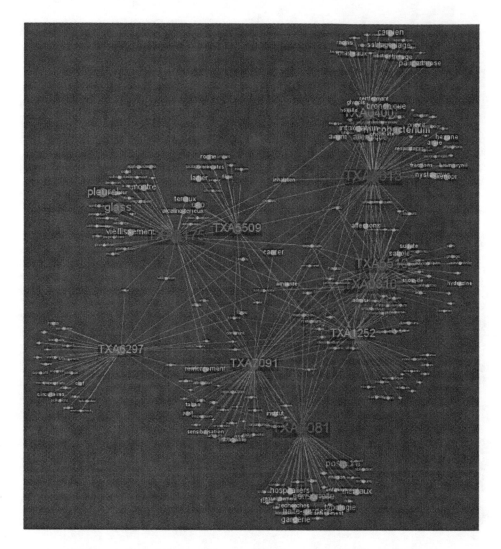

Fig. 4. Visualization of keywords networks - semantic weights

Figure 4, the same keywords network than in figure 3 is represented with a different nodes ranking, solely based on the semantic weights defined in equations (3), (4) and (5). It makes appear the most relevant texts, first, and the most relevant terms within each of these texts regarding the keywords. In this example, "TXA7175" and "TXA7081" are respectively ranked first and second most relevant texts regarding the topic *"amiante cancer"*, followed by 'TXA1613" and "TXA5510" and so on.

According to the experts, "TXA7175" is the latest important notice concerning the protection of population against the risks for health of the exposition to artificial and siliceous mineral fibers. It concerns the dispositions to be taken in all industries and institutions regarding the numerous new materials attempting to replace asbestos in its numerous applications. It alerts against the fact that epidemiological data are not in measure to evaluate the risk of pleural mesothelioma (*i.e.*, one of the main diseases resulting from the inhalation of asbestos) regarding most of the well-known new materials, such as ceramic fibers and micro fiberglass of type E or Glass-475.

"TXA7081" is a memorandum regarding the diagnostic of ceilings (*French tr. plafonds*) and insulation with asbestos in all sanitary and medical buildings welcoming young populations and/or ill populations. It concerns a hundred different types of institutions, such as hospitals, centers of actions against cancer, centers specialized against mental diseases, childcare centers and nurseries.

To their knowledge and after analyzing the "top five relevant texts" of the keywords network, specialized experts agree with the ranking of both the texts as "top relevant texts" concerning the studied topic and among the other results - also considered relevant. Based on the importance of anticipating and preventing new epidemiological issues due to new materials replacing asbestos in industry and building, they arbitrarily approve the proposed ranking of "TXA7175" as most important text, before and near "TXA7081". In the presented example, the weight of "TXA7175" is 373, the weight of "TXA7081" is 369, and the lowest node weight is around 14 for the word *cadre* (*tr.* frame or framework). We have repeated our empirical protocol with several other keywords networks and confirmed the robustness of our model, according to the experts.

4   Conclusion

We have defined a novel approach of knowledge engineering combining linguistic statistics and social networks analysis in a socio-semantic network structure named *keywords network*. According to our model, the predominance of key-texts regarding keywords is distributed depending (1) on similarity in-between texts and terms, and (2) on the topology of the keywords network. Therefore, our model defines relevance as a product of predominance and similarity, dynamically depending on the connections of each node with the other nodes, within keywords networks.

We have developed a software application of our model in line with the R&D project of a consultants' company specialized in Safety, Health and Hygiene. This application was experimented on a large dataset of official regulation texts, and evaluated by specialized experts so as to estimate the relevance and efficiency of our work. Our first results are approved by the experts who recognize their accuracy and relevance, and the application is considered as innovative and efficient tool for knowledge management and decision support. Outcomes are satisfying in theoretical and practical terms.

In the context of the presented work, official texts and terms representing these texts were processed in order to provide an innovative decision support system. In our current and future work, we are defining refinements of our model which enable the visualization and semantic analysis of social media and social networks, including opinions harvested on the Web. We aims at developing innovative software for decision-making and automatic recommendations, based on social content analysis and on social interactions in-between topics and people. We are currently involved in two recent projects oriented towards social and economical governance of touristic and digital territories (*e.g.*, smart cities), within international partnerships.

References

1. Rogers, D., Tanimoto, T.: A computer program for classifying plants. Science 132, 1115–1118 (1960)
2. Sparck Jones, K.: A statistical interpretation of term specificity and its application in retrieval. Journal of Documentation 28, 11–21 (1972)
3. Robertson, S.E., Sparck Jones, K.: Relevance weighting of search terms. Journal of the American Society for Information Science 27(3), 129–146 (1976)
4. Salton, G., MacGill, M.: 6 - Retrieval Refinements. In: Introduction to Modern Information Retrieval, pp. 201–215. McGraw-Hill Book Company (1986)
5. Aimé, X., Furst, F., Kuntz, P., Trichet, F.: Prototypicality gradient and similarity measure: a semiotic-based approach dedicated to ontology personalization. Journal of Intelligent Information Management 2(2), 65–158 (2010) ISSN: 2150-8194
6. Erétéo, G., Gandon, F., Buffa, M., Corby, O.: Semantic social network analysis. In: Proceedings of the WebSci 2009: Society On-Line, Athens, Greece, March 18-20 (2009)
7. Zhuhadar, L., Nasraoui, O., Wyatt, R., Yang, R.: Visual knowledge representation of conceptual semantic networks. Social Network Analysis and Mining Journal (SNAM) 3, 219–299 (2011)
8. Jaccard, P.: Distribution de la flore alpine dans le bassin des dranses et dans quelques régions voisines. Bulletin de la Société Vaudoise des Sciences Naturelles 37, 241–272 (1901)
9. Zaragoza, H., Craswell, N., Taylor, M., Saria, S., Robertson, S.E.: Microsoft Cambridge at trec-13: Web and hard tracks. In: Proceedings of Text REtrieval Conference, TREC 2004 (2004)
10. Freeman, L., Borgatti, S., White, D.: Centrality in valued graphs: A measure of betweenness based on network flow. Social Networks 13(2), 141–154 (1991)
11. Newman, M.: A measure of betweenness centrality based on random walks. Social Networks 27(1), 39–54 (2005)
12. Brandes, U., Fleischer, D.: Centrality measures based on current flow. In: Diekert, V., Durand, B. (eds.) STACS 2005. LNCS, vol. 3404, pp. 533–544. Springer, Heidelberg (2005)
13. Thovex, C., Trichet, F.: Static, Dynamic and Semantic Dimensions: From Social Networks Analysis to Predictive Knowledge Networks. In: Exploratory Analysis for Dynamic Social Networks, pp. 31–58. iConcept Press Ltd. (2012)
14. Brandes, U.: A faster algorithm for betweenness centrality. Journal of Mathematical Sociology 25, 163–177 (2001)

A Document Clustering Algorithm
Based on Semi-constrained Hierarchical
Latent Dirichlet Allocation

Jungang Xu, Shilong Zhou, Lin Qiu, Shengyuan Liu, and Pengfei Li

University of Chinese Academy of Sciences, Beijing, China
{xujg,lipengfei}@ucas.ac.cn,
{zhoushilong12,liushengyuan12}@mails.ucas.ac.cn,
qiulinster@gmail.com

Abstract. The bag-of-words model used for some clustering methods is often unsatisfactory as it ignores the relationship between the important terms that do not cooccur literally. In this paper, a document clustering algorithm based on semi-constrained Hierarchical Latent Dirichlet Allocation (HLDA) is proposed, the frequent itemsets is considered as the input of this algorithm, some keywords are extracted as the prior knowledge from the original corpus and each keyword is associated with an internal node, which is thought as a constrained node and adding constraint to the path sampling processing. Experimental results show that the semi-constrained HLDA algorithm outperforms the LDA, HLDA and semi-constrained LDA algorithms.

Keywords: Document clustering, HLDA, frequent itemsets, PMI, empirical likelihood.

1 Introduction

Document clustering has been studied intensively because of its wide applicability in many areas such as web mining, search engine, information retrieval, and topological analysis and etc. As a fundamental and enabling tool for efficient organization, navigation, retrieval, and summarization of huge volumes of text documents, document clustering has been paid more and more attentions to. With a good document clustering method, computers can automatically organize a document corpus into a meaningful cluster hierarchy, which enables an efficient browsing and navigation of the corpus.

The motivation of clustering a set of data is to find inherent structure in the data and to expose this structure as a set of groups. Zhao and Karypis [1] showed that the data objects within each group should exhibit a large degree of similarity while the similarity among different clusters should be minimized.

Unlike document classification, document clustering is one mean of no label learning. Merwe and Engelbrecht [2] showed that although standard clustering methods such as k-means can be applied to document clustering, they usually do not satisfy the

R. Buchmann et al. (Eds.): KSEM 2014, LNAI 8793, pp. 49–60, 2014.

special requirements for clustering documents: high dimensionality, high volume of data, easy for browsing, and meaningful cluster labels. In addition, many existing document clustering algorithms require the user to specify the number of clusters as an input parameter and are not robust enough to handle different types of document sets in a real world environment. For example, the cluster size in some document sets varies from few to thousands. This variation tremendously reduces the clustering accuracy of some of the state-of-the-art algorithms.

Jain et al [3] showed that document clustering methods can be mainly categorized into two types: document partitioning (flat clustering) and agglomerative clustering (bottom-up hierarchical). Although both types of methods have been extensively investigated for several decades, accurately clustering documents without domain-dependent background information, pre-defined document categories or a given list of topics is still a challenging task.

David et al [4] proposed a Hierarchical Latent Dirichlet Allocation (HLDA) algorithm for building topic model. Based on this algorithm, we study how to use the notion of frequent itemsets as prior knowledge into the HLDA algorithm, and propose to use semi-constrained HLDA to solve the problems described above.

The remainder of this paper is organized as follows. Section 2 describes the related work. Section 3 introduces the LDA and HLDA. Section 4 explains a document clustering algorithm based on semi-constrained HLDA. Section 5 presents the experimental results based on three different datasets. Finally, section 6 concludes the work.

2   Related Work

Generally, clustering methods can be categorized as agglomerative clustering and partitional clustering. Agglomerative clustering methods group the data points into a hierarchical tree structure or a dendrogram by bottom-up approach. The procedure starts by placing each data point into a distinct cluster and then iteratively merges the two most similar clusters into one parent cluster. Upon completion, the procedure automatically generates a hierarchical structure for the data set. The complexity of these algorithms is O(n2log n) where n is the number of data points in the data set. Because of the quadratic order of complexity, bottom-up agglomerative clustering methods could become computationally prohibitive for clustering tasks that deal with millions of data points. On the other hand, partitional clustering methods decompose a document corpus into a given number of disjoint clusters which are optimal in terms of some pre-defined criteria functions. Partitional clustering methods can also generate a hierarchical structure of the document corpus by iteratively partitioning a large cluster into smaller clusters. Typical methods in this category include K-means clustering, probabilistic clustering using the Naive Bayes or Gaussian mixture model, and so on. Willett [5] showed that K-means produces a cluster set that minimizes the sum of squared errors between the documents and the cluster centers, while Liu and Gong [6] showed that both the Naive Bayes and the Gaussian mixture models assign each document to the cluster that provides the maximum likelihood probability. The common drawback associated with these methods is that they all make harsh simplifying assumptions on the distribution of the document corpus to be clustered, K-Means assumes that each cluster in the document corpus has a compact shape, the Naive Bayes model assumes that all the dimensions of the feature space representing the document corpus are independent of each other, and the Gaussian mixture model assumes that the density of each cluster can be approximated by a Gaussian distribution. Obviously, these assumptions do not often hold true, and document clustering results could be terribly wrong with broken assumptions.

Deerwester et al [7] proposed a document clustering method using the Latent Semantic Indexing method (LSI). This method basically projects each document into the singular vector space through the Singular Value Decomposition (SVD), and then conducts document clustering using traditional data clustering algorithms (such as K-means) in the transformed space. Although it was claimed that each dimension of the singular vector space captures a base latent semantics of the document corpus, and that each document is jointly indexed by the base latent semantics in this space, negative values in some of the dimensions generated by the SVD, however, make the above explanation less meaningful.

In recent years, spectral clustering based on graph partitioning theories has emerged as one of the most effective document clustering tools. These methods model the given document set using an undirected graph, in which each node represents a document, and each edge (i, j) is assigned a weight w_{ij} to reflect the similarity between the document i and j. The document clustering task is accomplished by finding the best cuts of the graph that optimize certain pre-defined criterion functions. The optimization of the criterion functions usually leads to the computation of singular vectors or eigenvectors of certain graph affinity matrices, and the clustering result can be derived from the obtained eigenvector space. Chan et al [8] proposed one criterion functions named Average Cut; Shi and Malik [9] proposed two criterion functions named Average Association and Normalized Cut; Ding et al [10] proposed one criterion functions named Min-Max Cut. These criterion functions have been proposed along with the efficient algorithms for finding their optimal solutions. It can be proven that under certain conditions, the eigenvector spaces computed by these methods are equivalent to the latent semantic space derived by the LSI method. As spectral clustering methods do not make naive assumptions on data distributions, and the optimization accomplished by solving certain generalized eigenvalue systems theoretically guarantees globally optimal solutions, these methods are generally far more superior to traditional document clustering approaches. However, because of the use of singular vector or eigenvector spaces, all the methods of this category have the same problem as LSI, i.e., the eigenvectors computed from the graph affinity matrices usually do not correspond directly to individual clusters, and consequently, traditional data clustering methods such as K-means have to be applied in the eigenvector spaces to find the final document clusters.

3   Latent Dirichlet Allocation and Hierarchical Latent Dirichlet Allocation

3.1   Latent Dirichlet Allocation

In reference [11], David et al proposed Latent Dirichlet Allocation (LDA) model, which is a generative probabilistic model for processing collections of discrete data such as text corpus, and has quickly become one of the most popular probabilistic text modeling techniques. LDA uses the bag of words model, which considers each document as a word frequency vector. The graphical model of LDA is shown in Fig. 1.

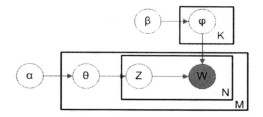

Fig. 1. The graphical model of LDA

LDA model could be described as follows:

- **Word:** a basic unit defined to be an item from a vocabulary of size W.
- **Document:** a sequence of N words denoted by $d = (w_1, ..., w_n)$, where w_n is the nth word in the sequence.
- **Corpus:** a collection of M documents denoted by $D = \{d_1, ..., d_m\}$.

Given corpus D is expressed over W unique words and T topics, LDA outputs the document-topic distribution θ and topic-word distribution ϕ. This distribution can be obtained by a probabilistic argument or by cancellation of terms in Equation 1.

$$p(z_i = j \mid z_{-i}, w) \propto \frac{n_{-i,j}^{(w_i)} + \beta}{\sum_{w'}^{W} n_{-i,j}^{(w')} + W\beta} \frac{n_{-i,j}^{(d_i)} + \alpha}{\sum_{j}^{T} n_{-i,j}^{(d_i)} + T\alpha} \qquad (1)$$

Where $\sum_{w'}^{W} n_{-i,j}^{(w')}$ is a count that does not include the current assignment of z_j. The first ratio denotes the probability of w_i under topic j, and the second ratio denotes the probability of topic j in document d_i. Critically, these counts are the only necessary information for computing the full conditional distribution, which allow the algorithm to be implemented efficiently by caching the relatively small set of nonzero counts. After several iterations for all the words in all the documents, the distribution θ and distribution ϕ are finally estimated using Equation 2 and 3.

$$\phi_j^{(w_i)} = \frac{n_j^{(w_i)} + \beta}{\sum_w^W n_j^{(w)} + W\beta} \tag{2}$$

$$\theta_j^{(d_i)} = \frac{n_j^{(d_i)} + \alpha}{\sum_j^T n_j^{(d_i)} + T\alpha} \tag{3}$$

Many applications are based on LDA model, such as clustering, deduction, forecast and so on. LDA for clustering and K-means are both unsupervised clustering methods that do not require any training corpus. Therefore, this method gets rid of the negative effect of the low-quality corpus. And on the other hand, we add a dictionary for the data set into the original segment system, which can improve the accuracy of segment results.

3.2   Hierarchical Latent Dirichlet Allocation

Since Topic Models are parametric methods which should make the topic number as the prior knowledge, furthermore, topic models such as LDA treat topics as a "flat" set of probability distributions, with no direct relationship between one topic and another. Then using a hierarchical tree, which the topic number is a nonparametric, to group multiple topics is the natural solution. Blei's Hierarchical Latent Dirichlet Allocation (HLDA) is one topic model of this kind.

Although the motivation for hierarchical topic modeling is provided in previous section as a bottom up process, the HLDA model presented by Blei is a top down process, and it is derived from a stochastic process known as Nested Chinese Restaurant Process (NCRP).In reference [12], Teh et al showed the details of Nested Chinese Restaurant Process.

NCRP is quite similar to CRP with the exception that a document is allocated to a topic each time, the document can spawn new topics within the previous topic. In other words, a document can spawn topic within topic, so the topic number can be changed dynamically. Fig. 2 briefly illustrates the HLDA initialization procedure which is based on the NCRP.

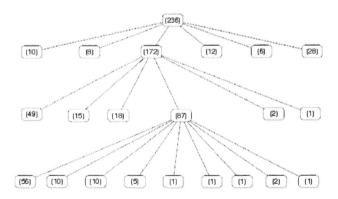

Fig. 2. Blei's Hierarchical Latent Dirichlet Allocation initialization procedure

1. Firstly, start with a root node at the top of the tree.
2. For each document, let initial node be root,
(a) Visit node.
(b) Compute the CRP probability P $(z_n = k)$ and P $(z_n = new)$ using Equation 4 and Equation 5.

$$P(z_n = k) = \frac{C_{n.k}}{N - 1 + \alpha} \tag{4}$$

$$P(z_n = new) = \frac{\alpha}{N - 1 + \alpha} \tag{5}$$

Where $C_{n.k}$ is the number of documents on the *kth* topic, N is the total number of documents and α is the parameter of NCRP, which is used to control the increment speed of new nodes.

(c) Sample and decide whether to visit an existing child node or spawn new child node using above probabilities.

(d) Once current level node has been chosen, using CRP to select the next level node until reaches a specified depth.

(e) The nodes visited in the order as specified above constitute a path in the tree.

(f) For each word in a document, randomly assign a node in the path which is chosen earlier to this word. In other words, the words are allocated to different levels in the tree.

Now that we have randomly created a tree, we shall proceed to perform the hierarchical topic modeling using Gibbs sampling.

Sampling Paths. Given the level allocation variables, the path associated with each document conditioned on all other paths and the observed words need to be sampled. In their seminal paper [13], Mao et al showed the sampling path calculation method as in Equation 6.

$$p(c_d \mid w, c_{-d}, z, \eta, \gamma) \propto p(c_d \mid c_{-d}, \gamma) p(c_d \mid c, w_{-d}, z, \eta) \tag{6}$$

Where $p(c_d \mid c_{-d}, \gamma)$ is the prior on paths implied by the nested CRP, $p(c_d \mid c, w_{-d}, z, \eta)$ is the probability of the data given a particular choice of path. Suppose that the sampled node is an internal node instead of a leaf node, then it means we spawn new leaf nodes until we reach the maximum depth as defined. Suppose for a given path c_m, the path has topic levels $1, \ldots, K$ and there are T number of word topic distributions. And the sampling path is described by Equation 7.

$$P(w_m \mid C_m, w_{-m}, z_m, \eta) = \frac{P(w \mid c_m, z_m, \eta)}{P(w_{-m} \mid c_m, z_m, \eta)} \tag{7}$$

And expressing Logarithm form is described as in Equation 8.

$$\log P(w_m \mid C_m, w_{-m}, z_m, \eta) =$$

$$\log[\prod_{t=1}^{T}[\frac{\Gamma(V_\eta + \sum_{v=1}^{V} g_{t,k,v})}{\prod_{v=1}^{V} \Gamma(\eta + g_{t,k,v})} \frac{\prod_{v=1}^{V} \Gamma(g_{t,k,v} + f_{t,k,v} + \eta)}{\Gamma(V_\eta + \sum_{v=1}^{V}(g_{t,k,v} + f_{t,k,v}))}] \quad (8)$$

When sampling, whether to branch off needs to be decided, the Equation 9 is used, it is pretty similar to the Equation 8 except that $g_{t,k,v}$ is always zero.

$$\log P(w_m \mid C_m, w_{-m}, z_m, \eta) =$$

$$\log[\prod_{t=1}^{T}[\frac{\Gamma(V_\eta + \sum_{v=1}^{V})}{\prod_{v=1}^{V} \Gamma(\eta)} \frac{\prod_{v=1}^{V} \Gamma(f_{t,k,v} + \eta)}{\Gamma(V_\eta + \sum_{v=1}^{V}(f_{t,k,v}))}] \quad (9)$$

Sampling Level Allocations. Given the current path assignments, the level allocation variable for each word in each document from its distribution given the current values of all other variables need to be sampled. Suppose that we have D documents and (d,M) words are in document d. For each word n in document d, the sampling of the topics in the path is described by the Equation 10.

$$P(z_{d,m} = k \mid V_1,...,V_k) = V_k \prod_{i=1}^{k-1}(1-V_i) \quad (10)$$

Let $e_{d,n}$ denote the counts of occurrence for each word n in document d, then derive the posterior distribution of $V_1, . . . , V_k$, hence,

$$P(z_{d,m} = k \mid z_{d,-m}, n, \pi) = E[V_k \prod_{i=1}^{k-1}(1-V_i)] \quad (11)$$

As to the word, suppose there are N number of words in the vocabulary, let $d_{k,n}$ be the number of times that word n is allocated to topic k, hence,

$$P(w_{d,m} = n \mid z, c, w_{d,-m}) = \frac{d_{k,n} + \eta}{\sum_{n=1}^{N} d_{k,n} + V\eta} \quad (12)$$

So, in order to sample a topic, the expression is described as in Equation 13.

$$P(z_{d,m} = k \mid z_{d,-m}, c, w, n, \pi, \eta) =$$

$$[\frac{n\pi + e_{d,n}}{\pi + \sum_{i=k}^{K} e_{d,i}} \prod_{i=1}^{k-1} \frac{(1-n)\pi + \sum_{j=i+1}^{K} e_{d,j}}{\pi + \sum_{j=i}^{K} e_{d,j}}] \frac{d_{k,n} + \eta}{\sum_{n=1}^{N} d_{k,n} + V\eta} \quad (13)$$

4   The Document Clustering Algorithm Based on Semi-constrained HLDA

4.1   Knowledge Extraction

As HLDA is a nonparametric model and it can obtain a more suitable topic number, we adopt it to construct our topic model to cluster the documents. As we know, the pre-existing knowledge can help us get more significant cluster results. In this study, we also adopt this strategy to incorporate the prior knowledge which is extracted automatically. The keywords are extracted as the prior knowledge from the original corpus and each keyword is associated with an internal node, which is thought as constraint node.

Han et al [14] used frequent pattern tree (FP-tree) algorithm to mine frequent patterns. Danushka et al [15] used point-wise mutual information (PMI) to calculate semantic similarity between words using web search engines. Here, in order to discovery appropriate keywords, the FP-tree algorithm and PMI are also applied. First, we extract the frequent items from the corpus as candidate keywords. Then we compute PMI of these words. If the PMI of two words is less than the given threshold, we add a cannot-link to these two words, which means that they should belong to different internal nodes. If the PMI of two words is greater than the given threshold, we add a must-link to these two words, which means that they should belong to the same internal node. Finally, we can get the final keyword set for all internal nodes.

4.2   Knowledge Incorporation

After knowledge extraction for internal nodes, we will build the underlying tree structure like as in Fig. 3. The node 0 is root node, the node 1 and node 2 are constraint nodes, which are related to the keywords. Therefore, we revise the path sampling as follows: (1) if the current document contains the keywords that belong to an internal node, we can deduce that the underlying path space of this document is no longer the tree, it should be the sub-tree of the corresponding internal node. Therefore, only sampling the nodes which are the child nodes of the corresponding internal node is enough. (2) If the current document does not contain the keywords, it may not contain important information. So it does not belong to the constraint nodes, and its underlying path space should be the root node and the sub-tree of the other unconstraint nodes.

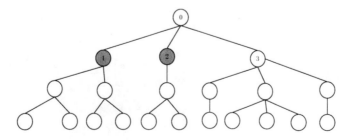

Fig. 3. Underlying tree structure

4.3   The Formulation of the Algorithm

The formulation of the algorithm is shown as in Fig. 4.

Algorithm 1. A document clustering algorithm based on semi-constrained HLDA

1: for each document d_i in document   set D do
2: $WS \leftarrow$ word segment for the document, and then choose the nouns, verbs, adjectives;
3: end for
4: $FI \leftarrow$ extract the frequent items as candidate keywords;
5: $NL \leftarrow$ add a cannot-link to these two words based on PMI, if PMI of two words is less than the given threshold;
6: $FK \leftarrow$ get the final keyword set for the internal nodes;
7: Cluster$\leftarrow$ document clustering based on semi-constrained HLDA.

Fig. 4. The formulation of the algorithm

Firstly, for each document in the data set, we need to do some pre-processing, such as word segmentation, part-of-speech (POS) tagging and feature clustering (Line 2). Except the prepositions, auxiliary words, etc. which are generally seen as stop words, nouns, verbs and adjectives may appear most frequent and include almost all information in the text, so we choose them. Then we get the frequent items as the candidate words, which consist of constrains (Line 4). Based on the PMI, we add the cannot-link to HLDA (Line 5). And we can get the final keyword set for the internal nodes (Line 6). And at last, we use the semi-constrained HLDA to conduct document clustering (Line 7).

5   Experiments

5.1   Datasets

In this paper, we adopt three different data sets: (1) 516 People's Daily editorials from year 2008 to 2010; (2) 540 academic papers of Chinese Journal of Computers (CJC) from year 2010 to 2012; (3) 500 Baidu encyclopedia texts.

5.2   Evaluation Method

In this paper, the empirical likelihood estimation method is used to estimate clustering algorithms. Empirical likelihood is an estimation method in statistics, which performs well even when the distribution is asymmetric or censored, and it is useful since it can easily incorporate constraints and prior information. Empirical likelihood estimation method requires few assumptions about the error distribution compared to similar method like maximum likelihood. In our Experiments, the empirical likelihood estimation can be calculated as follows.

First, the each dataset is divided into two categories, one is the training set called D_1, the other is the test set called D_2, where the proportion of these two sample data-sets are 20% and 80%. Second, one topic model is established through clustering algorithm based on D_1, and then D_2 is tested using the already established topic model. After each iteration, compare with the topic model established based on D_1. Finally, the Empirical Likelihood of each iteration can be obtained, its representation is shown as in Equation 14,

$$p\left(W_1^{held-out}, \dots, W_{D_2}^{held-out} \mid W_1^{obs}, \dots, W_{D_1}^{obs}, M\right) \tag{14}$$

Where M denotes topic model, including LDA, HLDA, semi-constrained LDA and semi-constrained HLDA. The expression of Empirical Likelihood of D_2 is shown as in Equation 15.

$$EL(D_2) = \log_2\left\{-\frac{\sum_{d=1}^{M}\sum_{m=1}^{N_d} \log_p w_{dm}}{\sum_{d=1}^{M} N_d}\right\} \tag{15}$$

Where D_2 denotes the training dataset, N_d denotes the length of the Document d, W_{dm} denotes the mth word in document d.

5.3   Experimental Results and Analysis

We compared LDA, HLDA, semi-constrained LDA, semi-constrained HLDA four clustering algorithm through the experiments. The clustering granularities are set to sentence, paragraph, word separately. The sentence granularity means we select one topic word from each sentence as a dimension of a document, the paragraph granu-larity means we select one topic word from each paragraph as a dimension of a docu-ment, and the word granularity means one word we select from a document is consi-dered as a dimension of a document. Experimental results on three different datasets are shown in Fig. 5, Fig. 6 and Fig. 7.

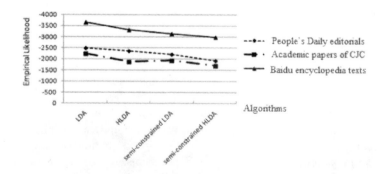

Fig. 5. Empirical likelihood of four clustering algorithms in sentence granularity

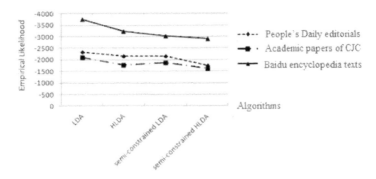

Fig. 6. Empirical likelihood of four clustering algorithms in paragraph granularity

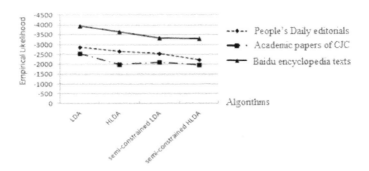

Fig. 7. Empirical likelihood of four clustering algorithms in word granularity

In this paper, the higher the empirical likelihood, the more concentrated the document's topic, which means the document with higher empirical likelihood has better quality. From Fig. 5, Fig. 6 and Fig. 7, we can see that semi-constrained HLDA can get the best empirical likelihood among four algorithms in all three different granularities. And we can also see that all four algorithms can get best empirical likelihood in paragraph granularity. In addition, we can deduce from the experimental results that the dataset of academic paper of CJC has the highest text quality among the three datasets, the dataset of People's Daily editorials takes the second place.

6   Conclusions

In this paper, one document clustering algorithm based on semi-constrained HLDA is proposed. We extract the frequent itemsets as candidate keywords through the FP-tree algorithm, and then compute the PMI of these words, add a cannot-link or must-link to each two word according to the PMI, finally, we can get the final keyword set for all internal nodes. After knowledge extraction for internal nodes, we will build the underlying tree structure of the document and cluster them. Based on three different datasets, we use the empirical likelihood estimation method to estimate LDA, HLDA, semi-constrained LDA and semi-constrained HLDA algorithms, experimental results show that the semi-constrained HLDA algorithm has the best empirical likelihood.

Acknowledgements. This work is supported in part by the National Natural Science Foundation of China under Grant no. 61372171 and the National Key Technology R&D Program of China under Contract no. 2012 BAH23B03.

References

1. Zhao, Y., Karypis, G.: Empirical and Theoretical Comparisons of Selected Criterion Functions for Document Clustering. Machine Learning 55(3), 311–331 (2004)
2. Merwe, V.D., Engelbrecht, A.P.: Data Clustering Using Particle Swarm Optimization. In: 2003 IEEE Congress on Evolutionary Computation, pp. 215–220. IEEE Press, New York (2003)
3. Jain, A.K., Murty, M.N., Flynn, P.J.: Data Clustering: A Review. ACM Computing Survey 31(3), 264–323 (1999)
4. David, B., Griffites, T., Tennbaum, J.: Hierarchical Topic Models and the Nested Chinese Restaurant Process. In: Advances in Neural Information Processing Systems, vol. 16, pp. 106–113 (2004)
5. Willett, P.: Document Clustering Using An Inverted File Approach. Journal of Information Science 2, 223–231 (1990)
6. Liu, X., Gong, Y.: Document Clustering with Cluster Refinement and Model Selection Capabilities. In: 25th Annual International ACM SIGIR Conference on Research and Development in Information Retrieval, pp. 117–122. ACM Press, New York (2002)
7. Deerwester, S.C., Dumais, S.T., Landauer, T.K.: Indexing by Latent Semantic Analysis. Journal of the American Society for Information Science 41(6), 391–407 (1990)
8. Chan, P.K., Schlag, D.F., Zien, J.Y.: Spectral K-way Ratio-cut Partitioning An Clustering. IEEE Council on Electronic Design Automation 13(9), 1088–1096 (1994)
9. Shi, J., Malik, J.: Normalized Cuts and Image Segmentation. IEEE Transactions on Pattern Analysis and Machine Intelligence 22(8), 888–905 (2000)
10. Ding, C., He, X., Zha, H., Simon, H.D.: A Min-max Cut Algorithm for Graph Partitioning and Data Clustering. In: 2001 IEEE International Conference on Data Mining, pp. 107–114. IEEE Press, New York (2001)
11. David, M.B., Andrew, Y.N., Michael, I.J.: Latent Dirichlet Allocation. Journal of Machine Learning 3, 993–1022 (2003)
12. Teh, Y.W., Jordan, M.I., David, M.B.: Hierarchical Dirichlet Processes. Journal of the American Statistical Association 101(476), 1566–1581 (2006)
13. Mao, X.L., Ming, Z.Y., Chua, T.S., Li, S., Yan, H.F., Li, X.M.: SSHLDA: A Semi-supervised Hierarchical Topic Model. In: 2012 Joint Conference on Empirical Methods in Natural Language Processing and Computational Natural Language Learning, pp. 800–809. ACL Press, Stroudsburg (2012)
14. Han, J.W., Pei, J., Yin, Y.W., Mao, R.Y.: Mining Frequent Patterns without Candidate Generation. In: 2000 ACM SIGMOD International Conference on Management of Data, pp. 1–12. ACM Press, New York (2000)
15. Danushka, B., Yutaka, M., Mitsuru, I.: Measuring Semantic Similarity between Words Using Web Search Engines. In: 16th International Conference on World Wide Web, pp. 757–766. ACM Press, New York (2007)

An Ordinal Multi-class Classification Method for Readability Assessment of Chinese Documents

Zhiwei Jiang, Gang Sun, Qing Gu*, and Daoxu Chen

State Key Laboratory for Novel Software Technology,
Nanjing University, Nanjing 210023, China
jiangzhiwei@outlook.com, sungangnju@163.com,
{guq,cdx}@nju.edu.cn

Abstract. Readability assessment is worthwhile in recommending suitable documents for the readers. In this paper, we propose an Ordinal Multi-class Classification with Voting (OMCV) method for estimating the reading levels of Chinese documents. Based on current achievements of natural language processing, we also design five groups of text features to explore the peculiarities of Chinese. We collect the Chinese primary school language textbook dataset, and conduct experiments to demonstrate the effectiveness of both the method and the features. Experimental results show that our method has potential in improving the performance of the state-of-the-art classification and regression models, and the designed features are valuable in readability assessment of Chinese documents.

Keywords: Readability Assessment, Classification Model, Text Feature, Chinese, Ordinal Classification.

1 Introduction

The readability of a document refers to the comprehending easiness or reading difficulty of the text in the document [1]. It is always helpful to give a reader documents which match his/her reading ability. For example, teachers often need to find documents suitable for the students, where readability is one of the most important factors. The readability assessment of documents is worthwhile in these situations.

Researches on readability assessment have a relatively long history from early 20th century [22]. In most cases, reading levels are used to quantify the readability of a document. In the early stage, many researchers make use of the surface features, e.g. average number of words per sentence or number of syllables per word, from a document to estimate its reading level [6]. By using linear regression on selected surface features, some well-known readability formulae such as FK [14] and SMOG [17] have been developed. Nowadays, the latest developments of natural language processing and the machine learning technologies have been successfully applied in readability assessment [19,5,10]. The former is used to develop new valuable text features, while the latter is often used to handle the assessment problem as a classification problem.

To our knowledge, readability assessment should be viewed as an ordinal classification problem, and ordinal classification techniques should be developed to handle it.

---

* Corresponding author.

R. Buchmann et al. (Eds.): KSEM 2014, LNAI 8793, pp. 61–72, 2014.
© Springer International Publishing Switzerland 2014

In this paper, we propose an Ordinal Multi-class Classification with Voting (OMCV) method for readability assessment of Chinese documents, which solves the problem by using multiple binary classifiers and voting the classification results to get an ordinal classification. We also design five groups of features to explore the peculiarities of Chinese, so that the performance of OMCV can be improved. Based on the Chinese primary school language textbook dataset, the experimental results demonstrate the usefulness of both OMCV and the designed text features.

The rest of the paper is organized as follows. Section 2 introduces the background of readability assessment, along with some related work. Section 3 gives our proposed method and the designed text features for Chinese documents. Section 4 presents the experimental studies, and finally Section 5 concludes the paper with the future work.

2 Background

2.1 Readability Assessment

In this paper, we define the readability assessment problem as how to determine the reading level of a document according to the grading levels available, e.g. in textbooks. Reading levels are convenient to rank the reading difficulty of documents [4], and the grading levels of textbooks [9] cover a sufficient large range of reading difficulties, e.g. in China, from primary school to postgraduate. To be more specific, the following gives the formal definition:

Given n ascending reading levels, $L_1, L_2, \ldots, L_n$, and a document whose reading level is unknown. The objective is to determine the most suitable reading level for it.

It should be noted that, assessing reading levels is a subjective judgement, since it should take into account the interaction between the documents and the readers. Above definition makes a trial to give an objective handling for readability assessment. To be more specific, judging a document to be in level L_i means the document is more difficult to read than those in levels less than L_i, but easier to read than those in levels greater than L_i. This implies that the assessment problem shall be viewed as an ordinal classification problem.

2.2 Related Work

The researches on readability assessment have a relatively long history from the beginning of the last century, and are mostly for the English documents. At an early stage, many readability formulae have been developed to measure the lexical and grammatical complexity of a document [6]. The Flesch-Kincaid formula [14] and the SMOG [17] are examples of the readability formulae still extensively used.

Recently, inspired by the achievements of natural language processing and machine learning techniques, many new methods have been developed for readability assessment. For example, Collins-Thompson and Callan [3] used statistical language models to estimate the reading level of a document based on the generation probabilities. Schwarm and Ostendorf [19], Feng et al. [5], and Chen et al. [2] built various SVMs (Support Vector Machines) to classify the readability of documents. Heilman et al. [11], and François [7] suggested that the measurement scale of reading levels should be ordinal [21] and used the proportional odds model, which could capture the ordinal relationship among the reading levels. Frank and Hall [8] transformed the k-class ordinal problem to $k - 1$ binary class problems, and determined the suitable class by computed probabilities. Li[16] used the ECOC (Error Correcting Output Coding) framework to handle conflicts among the results of multiple binary classifiers, in case no ordinal relationship existed among the multiple classes.

Besides the assessment methods, new valuable text features form another important research area for readability assessment. For example, Schwarm and Ostendorf [19] used the parse tree based features; Schwarm and Ostendorf [19], and Kate et al. [13] applied the language models to compute text features for a document; Feng et al. [5] used part of speech features; and Islam et al. [12] applied information theory to compute text features. It should be noted that, for documents in languages other than English, appropriate text features usually play an important role to improve the performance of the assessment. For example, Hancke et al. [10] developed text features that made use of the rich morphology of German; Sinha et al. [20] computed features to suit the Indian language; Islam et al. [12] developed the lexical features for the Bangla language; and for Chinese documents, Chen et al. [2] computed features to capture the semantic similarity among nouns of high frequent occurrence.

3 Our Method

3.1 An Ordinal Multi-class Classification Method

The readability assessment problem can be seen as an ordinal multi-class classification problem. We propose an ordinal classification method which can incorporate multiple binary classifiers, and vote the results to catch the ordinal relationships among the reading levels.

To fulfil the purpose, the Ordinal Multi-class Classification with Voting (OMCV) method is used. Given n ascending reading levels $L_1, L_2, \ldots, L_n$, $n - 1$ binary classifiers are trained first. Any classification technique can be used to build the classifiers, such as Naive Bayesian or Support Vector Machine. For the i-th classifier, the training set is divided into two classes: one contains the documents with reading levels less than or equal to L_i, relabelled as 0, the other contains documents with reading levels greater than L_i, relabelled as 1. Hence, by applying the $n - 1$ binary classifiers, each target document will get $n - 1$ classification results, denoted as $\{c_1, c_2, \ldots, c_{n-1}\}$. After that, a voter is designed to get the final result.

To get a proper reading level for the target document, and catch the ordinal nature of reading levels, we propose four voters (Voter1~Voter4) to vote the results of $n - 1$ classifiers. Figure 1 shows the decision processes of Voter1 $\sim$ 4 respectively.

Voter1: For each target document, make the choice along the $n-1$ results ascendingly from the 1-st to the $(n - 1)$-th. If the current i-th result is 0, then the decided reading level will be L_i, otherwise, transfer to the $(i + 1)$-th result. The final reading level of the document will be L_n if the result of the last classifier is 1.

Voter2: For each target document, make the choice along the $n - 1$ results decendingly from the $(n - 1)$-th to the 1-st. If the current i-th result is 1, then the decided

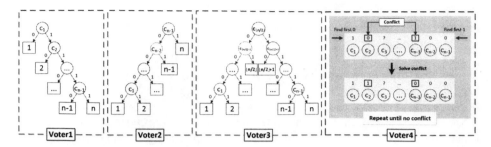

Fig. 1. The voter of the ordinal method

reading level will be L_{i+1}, otherwise, transfer to the $(i-1)$-th result. The final reading level of the document will be L_1 if the result of the first classifier is 0.

Voter3: For each target document, start from the median (the $n/2$-th for even n, and $(n-1)/2$-th for odd n), and make the choice along either side according to the median value. If the median value is 0, then make the choice along the left side decendingly like Voter2, otherwise, along the right side ascendingly like Voter1.

Voter4: Above three voters will get different outputs when inconsistency exists among the $n-1$ results, i.e. there exist i, j $(1 \leq i < j \leq n-1)$, where $c_i = 0$ and $c_j = 1$. Such a conflict is resolved by locating the first i and the last j, and then swap the two values. This procedure is repeated until no conflict can be identified. After that, make the choice just like Voter1.

3.2   Text Features for Chinese Documents

To our knowledge, available text features used for the readability assessment of Chinese documents are relatively few. In this paper, we employ features from other languages (e.g. English) to Chinese with necessary adaptation. We also design new features to fit the characteristics of Chinese. The text features for Chinese documents designed in this paper are divided into five groups: surface features, part of speech features, parse tree features, model based features and information theoretic features.

Surface Features. Surface features (denoted as SF) refer to the features which can be acquired by counting the grammatical units in a document, such as characters and words. In this paper, we implement the group of surface features suitable for Chinese, and calculating both the total numbers and the relative numbers of the grammatical units, e.g. per sentence. Examples include the "average number of strokes per character", the "complex character ratio", the "long word ratio", and the "unique character/word ratio" etc. Totally there are 21 surface features collected.

To be specific, the "stroke" which forms the "strokes per Chinese character" feature is a good alternative of the syllable in "syllables per English word", since it can represent the lexical complexity of a Chinese character just like what the syllable do in English. The "complex character" refers to a Chinese character of which the stroke number is more than 15. The "long word" refers to a word in which the number of characters is more than 3, and the "unique character/word" refers to a character/word which appears only once in a document.

Part of Speech Features. Part of speech features (denoted as POS) have been largely used in current researches on readability assessment [5]. It should be noted that part of speech features must be specifically designed for each language, since it may be defined semantically, e.g. in Chinese. As the reading level of a Chinese document increases, the ratio of decorative words usually increases too, hence counting the proportion of these words may be useful for readability assessment. In Chinese, the decorative words can be classified into dozens of distinct types, and the same words may belong to different types in different contexts.

In this paper, we use the ICTCLAS2011 tool[1] to identify the part of speech units (e.g. noun, verb, adjective, adverb, etc.) in a Chinese document, and count both the total numbers of occurrences and the average numbers per sentence as the part of speech features. ICTCLAS2011 tags 22 types of part of speech units, therefore we have 44 features collected in total.

Parse Tree Features. Parse tree features (denoted as PT) are extracted based on the parse tree of each sentence in a document. In this paper, we use the Stanford Parser Tool[15] to build parse trees for a Chinese document. We choose the parse tree nodes, noun phrases (NP), verb phrases (VP) and adjective/adverb phrases (ADJP/ADVP) as the basic types of units. For each type, we count the total number, the average number per sentence, and the relative number per word respectively. Moreover, since a sentence with the relatively high parse tree would have a much more complex syntax, we implement two extra features: the "total number of extra high tree" and the "ratio of extra high tree", where the threshold is 15. We also count the total and average tree heights. Totally we have 16 parse tree features collected.

Model Based Features. Model based features (denoted as MB) refer to those computed by available readability assessment models, such as the readability formula features [19], statistical language model features [19,13,10], and similarity model features [23]. For readability formula features, we use 4 well-known readability formulae, such as the Flesch-Kincaid measure; while the coefficients are re-computed by linear regression, since these formulae are not designed for Chinese. For statistical language model features, we build the smoothed unigram model [3] for each reading level, and for each document, compute its probability of being generated by the model of that level. For similarity model features, we calculate cosine similarity [23] of the target document, represented by a vector of <word, frequency>, to documents of known reading levels. These features are useful in that they may have a high correlation with the reading levels.

Information Theoretic Features. Islam et al. [12] have used the information theoretic features (denoted as IT) to estimate reading levels of documents in Bangla. The basic assumption of applying information theory to readability assessment is that a sentence which has more information will be more difficult to read and understand, and a document with higher ratio of those "difficult" sentences would be assigned a higher reading level. Inspired by Islam's research, we use the entropy based and the Kullback-Leibler divergence-based features for readability assessment of Chinese documents. In this paper, 5 random variables are considered: the character probability, word probability,

---

[1] http://ictclas.org/

character stroke probability, character frequency probability and word frequency probability. For each random variable, one entropy based feature and n Kullback-Leibler divergence-based features corresponding to the n readability levels are calculated.

4   Empirical Analysis

In this section, we conduct experiments based on the collected dataset to investigate the following two research questions:

RQ1: Whether the proposed method (i.e. OMCV) has potential in improving the performance of the state-of-the-art classification and regression models?

RQ2: Among the five groups of features, which group contributes the most to the classification performance for readability assessment of Chinese documents?

4.1   The Corpus Description

To our knowledge, there is no standard dataset available for readability assessment of Chinese documents, hence, we collect Chinese primary school language textbooks as the dataset, which contains 6 reading levels corresponding to 6 distinct grades. To assure the representativeness of the dataset, we combine two public editions of Chinese textbooks, which cover 90% of Chinese primary schools. By combining these two editions and removing the duplicates, we get the final dataset, the details of which are shown in table 1.

Table 1. Statistics of the collected dataset

Grade	Number of documents	Char/doc		Sent/doc	
		Average	Std. Dev.	Average	Std. Dev.
1	96	153.2	95.4	10.7	4.8
2	110	259.6	106.5	16.0	5.6
3	106	424.1	172.7	20.5	8.2
4	108	622.9	267.7	28.1	13.5
5	113	802.0	370.9	36.1	18.5
6	104	902.0	418.6	39.0	17.5

We select 80% of the documents from each grade (i.e. reading level) to build instances for the classification models, and use the rest to compute the model based features, which require extra data to train the models as described in Section 3.2.

In each experiment, cross-validation is used in which 10% randomly held-out instances are used as the test set, and the remaining 90% as the training set. The cross-validation are repeated 100 times to get statistically confident results.

4.2   Metrics

Three standard metrics which are commonly used in previous researches, are used to evaluate the performance of readability assessment. The three metrics are described as follows:

Accuracy (Acc): This metric is a measure of the ratio of the number of documents being predicted as the correct reading level to the total number of documents.

Adjacent Accuracy ($\pm$Acc): This metric is a measure of the ratio of the number of documents being predicted as the correct reading level or the adjacent levels within distance 1 to the total number of documents.

Pearson's Correlation Coefficient (ρ): This metric is a measure of how the trends of predicted levels match the trends of the documents' real level.

We measure the three metrics (Acc, $\pm$Acc and ρ) across the results of 100 validations, and consider both the mean value and the variance to count in the randomness.

4.3 Performance of the Ordinal Methods

To address RQ1, we select five base classifiers which are implemented in the scikit-learn library for Python[18], including support vector machine (SVM), decision tree (DT), nearest neighbor (NN), linear regression (LR) and logistic regression (LoR). Each of these base classifiers can be incorporated into our method (OMCV) to form the ordinal classifiers with one of the four voters, namely $OSVM\_V_i$, ODT_V_i, ONN_V_i, OLR_V_i and $OLoR\_V_i$, or $OMCV_i$ if the base classifier is not specific, where i can be $1, 2, 3, 4$ meaning Voter$i$ being used. To study the performance changes by applying OMCV, we run each of the classifiers respectively, and compare the performance measures among the base classifiers and the ordinal ones with different voters. The experimental results are shown in Table 2.

As shown in Table 2, values marked in bold refer to the maximum (best) measure in each row. It can be seen that OMCV usually improves the performance of the base classifiers on all three metrics. For example, for SVM, the mean Acc, $\pm$Acc and ρ measures are improved from 46.38%, 87.69% and 0.8149 to 47.94%, 89.27% and 0.8236 by $OSVM\_V_4$; and for NN, the three measures are improved from 45.48%, 86.73% and 0.8159 to 47.44%, 92.15% and 0.8452 by ONN_V_4. The only exception is the adjacent accuracy measure of LR. Among the four voters, the performance measures are nearly similar, e.g. identical in ONN. In most cases, $OMCV_4$ gets the best measure, which

Table 2. Performance changes provided by OMCV

Classifier	Metric	Base	OMCV$_1$	OMCV$_2$	OMCV$_3$	OMCV$_4$
	Acc	46.38	47.73	47.90	47.69	**47.94**
SVM	$\pm$Acc	87.69	88.92	89.21	89.08	**89.27**
	ρ	0.8149	0.8171	**0.8237**	0.8217	0.8236
	Acc	43.85	47.54	47.54	47.56	**48.04**
DT	$\pm$Acc	86.75	89.96	90.15	90.02	**90.40**
	ρ	0.8124	0.8298	0.8302	0.8307	**0.8332**
	Acc	45.48	47.44	47.44	47.44	**47.44**
NN	$\pm$Acc	86.73	92.15	92.15	92.15	**92.15**
	ρ	0.8159	0.8452	0.8452	0.8452	**0.8452**
	Acc	42.85	**46.25**	45.25	45.44	44.48
LR	$\pm$Acc	**89.71**	86.13	84.98	85.08	86.79
	ρ	0.7898	0.8063	0.7674	0.8066	**0.8265**
	Acc	46.08	48.81	49.04	49.06	**49.52**
LoR	$\pm$Acc	83.15	89.87	90.02	90.15	**90.42**
	ρ	0.8021	0.8480	0.8483	0.8506	**0.8540**

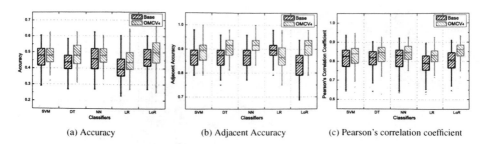

| (a) Accuracy | (b) Adjacent Accuracy | (c) Pearson's correlation coefficient |

Fig. 2. The box-plots of the three metrics measured from the 100 times cross-validations

means Voter4 is the one which has the greatest and relatively stable improvement on the performance of the base classifiers. During the experiments, among all the classifiers, according to the three metrics, the best base classifier is SVM, while the best ordinal one is $OLoR\_V_4$.

To observe the stability of the performance of classifiers, we draw box-plots on each of the three metrics (Acc, $\pm$Acc and ρ) measured from the 100-times cross-validations using each classifier. Due to space reasons, we only list the box-plots for the base classifiers and the corresponding $OMCV_4$s, which appear to be the best in Table 2.

As shown in Fig. 2-(a), the midhinge (i.e. the average of the first and third quartiles) of each $OMCV_4$ is greater than that of the corresponding base classifier. For example, the midhinge of DT improves from 44% to 48% by ODT_V_4. The interquartile range (i.e. the difference between the first and third quartiles) of each $OMCV_4$ is same as or smaller than that of the corresponding base classifier. For example, the interquartile range of NN improves from 11% to 8% by ONN_V_4. The only exception is in DT. We can say $OMCV_4$ brings a stable improvement on Acc to the base classifiers. The same result can be obtained by analyzing the other two metrics shown in Fig. 2-(b) and Fig. 2-(c). On $\pm$Acc, $OMCV_4$ can significantly improve the performance of the base classifiers, except in LR, where the interquartile range is worse. On Pearson's ρ, $OMCV_4$ makes the performance improvement more stable since all the interquartile ranges become smaller.

Besides OMCV, the simple ordinal classification method (SOC) and the proportional odds model (POM) are also ordinal classification methods. The former is proposed by Frank and Hall [8], while the latter was used for readability assessment by Heilman et al. [11], and François [7]. Both methods incorporate the idea of binary partition. For SOC, C4.5 decision tree is used as the base classifier since Frank and Hall[8] chose it in their work and it had good performances in our experiments, and for POM, the base classifier is logistic regression. OMCV can make use of multiple classifiers, and the voting mechanism is open for enhancement. To see which one can get better performance for readability assessment of Chinese documents, we perform the 100-times cross-validations using SOC^2, POM^3 and ODT_V_4, $OLoR\_V_4$ respectively. Table 3 shows the performance measure on the three metrics. From Table 3, on all the three metrics, ODT_V_4 outperforms SOC, while $OLoR\_V_4$ outperforms POM, although the base classifier is

---

[2] http://www.cs.waikato.ac.nz/ml/weka/

[3] http://cran.r-project.org/web/packages/MASS/index.html

Table 3. Performance comparison among SOC, POM and $OMCV_4$

Classifier	Acc	±Acc	ρ
SOC	43.02	86.46	0.7888
POM	47.65	88.31	0.8332
ODT_V_4	48.04	90.40	0.8332
$OLoR\_V_4$	49.52	90.42	0.8540

similar in either case. One possible reason is that C4.5 decision tree and logistic regression may not be the best choice of the base classifier for SOC and POM respectively, at least for readability assessment of Chinese.

In summary, for RQ1, we can conclude that OMCV has potential in improving the performance of the state-of-the-art classification and regression models, and can outperform the simple ordinal classification method and the proportional odds model for readability assessment of Chinese.

4.4   Contributions of the Features

For RQ2, we design experiments to evaluate the individual contribution of each of the five groups of features described in Section 3.2. By the way, to demonstrate the effects of the new or Chinese specific features developed in this paper (e.g. unique character ratio, number of strokes per character, similarity based features, etc.), we make them a new group, denoted as CS. Table 4 shows the number of features in each group respectively, together with the total number of features (denoted as All).

We compare the performance between two schemes of feature employment for each group: one uses the designated group of features by itself, and the other uses all the features except the designated group. Table 4 shows the mean values of the three metrics by running the 100-times validations for each scheme. The ordinal classifier used is $OLoR\_V_4$.

Table 4. Comparison of the performance resulted with or without each group of features

Group	Number of features	By itself			All but it		
		Acc	±Acc	ρ	Acc	±Acc	ρ
SF	21	49.21	91.35	0.8559	52.08	91.15	0.8583
POS	44	45.65	90.50	0.8267	51.42	91.38	0.8613
PT	16	45.92	89.08	0.8297	48.92	90.94	0.8564
MB	16	43.42	87.96	0.8286	49.23	89.77	0.8483
IT	35	47.63	91.25	0.8491	50.33	89.96	0.8495
CS	20	48.85	92.02	0.8456	51.33	91.19	0.8599
All	132	49.52	90.42	0.8540	–	–	–

In Table 4, the last row shows the values resulted from using all the features, which can be viewed as a benchmark. It can be seen that, with all the features used, the resulted performance is not always the best. Such phenomena suggest that some form of feature selection is necessary to improve the performance of readability assessment, which will be part of our future work. Among the five groups of features, the surface features (SF) perform the best on their own, but without them, the remaining features can do better than all (compared with the last row). Although the performance of the parse tree features (PT) is not good on their own, compared with other groups, such as the part of speech features (POS) and the information theoretic features (IT), removing them will decrease the performance on Acc. The same is true for the model based features (MB), where all the three metrics become worse when these features are removed. Moreover, the worst performance got by using MB only may be an under-estimate of the potential of these features, since the poor result may be caused by the insufficient training data used in our experiments.

In addition, the group of the Chinese specific features (i.e. CS) achieves the second best performance by itself according to Acc, and the best according to $\pm$Acc. However, by removing them, the performance is improved on all the three metrics (compared with "All"). This suggests that the features developed here for Chinese are far from enough, and deserve further study.

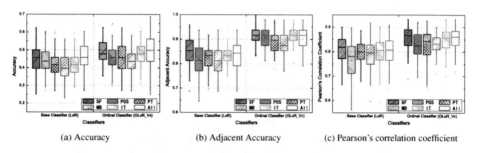

(a) Accuracy   (b) Adjacent Accuracy   (c) Pearson's correlation coefficient

Fig. 3. The box-plots of the three metrics measured by using each group of features

To further study the performance of each group of features, we use both the base classifier and $OMCV_4$ to check on each feature group. The logistic regression (LoR) is used as the base classifier to run the 100-times cross-validations. Figure 3 shows the results in box-plots.

As shown in Fig. 3, according to the midhinges, the surface features (SF) have the best overall performance on all the three metrics, while the model based features (MB) have the worst. The information theoretic features (IT) can also have good enough mid-hinges by using $OLoR\_V_4$. The part of speech features (POS) have smaller variability ranges by using $OLoR\_V_4$ than by LoR. The performance of the parse tree features (PT) is not good since the midhinges are usually less than those of the other features, but the stability is perfectly well by using LoR, since the interquartile ranges are sometimes the smallest.

In summary, for RQ2, we can conclude that all the five groups of features contribute to the performance of the classifiers for readability assessment. The group of surface features does play an important role in estimating the reading level of a document, and the information theoretic features perform the second best on their own. Although not effective by themselves, the parse tree and the part of speech features, which need be specifically designed for each language, cannot be ignored by the classifiers. The model based features may be useful when the standard dataset with sufficient documents of known levels is available, and they also suggest a way of combining the strength of different assessment models and techniques. In addition, the features developed for Chinese in this paper are proved useful, but require further exploration.

5 Conclusion and Future Work

In this paper, we propose an Ordinal Multi-class Classification with Voting (OMCV) method for readability assessment of Chinese documents, which solves the problem by using multiple binary classifiers and voting the classification results to get an ordinal classification. We also design five groups of text features to explore the peculiarities of Chinese documents, so that the performance of classification can be improved. We collect the Chinese primary school language textbook dataset, and conduct experiments to demonstrate the effectiveness of our method. The experimental results show that our method has potential in improving the performance of many state-of-the-art classification and regression models. We also design experiments to evaluate the individual contribution of each of the five groups of features. The results demonstrate that all the five groups of features contribute to the performance of the classifiers for readability assessment.

There are numerous other excellent and useful classifiers which are not discussed in this paper. During our future work, we plan to enhance our method using these classifiers, and expand our dataset to include senior school documents to explore whether our method and the developed feature set can handle a larger set of reading levels. Moreover, we plan to extend the experiments to involve documents of other languages, for example, English, to further demonstrate the potential of our method. For Chinese documents, we plan to develop new text features based on the achievements in natural language processing, and confirm their usefulness in readability assessment.

Acknowledgments. This work is supported by the National NSFC projects under Grant Nos. 61373012, 61321491, and 91218302.

References

1. Benjamin, R.G.: Reconstructing readability: Recent developments and recommendations in the analysis of text difficulty. Educational Psychology Review 24(1), 63–88 (2012)
2. Chen, Y.T., Chen, Y.H., Cheng, Y.C.: Assessing chinese readability using term frequency and lexical chain. In: Computational Linguistics and Chinese Language Processing (2013)
3. Collins-Thompson, K., Callan, J.P.: A language modeling approach to predicting reading difficulty. In: Proceedings of the Human Language Technology Conference of the North American Chapter of the Association for Computational Linguistics, pp. 193–200 (2004)
4. DuBay, W.: Smart language: Readers, readability, and the grading of text. BookSurge Publishing, Charleston (2007)
5. Feng, L., Jansche, M., Huenerfauth, M., Elhadad, N.: A comparison of features for automatic readability assessment. In: Proceedings of the 23rd International Conference on Computational Linguistics: Posters, pp. 276–284. Association for Computational Linguistics (2010)
6. Flesch, R.: A new readability yardstick. Journal of Applied Psychology 32(3), 221 (1948)

7. François, T.L.: Combining a statistical language model with logistic regression to predict the lexical and syntactic difficulty of texts for ffl. In: Proceedings of the 12th Conference of the European Chapter of the Association for Computational Linguistics: Student Research Workshop, pp. 19–27. Association for Computational Linguistics (2009)
8. Frank, E., Hall, M.: A simple approach to ordinal classification. In: Flach, P.A., De Raedt, L. (eds.) ECML 2001. LNCS (LNAI), vol. 2167, pp. 145–156. Springer, Heidelberg (2001)
9. Fry, E.: Readability versus leveling. The Reading Teacher 56(3), 286–291 (2002)
10. Hancke, J., Vajjala, S., Meurers, D.: Readability classification for german using lexical, syntactic, and morphological features. In: Proceedings of the 24th International Conference on Computational Linguistics, pp. 1063–1080 (2012)
11. Heilman, M., Collins-Thompson, K., Eskenazi, M.: An analysis of statistical models and features for reading difficulty prediction. In: Proceedings of the Third Workshop on Innovative Use of NLP for Building Educational Applications, pp. 71–79. Association for Computational Linguistics (2008)
12. Islam, Z., Mehler, A., Rahman, R.: Text readability classification of textbooks of a low-resource language. In: Proceedings of the 26th Pacific Asia Conference on Language, Information and Computation, pp. 545–553 (2012)
13. Kate, R.J., Luo, X., Patwardhan, S., Franz, M., Florian, R., Mooney, R.J., Roukos, S., Welty, C.: Learning to predict readability using diverse linguistic features. In: Proceedings of the 23rd International Conference on Computational Linguistics, pp. 546–554. Association for Computational Linguistics (2010)
14. Kincaid, J.P., Fishburne Jr., R.P., Rogers, R.L., Chissom, B.S.: Derivation of new readability formulas for navy enlisted personnel. Tech. rep., Research Branch Report 8-75 (1975)
15. Levy, R., Manning, C.: Is it harder to parse Chinese, or the Chinese treebank? In: Proceedings of the 41st Annual Meeting on Association for Computational Linguistics, vol. 1, pp. 439–446. Association for Computational Linguistics (2003)
16. Li, B., Vogel, C.: Improving multiclass text classification with error-correcting output coding and sub-class partitions. In: Farzindar, A., Kešelj, V. (eds.) Canadian AI 2010. LNCS, vol. 6085, pp. 4–15. Springer, Heidelberg (2010)
17. McLaughlin, G.H.: Smog grading: A new readability formula. Journal of Reading 12(8), 639–646 (1969)
18. Pedregosa, F., Varoquaux, G., Gramfort, A., Michel, V., Thirion, B., Grisel, O., Blondel, M., Prettenhofer, P., Weiss, R., Dubourg, V., Vanderplas, J., Passos, A., Cournapeau, D., Brucher, M., Perrot, M., Duchesnay, E.: Scikit-learn: Machine learning in Python. Journal of Machine Learning Research 12, 2825–2830 (2011)
19. Schwarm, S.E., Ostendorf, M.: Reading level assessment using support vector machines and statistical language models. In: Proceedings of the 43rd Annual Meeting on Association for Computational Linguistics, pp. 523–530. Association for Computational Linguistics (2005)
20. Sinha, M., Sharma, S., Dasgupta, T., Basu, A.: New readability measures for Bangla and Hindi texts. In: Proceedings of COLING 2012: Posters, pp. 1141–1150 (2012)
21. Stevens, S.S.: On the theory of scales of measurement. Science 103(2684), 677–680 (1946)
22. Vogel, M., Washburne, C.: An objective method of determining grade placement of children's reading material. The Elementary School Journal 28(5), 373–381 (1928)
23. Wan, X., Li, H., Xiao, J.: Eusum: extracting easy-to-understand English summaries for non-native readers. In: Proceedings of the 33rd International ACM SIGIR Conference on Research and Development in Information Retrieval, pp. 491–498. ACM (2010)

Applying Triadic FCA
in Studying Web Usage Behaviors

Sanda Dragoş, Diana Haliţă, Christian Săcărea, and Diana Troancă

Babeş-Bolyai University, Department of Computer Science, Cluj-Napoca, Romania
{sanda,dianat}@cs.ubbcluj.ro, diana.halita@ubbcluj.ro,
csacarea@math.ubbcluj.ro

Abstract. Formal Concept Analysis (FCA) is well known for its features addressing Knowledge Processing and Knowledge Representation as well as offering a reasoning support for understanding the structure of large collections of information and knowledge. This paper aims to introduce a triadic approach to the study of web usage behavior. User dynamics is captured in logs, containing a large variety of data. These logs are then studied using Triadic FCA, the knowledge content being expressed as a collection of triconcepts. Temporal aspects of web usage behavior are considered as conditions in tricontexts, being then expressed as modi in triconcepts. The gained knowledge is then visualized using CIRCOS, a software package for visualizing data and information in a circular layout. This circular layout emphasizes patterns of user dynamics.

1 Introduction

Investigating knowledge structures has a long tradition. In this paper, we propose an approach based on the Conceptual Knowledge Processing paradigm [1]. We use the idea of conceptual landscapes in order to highlight the visual part of organizing knowledge in a format which supports browsing and queries but also critical discourse. The implementation of such a system is thought to be a valuable help for the human expert, organizing knowledge in a way which supports human thinking and decision making.

Conceptual Knowledge Processing is an approach that underlies the constitutive role of thinking, arguing and communicating human being in dealing with knowledge and its processing. The term processing also underlines the fact that gaining or approximating knowledge is a process which should always be conceptual in the above sense. The methods of Conceptual Knowledge Processing have been introduced and discussed by R. Wille in [2], based on the pragmatic philosophy of Ch. S. Peirce, continued by K.-O. Apel and J. Habermas.

The mathematical theory underlying the methods of Conceptual Knowledge Processing is Formal Concept Analysis (FCA), providing a powerful mathematical tool to understand and investigate knowledge, based on a set-theoretical semantics, developing methods for representation, acquisition, and retrieval of knowledge. FCA provides a formalization of the classical understanding of a concept. Knowledge is organized in conceptual structures which are then graphically

R. Buchmann et al. (Eds.): KSEM 2014, LNAI 8793, pp. 73–80, 2014.

represented. These graphical representations are forming the basis for further investigation and reasoning.

In this paper, we apply Formal Concept Analysis to investigate web usage behavior. This study is conducted within a previously built conceptual information system (see Section 3). Herefrom, we select triadic data and compute a set of triadic concepts. These triadic concepts contain all relevant information related to knowledge structures encoded in the selected data set. We also use some derivation operators to process data for our web usage behavior study (see Section 4). In the last part of this paper, we focus on emphasizing patterns of user dynamics and their temporal behaviour using a circular visualization tool.

2   Prerequisites: Triadic Formal Concept Analysis

In the following, we briefly recall some definitions. For more, please refer to [3] and [4]. The fundamental data structure triadic FCA uses is *a tricontext*, which exploits the fact that data might be represented in 3D tables of objects, attributes, and conditions. Hence, a *tricontext* is a quadruple $\mathbb{K} := (K_1, K_2, K_3, Y)$ where K_1, K_2 and K_3 are sets, and Y is a ternary relation between them, i. e., $Y \subseteq K_1 \times K_2 \times K_3$. The elements of K_1, K_2 and K_3 are called (formal) objects, attributes, and conditions, respectively. An element $(g, m, b) \in Y$ is read *object g has attribute m under condition b*. Every tricontext has an underlying conceptual structure reflecting the knowledge which is encoded in the triadic data set. This conceptual structure is described by the so-called triconcepts. In order to mine them, derivation operators are defined ([4]). Every triconcept is a maximal triple (A, B, C), where $A \subseteq K_1, B \subseteq K_2, C \subseteq K_3$ having the property that for every $a \in A, b \in B, c \in C$, we have $(a, b, c) \in Y$.

3   Web Usage Mining and Previous Work

A large amount of collateral information about web usage information is stored in databases or web server logs. Statistics and/or data mining techniques are used to discover and extract useful information from these logs [5].

Web analytics tools are based on some web analytics metrics. They prove to be a proper method to give a rough insight about the analysed web site, especially if it is a commercial site. However, the purpose of an e-commerce site is to sell products, while the purpose of an e-learning site is to offer information. Therefore, a visit on an educational site does not apply to the heuristics used by most analytics instruments [6].

Web usage mining focuses mainly on research, like pattern analysis, system improvement, business intelligence, and usage profiles [7,8]. The process of web usage mining undergoes three phases: preprocessing, pattern discovery, and pattern analysis. At the preprocessing phase data is cleaned, the users and the sessions are identified and the data is prepared for pattern discovery. Such usage analysis constitutes an important feedback for website optimization, but they are also used for web personalization [9,10] and predictions [11].

The web site used for collecting the usage/access data is an e-learning portal called PULSE [12]. The web usage data collected from PULSE was already processed by using Formal Concept Analysis [13], where a detailed description of using ToscanaJ to build a conceptual information system for a previous version of PULSE is given.

The analysis is performed on the data collected from the second semester of the academic year 2012-2013 (i.e., from the beginning of February 2013 to the end of July 2013). A log system records all PULSE accesses into a MySQL database. For the analysed time interval there were 40768 PULSE accesses. The data fields from the collected information used in the current investigation are Full request-URI, Referrer URL, Login ID and Timestamp.

The data to be analysed contains 751 distinct request-URIs (i.e., access files), 471 distinct referrers, 130 distinct login IDs and 25798 distinct timestamps. Using the same methodology as in [13], a ToscanaJ conceptual information system has been built over the PULSE log files. Each data field has been scaled and a conceptual scale has been created. The datasets are considered many-valued contexts, the semantics of attributes being expressed by *conceptual scales*.

For this research, we are interested in investigating temporal patterns of web usage behavior within PULSE. Hence, we will restrict our focus only to the access file classes, the referrer classes and the timestamps of the system. This is a natural triadic structure whereof we can extract user dynamics related knowledge structures in form of triadic concepts.

3.1 Access File Classes

The request-URI represents the address of the accessed webpage along with all query information used for that actual request. Although we value the information contained by this field, the granularity of the accessed web pages is too fine for our intent (i.e., there are 751 distinct access file entries in the database). Thus, the accessed webpages have been divided into 9 classes (see Figure 1(a)).

PULSE portal was intended to be used mainly during laboratory sessions for students to access appointed assignments with the related theoretical support or to consult lecture related content. Each user enters PULSE through the HOME page which contains general information such as: lab attendances, marks, evaluation remarks and current announcements. All webpages related to the content described above are grouped into the three main classes named: **Lab**, **Lecture** and **HOME** according with their purpose.

The other 6 classes represent administrative utilities for the teacher grouped into the **TeacherAdm** class and informative sections for the students, such as **FAQ** (i.e., frequently asked questions), **Feedback**, **News**, **Logout** or navigation through the content of the course from previous years of study (**Change**).

3.2 Referrer Classes

Referrer URLs represent the webpage/site from which the current webpage/access file was accessed. The referrers which are outside of PULSE are not used in this current research. The referrers which represent PULSE webpages fall into the same classes as access files do. The access file classes and referrer classes have been scaled nominally and visualized with ToscanaJ in Figure 1(b).

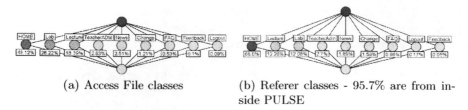

(a) Access File classes

(b) Referer classes - 95.7% are from inside PULSE

Fig. 1. Nominal scales of Access File and Referrer classes

4   Web Usage Mining with Triadic FCA

The extension Toscana2Trias allows the selection of triadic data starting from a given set of scales, if the data has been preprocessed with ToscanaJ. From the conceptual schema file, we have selected the scales presented above and obtained a triadic data set using the pairs Referrer class-Access File class (R_class-AF_class) as attribute set, timestamps as conditions and students Login as object set. Then, we have generated all triconcepts using the Trias algorithm [14].

The problem of visualizing triadic data has not been yet satisfactory solved. Triadic conceptual structures have been visualized for instance using trilattices or graphs. Circos as a visualizing tool has been developed to investigate structural patterns arising in bioinformatics. In this section, we present a proof of concepts in order to show possible applications of using Circos to visualize triadic content.

4.1   Interpreting Triadic FCA Results with Circos

Circos is a software package for visualizing data and information in a circular layout. This circular layout emphasizes patterns in the dataset, showing connections between represented data [15].

The input data for Circos is obtained from the tricontext using a derivation operator. We implemented an algorithm that analyzes the XML output of Trias and transforms it into a valid input for Circos.

The XML file that results as an output from Trias contains all triconcepts which can be derived from the tricontext defined over the data set using Toscana2Trias. Each of them is defined by an extent, an intent and a modus. The valid input data set for Circos is a bidimensional table $R \times C$, with numerical values, hence we have to derive these tables from the tricontext.

Starting from the tricontext (G, M, B, Y), we first build a dyadic projection $\mathbb{K}_{32} := (G, (B, M), I)$, where $(g, (b, m)) \in I \Leftrightarrow (g, m, b) \in Y$. Then, for each pair (b, m) we compute the corresponding attribute concept $\mu_{\mathbb{K}_{32}}$ and determine the cardinality of its extent $(b, m)'$.

The set of column indicators, denoted C, is the set obtained by projecting the ternary incidence relation Y on M, $\mathrm{pr}_M(Y) := \{m \in M \mid \exists(g,b) \in G \times C. \ (g,m,b) \in Y\}$. Similarly, the set of row indicators, denoted R, is the set obtaining by projecting Y on the set of conditions B.

The algorithm we have implemented builds a table having these sets as column and row indicators and calculates the numerical values of the table as follows. For each pair $(c,r) \in C \times R$, the cardinality of the extent $(c,r)'$ in $\mathbb{K}_{32}$ is computed directly from the XML output of Trias. This cardinality represents the numerical value of the cell corresponding to the column c and the row r.

As a final step, we visualize our data by running Circos and obtain an output in png or svg format. Figure 2 presents web usage of the student group "ar" on the 10th week as described in Section 5. Each ribbon corresponds to a pair (Referrer class, Access File class). Because the set of referrer classes and the set of access file classes have elements in common, the sets C and R are not disjoint.

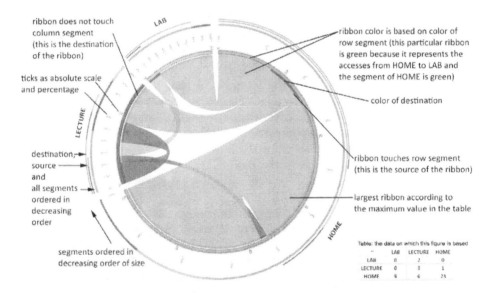

Fig. 2. Results for the "ar" students on the 10th week of school

5   Test Results

The data used was gathered during an entire semester and because of the high volume of data, one single circular representation did not reveal any useful patterns. Therefore, we reduced the volume of data and aggregated the object set (i.e., individual student login) into the set of student groups, as distributed according to their curricula. We continued our tests treating each group of students separately, considering R_classes as objects set, AF_classes as attribute set and timestamps as conditions. In order to investigate the temporal behaviour of students during one semester, we analysed the data on time intervals.

The first time granule we have considered was approximately one third of a semester (beginning, middle and the end). Such time intervals do not provide any significant patterns, and therefore we fine-tuned the time granule to a week.

The PULSE portal recorded data on two courses for the semester we have considered: Operating Systems (SO1), which is a compulsory course and Web Design Optimisation (WDO), which is an elective course. Two student groups enrolled in the SO1 course, denoted "ar" and "ri". For the WDO course, students from five different groups enrolled. WDO being an elective course, some of the student groups were poorly represented, hence we studied the behaviour of two of these student groups, denoted "ei" and "ie".

The entire set of results are posted at http://www.cs.ubbcluj.ro/~fca/ksemtests-2014/. We observed from these results that there are three types of behavior, which we named: relaxed, intense and normal.

The relaxed behaviour occurs mainly during holiday (e.g., the 10th week). The pattern for this type of behavior is depicted in Figure 2 and can be distinguished by the fewer accesses and the reduced number of Access File classes visited (e.g., usually only the main classes). The navigational patterns observed during this week were really simple. For instance for the "ar" group of students went from HOME to either LAB or LECTURE; from LAB they went to LECTURE, and from LECTURE they went back to HOME. For the elective course (i.e., WDO) the results show more relaxed patterns due to the fact the this type of course implies personal research. Therefore, the teaching material provided is less visited than in the case of the compulsory course (i.e., SO1). This type of behavior occurs also after final exams or between the final exam and the re-examination: 18th and 20th week for group "ar", 18th and 19th week for group "ri" and after the 14th week for groups "ei" and "ie".

The intense behavior occurs during examination periods. The pattern depicted in Figure 3(a) shows an increase number of accesses. The pattern can be observed even in the weeks before the examination, its peek occurring during the week of the exam. It is the case of the 17th week for the "ar" and "ri" groups for the final exam, and the week 19th for "ar" group and week 20th for the "ri" group for re-examination. For the elective course (i.e., groups "ie" and "ei") there are three evaluation dates during the weeks 7th, 9th and 13th.

The normal behavior occurs during the semester when there is no examination period or holiday. The pattern for this type of behaviour, as show in Figure 3(b), is that almost all Access File classes are visited. The three main classes contain the most visited pages. The next most visited class is News. These results are to be expected as PULSE is mainly intented to provide support for laboratory, lectures and to post news.

Although in Figure 3 the two behaviors can appear similar, there are some important differences. The main difference is the fact that during the intense period the webpages from the Lecture class are more visited as the students prepare for the examination, while in the normal period the webpages in the Lab

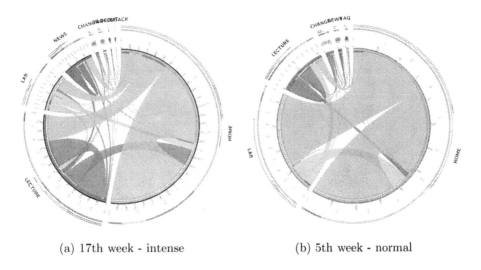

(a) 17th week - intense   (b) 5th week - normal

Fig. 3. Comparative behaviors on "ar" group: intense versus normal

class are most visited as students solve their lab assignments. Another difference is the fact that even if HOME is the most visited class, during the intense period it looks like it represents merely a connection to the other PULSE facilities (i.e., Lecture, Lab, News). The number of accesses is another difference as there are much more accesses during the intense periods.

The triadic conceptual landscape as computed by TRIAS provides a large amount of information that is suitable for a large variety of interpretation/visualization. Histograms are also provided at http://www.cs.ubbcluj.ro/~fca/ksemtests-2014/. This representation however, presents only the quantitative aspect of the navigation meaning the number of accesses. The circular visualization presented so far provides a more qualitative view on the navigational pattern, comprising more details about how and where students navigate.

6   Conclusion and Future Work

The research conducted so far and the previous related work, shows how triadic conceptual landscapes can be used for web usage mining and representation of user dynamics patterns. The main advantage of using conceptual landscape versus different interrogations rely on the completeness of the information clustered in a concept, or determined by a derivation, respectively. Circular visualization tools can be applied to any quantitative data, the qualitative interpretation comes from the conceptual preprocessing.

For further research, we will apply the methods of Temporal Concept Analysis and develop them for the triadic setting with the aim to describe user trails, life tracks and bundles of trails and tracks.

References

1. Wille, R.: Conceptual Landscapes of Knowledge: a Pragmatic Paradigm for Knowledge Processing. In: Gaul, W., Locarek-Junge, H. (eds.) Classification in the Information Age, Proceedings of the 22nd Annual Gfki Conference, Dresden, March 4-6, pp. 344–356 (1998)
2. Wille, R.: Methods of Conceptual Knowledge Processing. In: Missaoui, R., Schmidt, J. (eds.) ICFCA 2006. LNCS (LNAI), vol. 3874, pp. 1–29. Springer, Heidelberg (2006)
3. Ganter, B., Wille, R.: Formal Concept Analysis. Mathematical Foundations. Springer, Heidelberg (1999)
4. Lehmann, F., Wille, R.: A Triadic Approach to Formal Concept Analysis. In: Ellis, G., Rich, W., Levinson, R., Sowa, J.F. (eds.) ICCS 1995. LNCS, vol. 954, pp. 32–43. Springer, Heidelberg (1995)
5. Kosala, R., Blockeel, H.: Web Mining Research: A survey. ACM SIGKKD Explorations Newsletter 2, 1–15 (2000)
6. Dragoş S.: Why Google Analytics Can Not Be Used For Educational Web Content. In: Abraham A. et al (eds.) 7th International Conference on Next Generation Web Services Practices (NWeSP), pp. 113–118. IEEE (2011)
7. Spiliopoulou, M., Faulstich, L.C.: WUM: A Tool for Web Utilization Analysis. In: Atzeni, P., Mendelzon, A.O., Mecca, G. (eds.) WebDB 1998. LNCS, vol. 1590, pp. 184–203. Springer, Heidelberg (1999)
8. Srivastava, J., Cooley, R., Deshpande, M., Pang-Ning, T.: Web Usage Mining: Discovery and Applications of Usage Patterns from Web Data. ACM SIGKDD Explorations Newsletter 1(2), 12–23 (2000)
9. Eirinaki, M., Vazirgiannis, M.: Web Mining for Web Personalization. ACM Transactions on Internet Technology (TOIT) 3(1), 1–27 (2003)
10. Romero, C., Ventura, S., Zafra, A., Bra, P.D.: Applying Web Usage Mining for Personalizing Hyperlinks in Web-Based Adaptive Educational Systems. Computers & Education 53, 828–840 (2009)
11. Romero, C., Espejo, P.G., Zafra, A., Romero, J.R., Ventura, S.: Web Usage Mining for Predicting Final Marks of Students That Use Moodle Courses. Computer Applications in Engineering Education 21, 135–146 (2013)
12. Dragoş, S.: PULSE Extended. In: The Fourth International Conference on Internet and Web Applications and Services, Venice/Mestre, Italy, pp. 510–515. IEEE Computer Society (May 2009)
13. Dragoş, S., Săcărea, C.: Analysing the Usage of Pulse Portal with Formal Concept Analysis. In: Studia Universitatis Babes-Bolyai Series Informatica, vol. LVII, pp. 65–75 (2012)
14. Jaeschke, R., Hotho, A., Schmitz, C., Ganter, B., Stumme, G.: Trias - An Algorithm for Mining Iceberg Trilattices. In: Proceedings of the IEEE International Conference on Data Mining, Hong Kong, pp. 907–911. IEEE Computer Society (2006)
15. Circos, a circular visualization tool, http://www.circos.ca

Verification Based on Hyponymy Hierarchical Characteristics for Web-Based Hyponymy Discovery

Lili Mou, Ge Li*, Zhi Jin, and Lu Zhang

Software Institute, School of EECS, Peking University, Beijing, P.R. China
Key Laboratory of High Confidence Software Technologies, Peking University,
Ministry of Education, Beijing, P.R. China
{moull12,lige,zhijin,zhanglu}@sei.pku.edu.cn

Abstract. Hyponymy relations are the skeleton of an ontology, which is widely used in information retrieval, natural language processing, etc. Traditional hyponymy construction by domain experts is labor-consuming, and may also suffer from sparseness. With the rapid development of the Internet, automatic hyponymy acquisition from the web has become a hot research topic. However, due to the polysemous terms and casual expressions on the web, a large number of irrelevant or incorrect terms will be inevitably extracted and introduced to the results during the automatic discovering process. Thus the automatic web-based methods will probably fail because of the large number of irrelevant terms. This paper presents a novel approach of web-based hyponymy discovery, where we propose a term verification method based on hyponymy hierarchical characteristics. In this way, irrelevant and incorrect terms can be rejected effectively. The experimental results show that our approach can discover large number of cohesive relations automatically with high precision.

Keywords: Ontology, hyponymy learning, term verification.

1 Introduction

Ontologies are a formal model to represent domain knowledge, which is widely used in information retrieval, natural language processing and various other applications [1]. Hyponymy is the skeleton of a domain ontology, expressing "is-a-kind-of" or "is-a-subclass-of" relations [2]. For example, "Java" is a kind of "programming language" (PL), denoted as "Java $\sqsubseteq$ PL". "Java" is the hyponym while "PL" is the hypernym.

Traditional hyponymy relations are often constructed by domain experts, which is a labor- and time- consuming task. Moreover, manually-constructed knowledge is typically suffering from severe data sparseness. As reported in Section 6, the hyponymy relations in WordNet under certain root concepts cover only 3%-5% of what we have discovered.

In 1992, Hearst proposed an automatic hyponymy acquisition approach [3]. Given a large corpus, lexical-syntactic patterns (e.g. "A such as B, C") are used to extract hyponymy $B, C \sqsubseteq A$. Based on Hearst's work, many other researchers proposed approaches to discover more and more hyponyms in a bootstrapping framework [4]. Pattern learning approaches are also proposed to discover lexical-syntactic patterns as well

---

* Corresponding author.

R. Buchmann et al. (Eds.): KSEM 2014, LNAI 8793, pp. 81–92, 2014.

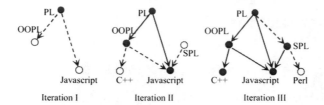

Fig. 1. The iterative hyponymy discovering process. Black dots and solid lines are existing terms and relations; blank dots and dashed lines are newly selected terms and relations in each iteration.

as hyponyms [5]. However, the hyponymy relations acquired in this fashion are typically loose and low-coupling because they are not confined in a certain domain, which fail to form the skeleton of a domain ontology.

To overcome this problem, a naïve and direct approach is to search on the Internet and discover hyponymy iteratively from a starting concept. Figure 1 illustrates the iterative searching process. Given the root concept, say programming language (PL), we search on the search engine, and use lexical-syntactic patterns to extract hyponyms of PL, e.g., "javscript", "object-oriented PL" (OOPL). Then we extract both hyponyms and hypernyms of the newly extracted concepts and we obtain "C++", "scripting programming language". So on and so forth, we acquire the "cohesive" hyponymy relation under PL, which form the skeleton of the domain ontology.

However, in such an iterative process, "irrelevant term explosion" is a common problem, where the number of irrelevant terms grows exponentially. When extracting hyponyms and hypernyms, irrelevant terms might be included because of polysemous terms and casual expressions. Even if very few mis-matched or irrelevant terms are extracted during one iteration, they will introduce even more other irrelevant terms in the next iteration, so on and so forth, so that the system will soon break down within one or two iterations.

In this paper, we propose a term verification method based on *Hyponymy Hierarchical Characteristics* (HHC). As we know, hyponymy is a specific kind of relation between two terms. A complete, well-formed hyponymy hierarchy has some inherent characteristics, e.g., hyponymy transitivity. These characteristics can be quantified and used to evaluate the degree to which a hyponymy hierarchy is well-formed. For a growing hyponymy hierarchy, its characteristics tend to be satisfied to a large degree if the newly discovered terms are probably correct. On the contrary, if an irrelevant or incorrect term is introduced, the characteristics are likely violated to a large degree. So, by quantifying these characteristics and using some appropriate thresholds, we can judge whether a newly discovered term should be included or not. Therefore, irrelevant terms can be removed as soon as possible during the iterative searching process.

2   Related Work

Hearst proposed the "such as" pattern to extract hyponymy in [3]. Z. Kozareva raised the doubly anchored pattern (DAP) for instance discovery in [6]. In the pattern "NP such as NP_1 and X", they specify at least one instance NP_1 in advance. Then they discover other instances by applying the pattern iteratively. H. Eduard et al. use the backward doubly-anchored pattern (DAP^{-1}) [4] to iteratively extract instances and intermediate concepts. Sanchez extracts immediate anterior word as an classification method [7], e.g., lung cancer $\sqsubseteq$ cancer. R. Snow et al. discover lexical-patterns and hyponymy relations iteratively in [5]. They use patterns to extract relations, and relations to extract patterns iteratively.

Many researchers focus on removing irrelevant terms during their hyponymy discovery. Co-occurrence is a common technique to judge the relevance between two terms [8,9,7]. [10] uses content words for domain filtering. [11] extracts instances from free texts coupled with semi-structured html pages. In [12,5], a group of patterns are used to determine whether a relation holds.

In our previous work [13], we proposed a term verification method based on text classification, where a training corpus should be provided in advance. In [14], we considered hyponymy transitivity. In this paper, we extend the method in [14] and semantically verify the candidate results by three hyponymy hierarchical characteristics. With irrelevant terms removed effectively, the system can search iteratively and automatically.

3   Overview of Our Approach

Our approach of "cohesive" hyponymy discovery is an iterative process. Given a root concept, we do web searching and pattern matching to extract candidate hyponyms and/or hypernyms. Then HHCs are used to judge whether the candidates are relevant to the root. Selected candidates are then searched iteratively. Figure 2 illustrates the overview of our approach and the main steps are listed as follows.

Step 1 We first specify a root term r.

Step 2 We construct a search query by a lexical-syntactic pattern and r. The query is sent to a search engine. Web snippets returned by the search engine are collected.

Step 3 By matching the lexical-syntactic pattern "A is a|an B", noun phrases A, B are extracted as candidate hyponyms.

Step 4 Since a noun phrase is not necessarily a term which represents a common concept in the human mind (e.g., "a well-designed PL"), we verify whether the noun phrases extracted in Step 3 are terms with a pragmatic method. If A appears in a pattern "A is a|an", then A is regarded as at term.

Step 5 Three types of hyponymy hierarchical characteristics are used to verify whether a term is relevant to r. Irrelevant terms are removed.

Step 6 We search the newly selected terms for both hyponyms and hypernyms on the web in the next iteration. (Go to Step 2.)

In Section 4, we introduce HHCs in detail. Based on these characteristics we design our candidate hyponymy verifying method in Section 5.

4   Hyponymy Hierarchical Characteristics

As mentioned in Section 1, the cohesive hyponymy relations under a root term have some inherent characteristics which can be for term verification. In this paper, we

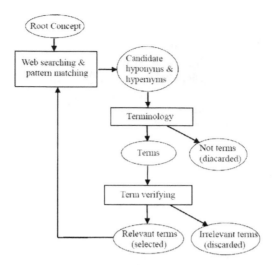

Fig. 2. Overview of our approach

propose three types of *Hyponymy Hierarchical Characteristics* (HHC), namely (1) *hyponymy transitivity characteristic*, (2) *hyponym-hypernym ratio decreasing characteristic* and, (3) *hyponym overlapping characteristic*.

A complete, well-formed hyponymy hierarchy will exhibit these hyponymy hierarchical characteristics, which can be used to evaluate the degree to which a growing hyponymy hierarchy is well-formed. The characteristics of a growing hyponymy hierarchy tends to be satisfied if the newly discovered term is probably correct. On the contrary, if an incorrect or irrelevant term is introduced to the growing hyponymy hierarchy, these characteristics tend to be violated to a large degree. We quantify these characteristics, and use thresholds to judge whether a candidate term should be included or not, introduced in detail in the rest part of this section.

4.1 Hyponymy Transitivity Characteristic

As we know, hyponymy is a transitive relation. For terms a, b, c, if $a \sqsubseteq b$ and $b \sqsubseteq c$, it can be inferred that $a \sqsubseteq c$. We call this *Hyponymy Transitivity Characteristic* (HTC).

Since the web contains massive information, these semantically redundant expressions all may stated explicitly on the web, and they can be cross-validated by each other.

In particular, if we want to get hyponyms or hypernyms of a term t, we first get candidate hyponyms and hypernyms of t. For each candidate c (either $c \sqsubseteq t$ or $t \sqsubseteq c$), we search potential hyponyms and hypernyms of c. In this case, hyponymy transitivity can be used to verify c. There are four typical scenarios when we verify c. Whether c is *cross-validated* by HTC in each scenario is shown in Figure 3 and explained as follows.

Scenario 1: If c is a candidate hyponym of t, and if c is the hyponym or hypernym of any one of other already selected terms (t excluded), then c is regarded as a relevant term since it is cross-validated by other hyponymy relations.

Scenario 2: If c is a candidate hypernym of t, and c is a hyponym of some other select terms, then c is also cross-validated.

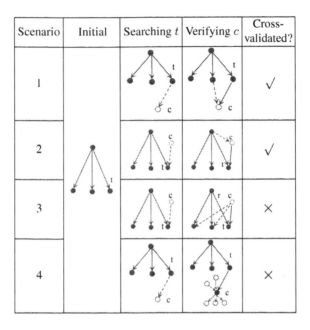

Scenario	Initial	Searching t	Verifying c	Cross-validated?
1				✓
2				✓
3				✗
4				✗

Fig. 3. Scenarios when verifying a candidate term by HTC

Scenario 3: In this scenario, c is a candidate hypernym of t, and some other selected terms are the hyponym of c. If no other hyponymy relations can infer $c \sqsubseteq r$, then c is not cross-validated.

Scenario 4: If no hyponyms and hypernyms of c have relations with any selected term (t excluded), then c is not cross-validated.

4.2 Hyponymy-Hypernym Ratio Decreasing Characteristic (R_{HH}DC)

We define *Hyponym-Hypernym Ratio* (R_{HH}) as follows[1].

$$R_{HH}(t) = \frac{\#\text{Hyponym}(t) + 1}{\#\text{Hypernym}(t) + 1}$$

where adding one in the equation is a widely used smoothing technique used in many NLP tasks [15].

For two terms a, b, If $a \sqsubseteq b$, then $R_{HH}(a) < R_{HH}(b)$. This is called *Hyponymy-Hypernym Ratio Decreasing Characteristic* (R_{HH}DC).

The above fact can be proven by the definition of hyponymy relation. If $a \sqsubseteq b$, b inherits all the hyponyms of a. So b has more hyponyms than a. Likewise, a has more hypernyms than b. Therefore $R_{HH}(a) < R_{HH}(b)$.

Though we cannot get the entire hyponymy relations, the phenomenon still holds true that when we search a term, a higher level term tends to have more hyponyms and less

---

[1] Hyponym(t) is the hyponyms of t, Hypernym(t) is the hypernyms of t. # refers to the number of elements in a set.

hypernyms, and thus has a higher R_{HH} value. A preset threshold is used to determine whether a candidate term satisfies R_{HH}DC.

Formally, if $c \sqsubseteq t$ and $R_{HH}(c) > t_1 \cdot R_{HH}(t)$, then we say R_{HH}DC is violated, which indicates an anomaly. A typical cause of R_{HH}DC violation is the presence of polysemous terms. For example, "scheme" is a kind of PL. Meanwhile, "scheme" has many different meanings, one of which refers to a schema or an outline. Therefore R_{HH}(scheme) may be much larger than R_{HH}(PL). Terms that violate R_{HH}DC cannot be used for verifying other terms with hyponymy transitivity (introduced in Section 4.1). Otherwise, a large number of irrelevant terms might be mis-validated.

4.3   Hyponym Overlapping Characteristic (HOC)

We define *Hyponym Overlapping Proportion* (HOP) of terms a, b as follows.

$$P_{HO}(a, b) = \frac{\#\{\text{Hyponym}(a) \cap \text{Hyponym}(b)\}}{\#\text{Hyponym}(a)}$$

In a complete, well-formed hyponymy hierarchy, if $a \sqsubseteq b$, then $P_{HO}(a, b)=1$, because Hyponym(a) $\subseteq$ Hyponym(b). Though we cannot extract the complete hyponyms of a and b, $P_{HO}(a, b)$ is probably high if $a \sqsubseteq b$. This is called *Hyponym Overlapping Characteristic* (HOC). A threshold is also used to determine whether HOC is satisfied.

When discovering hyponymy relations under the root term r by iterative searching, we calculate $P_{HO}(c, r)$ to verify a candidate term c. If $P_{HO}(c, r)$ is greater than a preset threshold t_2, HOC is satisfied, which indicates that t might be a relevant term.

However, because we add new terms to hyponyms of r in each iteration, Hyponym(t) $\cap$ Hyponym(r) grows larger naturally during the iterations. P_{HO} is not stable in different iterations. Therefore, we multiply a heuristic regularization factor and extend $P_{HO}(c, r)$ to the following equation.

$$P_{HO}(c, r) = \frac{(N + 1) \ln(N + 1)}{(\#\text{Hyponym}(c) + 1) \ln(S + 1)}$$

Here $N = \#\{\text{Hyponym}(c) \cap \text{Hyponym}(r)\}$, S is the size of the entire selected hyponyms of r. Adding-one smoothing is also applied for P_{HO}.

Hyponymy overlapping characteristic can deal with the situation where those hyponymy relations are not extracted explicitly from the web. For example, "Turing complete PL (TCPL) $\sqsubseteq$ PL" is not stated explicitly on the web. Therefore TCPL cannot be cross-validated by HTC because none of the known terms is stated explicitly as the hypernym of TCPL. However, when we examine the hyponyms of TCPL, we find the majority of them are known hyponyms of PL, e.g., javascript, ruby, matlab etc. So, P_{HO}(TCPL, PL) is large and TCPL is probably a relevant term to PL.

5   Verifying Candidate Terms Based on Hyponymy Hierarchical Characteristics

In this part, we explain our method of candidate term verification based on the three types of HHCs introduced in Section 4.

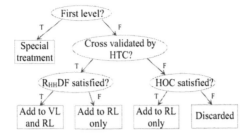

Fig. 4. The decision process of verifying a candidate term based on hyponymy hierarchical characteristics

To implement our semantically verification approach, we need to maintain two lists of terms: Result List (RL) and Cross Validate List (VL). RL is the hyponyms obtained as the result; VL is a subset of RL for which we have higher confidence, and VL is used for cross-validation. These two lists are updated in two iterations.

Figure 4 gives the decision process of candidate term verification. For a candidate term c, if c is cross-validated by HTC with VL (Scenario 1 and Scenario 2 in Subsection 4.1), c is added to RL. Furthermore, if $R_{HH}DC$ constraint of c is satisfied, i.e., $R_{HH}(c) < t_1 \cdot R_{HH}(c$'s hypernym), then c is considered as a relevant term with high confidence, and is added to VL. If $R_{HH}DC$ constraint of c is not satisfied, c is added to RL only since it may indicate that c is a polysemous term like "scheme".

If c is not cross-validated by HTC, but after we obtain the hyponyms of c, if the hyponym overlapping proportion is large, i.e., $HOP(c, r) > t_2$, this may indicate the absence of some hyponymy relations. Therefore we regard c as a relevant term and add c to RL. However, to be on the safe side, c is not added to VL.

For those terms that are neither cross-validated by HTC nor satisfying HOC constraint, they are discarded.

There still remains one problem to solve. Since there are no selected terms in the first iteration, how can we validate candidate terms at the beginning? In our approach, given the root term r, we get candidate hyponym c, and search c on the web. If r is extracted as the hypernym of c, then c is selected. This seems trivial, but can effectively remove some noises caused by html parsing errors, sentence segmenting errors, etc.

6   Experimental Results

We carried out two experiments under the root terms "programming language" and "algorithm". The root terms are assigned respectively by human in advance. Then the system works automatically by our approach.

Our approach needs two parameters t_1 and t_2, which are set to 3 and 0.1 empirically. These two parameters are fixed between the two experiments. The number of iterations is limited to 3.

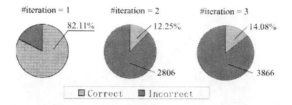

Fig. 5. Percentage of correct candidate terms and number of incorrect terms in each iteration in Experiment I before pruning by HHCs

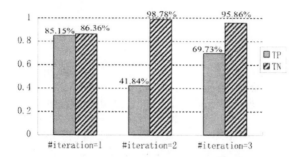

Fig. 6. TP and TN of term verification in each iteration in Experiment I

6.1 Experiment I

In the first experiment, we discovered hyponymy under the root term "programming language." The system has totally accepted 975 terms and discarded 7002 terms. Full sized evaluation is executed on the system-accepted terms; for those terms that are discarded by our system, we randomly sample about 100 terms and report the sampling statistics. Each term is annotated by human, which falls into three categories, namely CORRECT, INCORRECT and BORDERLINE. Those in BORDERLINE category are not counted in our result.

Accuracy of Term Verification. In this part, we evaluate our candidate term verification approach based on hyponymy hierarchical characteristics.

In Figure 5, we present the percentage of the actual correct terms and the number of incorrect terms in each iteration. In the first iteration, it achieves 82.11% accuracy for the candidate terms that are extracted. In the second and third iteration, the accuracy among the 3000-4000 total candidates is only about 13%. Therefore, the "irrelevant term explosion" problem is severe in the iterative hyponymy discovery framework. Selecting correct terms is like gold mining — finding small quantities of the wanted among massive unwanted.

In our approach, candidate terms are verified semantically by hyponymy hierarchical characteristics. True positive rate (TP) and true negative rate (TN) are used to evaluate accuracy. Figure 6 shows TP and TN in the three iterations. Except for the special treatment in the first iteration, TN rate is high (over 95%) in the remaining iterations,

Table 1. Recall against Wikipedia list and TIOBE index

	Our results	Standard			
		Wikipedia List	TIOBE Top 20	TIOBE Top 50	TIOBE Top 100
#	693	645	20	50	100
Recall	-	30.43%	100%	92%	79%

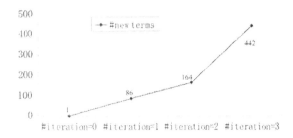

Fig. 7. Number of terms we get in each iteration in Experiment I

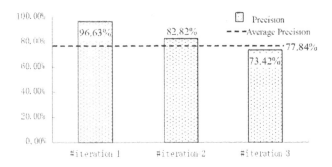

Fig. 8. Precision in each iteration in Experiment I

which ensures the system can work automatically and effectively. TP rate is relatively low at the first glance, since our candidate verifying approach is very strict. One reason for the low TP rate is that some programming languages are only mentioned in one or a few scientific papers, and therefore it cannot be cross-validated by either HTC or HOC. However, on second thought, it seems reasonable since we as human beings can hardly confirm a hyponym of PL if it is mentioned only once on the web. Despite the relatively low TP, the number of selected terms is large and the recall of commonly used PL is also high.

As seen, the iterative process will introduce a huge number of irrelevant terms as candidates. Our term verification method can effectively remove the irrelevant terms before they grow exponentially.

Precision and Recall. In this part, we evaluate precision and recall of our result.

Figure 7 presents the number of new terms in each iteration. Totally, we get 692 correct hyponym terms under the term PL. We compare the number to two open hyponymy

Table 2. Number of hyponyms under "programming language" in WordNet, WordNet+40k, and our result

WordNet	WordNet+40k	Our result
34	77	692

Table 3. Number of hyponyms under "algorithm" in WordNet, WordNet+40k, and our result

WordNet	WordNet+40k	Our result
3	8	223

lexicons. In particular, one is WordNet 2.1[2] ; the other is an automatic enlarging of WordNet by machine learning approaches [5][3], which adds up to 40,000 concepts to WordNet. As shown in Table 2, the number of hyponyms we get is much more significant than the other two related works.

Figure 8 shows the precision during the iterative searching process. The overall precision reaches 77.89%.[4]

Calculating recall is difficult because we can hardly find a complete, authentic gold standard. WordNet is not suitable to be the gold standard since it suffers from severe term sparseness. We find two lists that are suitable to compare our results to.

1. Wikipedia list[5]. Wikipedia reflects the collective wisdom of the online community. Everyone can edit and contribute to Wikipedia so that this list is comparatively complete, containing 645 instances of PL.
2. TIOBE index[6], which lists top 20/50/100 PLs indexed by TIOBE programming community. Comparing to TIOBE index, we can see the recall on common programming languages.

Table 1 shows the number of concepts in the golden standards, and the recall calculated against these standards. As we can see, more than 30% instances of Wikipedia lists are recalled; for the TIOBE top 20/50/100 PL index, our recall is 100%, 92% and 79% respectively, which is also high.

This experiment shows our approach has acquired both high precision (77.89% on average) and high recall against Wikipedia list and TIOBE index.

6.2 Experiment II

In the second experiment, we extract hyponymy relations under the root term "algorithm". The parameters of our algorithm remains unchanged.

Figure 9 presents the TP and TN of our term verification approach. As shown in Figure 10, we totally obtain 379 terms under "algorithm" while the average precision

---

[2] WordNet is available on http://wordnet.princeton.edu/

[3] This work is available on http://ai.stanford.edu/~rion/swn/

[4] Those terms annotated as BOARDERLINE are not counted.

[5] http://en.wikipedia.org/wiki/List_of_programming_languages

[6] http://www.tiobe.com/index.php/content/paperinfo/tpci/
index.html

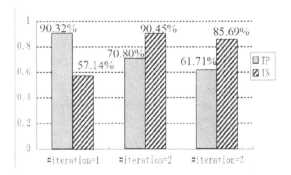

Fig. 9. TP and TR of term verification in each iteration in Experiment II

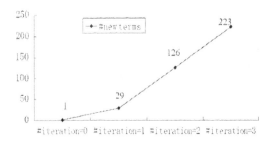

Fig. 10. Number of terms we get in each iteration in Experiment II

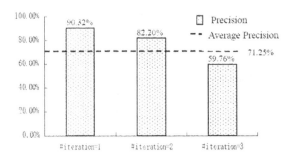

Fig. 11. TP and TN of term relevance verifying in each iteration in Experiment II

is 71.25%, which is also acceptable. The number of hyponyms is also more significant than WordNet and WordNet+40k (See Figure3).

This experiment shows that our approach is general applicable since we do not change the parameters.

From the two experiments, we can see that our approach has extracted a large number of terms with high recall. Incorrect and irrelevant terms can be removed effectively, so that the system can work iteratively and automatically.

7   Conclusion

In this paper, we propose a domain hyponymy discovery approach with term verification based on hyponymy hierarchy characteristics.

Our approach needs very few human supervision. Only a root concept should be given in advance. Then the system can search for domain hyponymy relations automatically and unsupervisedly. At the end of the process, we gain the "cohesive" hyponymy with relatively high precision and recall, which forms the skeleton of the domain ontology.

Acknowledgment. This research is supported by the National Natural Science Foundation of China under Grant Nos. 61232015 and 91318301.

References

1. Clark, M., Kim, M., Kruschwitz, U., Song, D., Albakour, D., Dignum, S., Beresi, U., Fasli, M., Roeck, A.: Automatically structuring domain knowledge from text: An overview of current research. Information Processing and Management 48(3), 552–568 (2012)
2. Buitelaar, P., Cimiano, P., Magnini, B.: Ontology learning from text: methods, evaluation and applications. IOS press (2005)
3. Hearst, M.A.: Automatic acquisition of hyponyms from large text corpora. In: Proceedings of the 14th Conference on Computational Linguistics, vol. 2 (1992)
4. Hovy, E., Kozareva, Z., Riloff, E.: Toward completeness in concept extraction and classification. In: Proceedings of the 2009 Conference on Empirical Methods in Natural Language Processing, vol. 2 (2009)
5. Snow, R., Jurafsky, D., Ng, A.Y.: Learning syntactic patterns for automatic hypernym discovery. In: Advances in Neural Information Processing Systems (2004)
6. Kozareva, Z., Riloff, E., Hovy, E.H.: Semantic class learning from the web with hyponym pattern linkage graphs. In: Proceedings of ACL 2008: HLT (2008)
7. Sanchez, D., Moreno, A.: A methodology for knowledge acquisition from the web. International Journal of Knowledge-Based and Intelligent Engineering Systems (2006)
8. Cimino, P., Steffen, S.: Learning by googling. SIGKDD Explor. Newsl. 6(2) (2004)
9. Sanchez, D.: Learning non-taxonomic relationships from web documents for domain ontology constructions. In: Data and Knowledge Engineering (2008)
10. Navigli, R., Paola, V., Stefano, F.: A graph-based algorithm for inducing lexical tax- onomies from scratch. In: Proceedings of the Twenty-Second International Joint Conference on Artificial Intelligence, vol. 3, pp. 1872–1877 (2011)
11. Andrew, C., Justin, B., Richard, W., Hruschka, J., Estevam, R., Mitchell, T.: Coupled semi-supervised learning for information extraction. In: Proceedings of the third ACM International Conference on Web Search and Data Mining, pp. 101–110 (2010)
12. Pantel, P., Pennacchiotti, M.: Espresso: leveraging generic patterns for automatically harvesting semantic relations. In: Proceedings of the 21st International Conference on Computational Linguistics and the 44th Annual Meeting of the Association for Computational Linguistics (2006)
13. Mou, L.L., Li, G., Jin, Z., Lu, Y.Y., Hao, Y.Y.: Discovering domain concepts and hyponymy relations by text relevance classifying based iterative web searching. In: Proceedings of the 2012 Asian Parcific Software Engineering Conference (2012)
14. Mou, L.L., Li, G., Jin, Z.: Domain hyponymy hierarchy discovery by iterative web searching and inferable semantics based concept selecting. In: Proceedings of the 2013 IEEE 37th Annual Computer Software and Applications Conference, pp. 387–392 (2013)
15. Jurafsky, D., Martin, J.: Speech And Language Processing: An Introduction to Natural Language Processing, Computational Linguistics, and Speech Recognition. Prentice Hall (2009)

Using Standardized Lexical Semantic Knowledge to Measure Similarity

Wafa Wali[1], Bilel Gargouri[1], and Abdelmajid Ben Hamadou[2]

[1] MIR@CL Laboratory, FSEGS, Sfax, Tunisia
{wafa.wali,bilel.gargouri}@fsegs.rnu.tn
[2] MIR@CL Laboratory, ISIMS, Sfax, Tunisia
abdelmajid.benhamadou@isimsf.rnu.tn

Abstract. The issue of sentence semantic similarity is important and essential to many applications of Natural Language Processing. This issue was treated in some frameworks dealing with the similarity between short texts especially with the similarity between sentence pairs. However, the semantic component was paradoxically weak in the proposed methods. In order to address this weakness, we propose in this paper a new method to estimate the semantic sentence similarity based on the LMF ISO-24613 standard. Indeed, LMF provides a fine structure and incorporates an abundance of lexical knowledge which is interconnected together, notably sense knowledge such as semantic predicates, semantic classes, thematic roles and various sense relations. Our method proved to be effective through the applications carried out on the Arabic language. The main reason behind this choice is that an Arabic dictionary which conforms to the LMF standard is at hand within our research team. Experiments on a set of selected sentence pairs demonstrate that the proposed method provides a similarity measure that coincides with human intuition.

Keywords: Lexical semantic knowledge, ISO standard, similarity measure, sense relations, semantic classes, thematic roles.

1 Introduction

Determining semantic similarity measure between sentences is one of the most fundamental and important tasks in the Natural Language Processing (NLP) domain and has a wide use in many text applications such as automatic translation, question answering systems, information extraction, and knowledge acquisition.

Two sentences are considered to be similar if they are a paraphrase of each other; that is, they talk about the same event or idea judging from the common principal actors and actions, or if one sentence is a superset of the other.

As sentence semantic similarity measure is increasingly in demand for a variety of applications, enormous achievements have been made recently in this area; which can be classified into three major groups: syntactic based methods [1], semantic based methods[2, 3] or hybrid methods [4, 5, 6]. These methods, which were proposed previously, typically compute sentence similarity based on the frequency of a word's

R. Buchmann et al. (Eds.): KSEM 2014, LNAI 8793, pp. 93–104, 2014.

occurrence or the co-occurrence between words. Although these methods benefit from the statistical information derived from the corpus, this statistical information is closer to syntactic representation than to semantic representation.

Some authors have used knowledge bases such as WordNet [7] to compute the similarity between two sentences based on synsets. Other authors have highlighted syntactic information -i.e. grammar dependency, common part of speech and word order -to calculate the semantic similarity between sentences. Most of these methods build on statistics corpuses, or lexical databases, such as WordNet, which provide that semantic information.

The main interest of this paper is to propose a novel method for measuring semantic similarity between sentences which takes into account lexical and semantic information. The method measures sentence semantic similarity via the LMF (Lexical Markup Framework) standard [8]. In fact, the LMF incorporates diversified lexical knowledge at the morphological, syntactic and semantic levels. Furthermore, it is finely structured, which facilitates access to information. These lexical pieces of information are interconnected by relationships. For example, the senses can be related by semantic relationships such as synonymy or antonymy. Also, the LMF standard identifies the syntactic-semantic relation. The originality of our approach is the highlighting of synonymy relations, semantic class and thematic role extracted from LMF to compute the sentence similarities.

It is through the applications carried out on the Arabic language that our approach proved to be reliable. Two main reasons are behind this choice. The first one is that research on measuring sentence similarity for the Arabic language is often deficient and the second is that an Arabic dictionary which conforms to the LMF standard is at hand within our research team [9].

The paper is organized as follows. First, we review the works related to our study. Next, we describe the new method for measuring sentence similarity. In section 4, we explain the experimental evaluation, including a presentation of the Arabic LMF standardized dictionary used in this study. Illustrative examples provide experimentation and experiment results that coincide with human perceptions. We summarize our research, draw some conclusions and propose future related works in section 5.

2   Related Works

During the last decade, there has been extensive literature on measuring the semantic similarity between short texts or sentences. In this section, we examine some related works in order to explore the advantages and the limitations of previous methods. To compare the sentences similarity, the semantic and syntactic information is making contributions to the meaning of a sentence. Related works usually consider semantic, Part Of Speech (POS) and syntactic (word order) information or all of them combined to give an overall similarity of two compared sentences.

Mandreoli et al. [1] proposed a method, adopting the Edit Distance as similarity measure between the parts of sentences; it analyzed the sentence contents in order to find similar parts. The disadvantage is that the method essentially focuses attention on the similarity of syntactic structure.

Hatzivassiloglou et al. [2] presented a composite similarity metric of short passages which uses only semantic information. The authors measured the semantic distance between pair of small textual units from multiple linguistic indicators such as word co-occurrence, matching noun phrases, WordNet synonyms, common semantic class for verbs and shared proper nouns.

Mihalcea et al. [3] developed a method to score the semantic similarity of sentences by exploiting the information that can be drawn from the similarity of the component words. Specifically, they used two corpus-based and six knowledge-based measures of word semantic similarity, and combined the results to show how these measures can be used to derive a text-to-text similarity metric, but the syntactic structure was ignored.

Yuhua et al. [4] presented another hybrid method that combines semantic and word order information. The proposed method dynamically forms a joint word set using all the distinct words in the sentences. For each sentence, a raw semantic vector is derived using the WordNet lexical database [7]. Again, a word order vector is associated with each sentence, with the assistance of the lexical database. Since each word in a sentence contributes differently to the meaning of the whole sentence, the significance of a word is weighted by using information content derived from a corpus. By combining the raw semantic vector with the information content from the corpus, a semantic vector is obtained for each of the two sentences. The Semantic similarity is calculated based on the two semantic vectors. An order similarity is calculated using the two order vectors. Finally, the sentence similarity is derived by combining semantic similarity and order similarity.

In [5], Xiaoying et al. measured semantic similarity between sentences with the Dynamic Time Warping (DTW) technique, which takes into account the semantic information, word order and the contribution of different parts of speech in a sentence.

Islam et al. [6] presented a method for measuring the semantic similarity of two texts from the semantic and syntactic information (in terms of common-word order) that they contain. The authors used a corpus-based measure of semantic word similarity and a normalized and modified version of the Longest Common Subsequence (LCS) string matching algorithm. But, the judgment of similarity is situational and depends on time (the information collected in the corpus may not be relevant to the present).

We can see that all of the methods described above exploit insufficiently the sentence information and are not suitable for some applications. However, none of the researchers considered the semantic predicate (semantic arguments) of a sentence as the factors to be considered to calculate sentence similarity. Furthermore, understanding a sentence depends on assigning its constituents to their proper thematic roles. Thus, the idea is to take advantage of the ISO standards proposed in this area, namely LMF. (ISO 24613) [8].

3   The Proposed Method

The proposed method consists essentially to measure semantic similarity between sentences based on lexical and semantic information using LMF [8] in order to have a semantic similarity score total (see Figure1). Before the measuring semantic similarity step between sentence S1 and S2, we perform a pre-processing step which eliminates punctuation signs in order to compare the important lexical units. Also, we should lemmatize the words of the sentence by reducing the words to their canonical forms (lemmas) by deleting all the grammatical inflections and derivations.

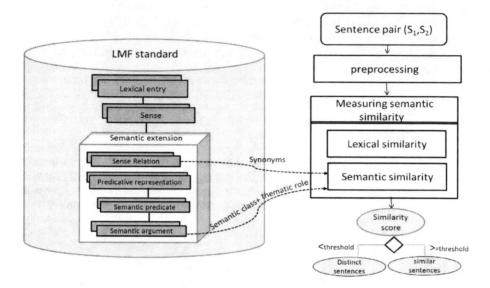

Fig. 1. Sentence semantic similarity computation diagram

Our semantic sentence similarity measure is based on two similarity levels, namely the lexical and the semantic ones using the LMF standard [8]. At the lexical level, we compare the lexical units constituting the sentences in order to extract lexically similar words. To compare the sentences lexically, we relied on the number of common terms between the sentences to compare. To calculate the degree of lexical similarity, which we call $SL_{(S1,S2)}$, using the Jaccard coefficient [10], we used the following formula:

$$SL_{(S1,S2)} = M_C / (M_{S1} + M_{S2} - M_C) \qquad (1)$$

Where:

M_C: the number of common words between the two sentences
M_{S1}: the number of words contained in the sentence S1
M_{S2}: the number of words contained in the sentence S2

In the second similarity level, our method dynamically forms the Semantic Vectors (SV) based on the compared sentences. The number of entries of SV is equal to the number of distinct words in the treated sentences. In fact, we intend to present all the distinct words from S1and S2 in a list called T. For example, if we have the sentences: S1: "RAM keeps things being worked with"; S2: "The CPU uses RAM as a short-term memory store", then we will have after the pre-processing step T = {RAM, to keep, thing, being, to work, CPU, to use, short-term, memory, store}. The process of deriving a semantic vector for S1and S2 is explained as follows: each entry of the semantic vector $SV_i$is determined by the semantic similarity of the corresponding word in T to a word in the sentence. The semantic similarity between words can be determined from one of two cases taking S_1 as an example:

Case1:   if W_i(the word at the position i in T)appears in the sentence S1 then SV_i is set to 1.

Case2: if W_i is not contained in S_1, a semantic similarity score is computed between W_i and each word in S_2, using the LMF standard. Indeed, the LMF model defines many types of semantic relationships (e.g., synonymy, antonymy, etc.) between the senses of two or several lexical entries by means of the SenseRelation class (see Figure2)as opposed to WordNet that defines the semantic relationships between concepts.

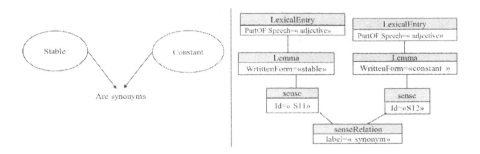

Fig. 2. Synonymy relation between two word senses

Given two words W1 and W2, we need to find the semantic similarity $Sim_{(W1,W2)}$. We can do this by analyzing the relations between word meanings as follows: words are linked into a semantic relation in the LMF standard (in this paper, we are interested in the synonymy relation) and with relation pointers to other synsets. One direct method for words similarity calculation is to find the synonymy set of each word so as to detect the common synonyms between the two words. For example, the common synonyms between the words "stable' and "constant" are "steady" and "firm" as the synonyms of "stable" are {steady, constant, enduring, firm, stabile} while the synonyms of "constant" are {steady, abiding, firm, perpetual, hourly}.

Once the two sets of synonyms for each word have been collected, we calculate the degree of similarity between them using the Jaccard coefficient [10]:

$$Sim_{(W1,W2)}=M_C/(M_{W1}+M_{W2}-M_C) \qquad (2)$$

Where:

M_C: the number of common words between the two synonym sets
M_{W1}: the number of words contained in the W_1 synonym set
M_{W2}: the number of words contained in the W_2 synonym set

Thus, the most similar word in sentence i to the W_i is one that has the highest similarity score δ. If δ exceeds a preset threshold, then S_i=δ; otherwise S_i=0.

From the semantic vectors generated as described above, we compute the degree of semantic similarity that we note $SM_{(S1,S2)}$ using the Cosine similarity.

$$SM_{(S1,S2)}= V1.V2/(\|V1\|*\|V2\|) \qquad (3)$$

Where:
V_1: the semantic vector of sentence S_1
V_2: the semantic vector of sentence S_2

On the other hand, the semantic class and the thematic role for each semantic argument of sentence provide information about the relationships between words. Thereby, the two semantic pieces of information play a role in conveying the meaning of sentences. For example, in the sentence "Susan ate an apple", "Susan" is the doer of the eating, so she is an agent as thematic role and human as semantic class; the apple is the item that is eaten, so it is a patient as thematic role and vegetal as semantic class.   Also, there is a variety of semantic classes and thematic roles described in linguistics. In our method, these bits of information are extracted from the semantic argument class of LMF associated to a semantic predicate (see Figure3).

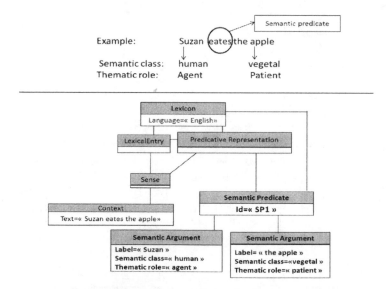

Fig. 3. Illustration of the semantic predicates and the semantic arguments

Indeed, LMF standard presents for each sense many semantic predicates that are interconnected at the syntactic level via the Predicative Representation class. In our method, these semantic predicates are provided by an expert according to the LMF structure. The pairs of semantic arguments are considered similar if they have the same attributes like thematic role and semantic class.

At this level, we calculate the degree of similarity between the two sentences S1 and S2 from the common semantic arguments between the pair of sentences, which we call $SSM_{(S1,S2)}$, using the Jaccard coefficient [10]:

$$SSM_{(S1,S2)} = AS_C/(AS_{S1} + AS_{S2} - AS_C) \tag{4}$$

Where:

AS_C: the number of common semantic arguments between the two sentences
AS_{S1}: the number of semantic arguments contained in the sentence S_1
AS_{S2}: the number of semantic arguments contained in the sentence S_2

The combined sentences similarity represents the overall sentence similarity, which is the sum of the lexical and semantic similarities calculated as follows:

$$Sim(S1,S2) = \lambda * SL_{(S1,S2)} + \beta * SM_{(S1,S2)} + \gamma * SSM_{(S1,S2)} \tag{5}$$

Where λ, β, $\gamma < 1$ decides the relative contributions of the lexical and semantic information to the overall similarity computation. Since the lexical similarity plays a subordinate role for semantic similarity processing of sentences, β and γ should be a value greater than λ. These coefficients are fixed by an expert.

After the calculation of the similarity score, we will fix a threshold for judging the similarity between sentences. Besides, the threshold varies from one context to another and $\in [0, 1]$. The score similarity between two sentences, S1, S2, can be resolved from one of two cases:

Case1: if (Sim(S1,S2)<threshold) then the sentences are distinct.
Case2: if (Sim(S1,S2)>=threshold) then the sentences are similar.

4 Case Study on the Arabic Language

In the experimentation of the proposed method, we used the Arabic LMF standardized dictionary [8]. In the following subsections, we will first present the LMF standardized Arabic dictionary. Then, we will show the results of applying the three stages of our approach to the chosen two example sentences. And finally, we will evaluate our similarity algorithm experimentally on a set of sentence pairs and compare it with human perceptions and we will exhibit the performance of the prototype realized.

4.1 General Features of the LMF Arabic Dictionary

The Arabic LMF standardized dictionary was developed within our research team MIR@CL [9]. It is currently composed of 38423 lexical entries, 28786 semantic relationships and 8278 semantic predicates. In addition, thanks to the LMF meta-model, our dictionary would certainly be finely structured, cover lexical knowledge at the morphological, syntactic and semantic level and be an extendable resource that could be incremented with entries and lexical properties, extracted from other sources (e.g., Arabic lexicons, text corpora, etc.). Figure 4 below shows the principal classes and attributes of the Arabic LMF dictionary.

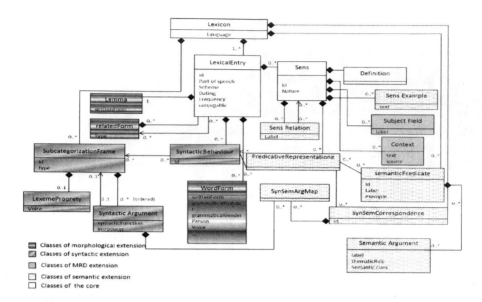

Fig. 4. LMF Arabic standardized dictionary model

4.2   Experimentation

To illustrate the proposed method on an Arabic sentence pair, we provide below a detailed description of two examplesof sentence.

S1: أحسُّ بألم في بَطني(>uHis¨u bi >alamN fi bat°ny) [I feel pain in my belly]

S2:أحَسَّ بالوَجَع (>aHas¨a bi >alwajaEi) [He feels of the ache]

For the step of pre-processing the sentence pair, we give the set of words as follows: for S1{ أحَسَّ، ب، ألمّ ، في ، بَطنْ} and for S2 { وجَعْ ، ب ،أحَسَّ}.

From S1 and S2, the lexical similarity between the two sentences is SL$_{(S1,S2)}$= 2/6.

In the second step, the semantic similarity which is derived from the semantic vectors for S1 and S2 is shown in Table 1.

Table 1. Process for deriving the semantic vectors

	وجَعْ	بَطنْ	في	ألمّ	ب	أحَسَّ
V1	2/5	1	1	1	1	1
V2	1	0	0	2/5	1	1

In Table1, the first row lists the distinct words T= { أحَسَّ، ب، ألمّ ، في ، بَطنْ،وجَعْ}. For each word in T, if the same word exists in S1 or S2, the value of similarity is 1. Otherwise, the value of semantic vector cell is the highest similarity value. For example, the word "وجَعْ" [ache] is not in S$_1$, having the following synonym list {ألم(>alamN)،عَذاب (Ea*AbN)، وَصَب (waSabN)، تَأوُّه (ta>aw¨uhN)}, the most similar word is "ألمّ" [pain] which is the pursuant list synonym { وجَع(wajaEN)، عَذاب (Ea*AbN)، وَصَب (wasSabN)}, with a similarity of 2/5.

From S1 and S2, the semantic similarity between the two sentences is $SM_{(S1,S2)}=0.65$.

Besides, the similarity between S1 and S2is derived from common semantic arguments. The syntactic-semantic similarity $SSM_{(S1,S2)}$ is 2/3 as the semantic arguments for the first sentence are {{agent, human}, {Force, inanimate}, {location, inanimate}} and for the second sentence are {{agent, human}, {Force, inanimate}}.

Finally, the similarity between the sentences "أَحَسَّ بِالوَجَع" and "أحسُ بألم في بَطني" is 0.624, using the coefficient 0.2for λ and 0.4 for β and γ.

This pair of sentences has only one co-occurrence word, "أَحَسّ", but the meaning of the sentences is similar. Thus, the proposed method gives a relatively high similarity. This example demonstrates that the proposed method can capture the meaning of the sentence independently of the co-occurrence of words.

4.3   Experiment Results

In our experiments, we used the Arabic normalized dictionary [9] as the main semantic and syntactic knowledge base to get the synonymy relation between words, semantic class and thematic role of semantic arguments which are linked to semantic predicate that is specified to a noun or a verb. These attributes are considered in order to compute the degree of semantic similarity.

In order to evaluate our similarity measure, we constructed a data set of about 1350 sentence pairs using Arabic dictionaries such as AlWasit [11], AlMuhit [12] and Lissan AlArab [13]. The selected pairs of sentences are categorized as follows: definitions related to one word sense, definitions related to different senses for the same word, examples related to one word sense, examples related to different definitions of one word, sentences having the same verb or subject or object complement with different contexts and presented respectively in Table 2 below. For each category we will give two exemplars.

Moreover, the threshold of similarity between the selected sentences varies from one context to another and should have a value greater than 0.5. In this way, we empirically found a threshold 0.6 for definitions and examples related to one sense, and for definitions and examples related to different senses. Also, we have fixed the coefficient 0.2 for lexical similarity and 0.4 for semantic similarity.

An initial experiment on this data illustrates that the proposed method provides similarity measures that are fairly consistent with human knowledge as shown below in Table2.

Our algorithm's semantic similarity measure achieved reasonable results that were measured using recall (6) and precision (7) metrics.

Recall = Number of similar sentences identified /Total number of similar sentences (6)

Precision= Number of similar sentences identified/ Total number of sentences   (7)

Table 2. Semantic similarities between Arabic selected sentence pairs

Sentence pair	English translation	Our Measure	Human-similarity (yes/no)
الوَقتُ المُمتَدُّ مِنَ الفَجْرِ إلى غُرُوبِ الشَّمْس [alwaqotualmumotadu mina alfajri<lYgurubial$amsi] زَمَنٌ مِقدَارَهُ مِنْ طُلوع الشَّمْس إلى غُرُوبهَا [zamanNmiqodArahu min TuluEial$amsi<lYgurubihA]	The time from dawn to sunset Time from sunrise to sunset	0.76	Yes
أكلَ غَدَاءَهُ بِشَهِيَّةٍ [>akalagadA'ahubi$ahiyatK] أكلتِ النَّارُ الحَطبَ [>akalatialnArualHaTaba]	He ate his lunch with appetite The fire destroyed the wood	0.17	No
حَلأ قلانٌ دِرْهَمًا [Hala>a fulAnNdirhamF] حَلأَهُ دِرْهَمًا [Hala>ahudirhamF]	Someone gave money He gave him money	0.63	Yes
أجَّلَ مَوْعِدَ الاجتِمَاع إلى يَوْم آخرَ [>aj¨lamawoEda al<jtimAi<lYyawomK lxra] لا تُؤَجِّلْ عَمَلَ اليَوْم إلى غدٍ [La tu&aj¨loEamalaalyawomi<lYgd K]	The date of the meeting has been delayed to another day. Do not delay today's work until tomorrow	0.4	No
مَالَ عَن الطَّريق المُشتَّقيم [mAlaEanialTariqialmustaqimi] مَالتِ السيَّارَهُ عن الطَّريق [mAlatoalsay¨AratuEanialTariqi]	He deviated from the right way The car deviated from the road	0.4	No
كتَب له الأرض [katabalahu al>aroDa] كتَب له رسَالة [katabalahurisAlatF]	He gave him the land He wrote him a letter	0.46	No

Table 3 shows respectively the precision and the recall obtained by a prototype that is implemented in Java and whose performance was evaluated on a data test set of 1350 Arabic sentences.

Table 3. The developed prototype performance

Recall	0.812
Precision	0.722
F-measure	0.764

The results indicate that the prototype has the capability to detect precisely the semantic similarity between sentences. In addition, the performance of our prototype is dependent on the stemming system, syntactical analyzer, synonyms and semantic predicates retrieved from Arabic LMF dictionary [9].According to a comparative evaluation study of Arabic language stemmers [14] and syntactical analyzers, MADA [15] and the Stanford parser [16] achieves the highest accuracy. So, we do not expect to increase the performance of our prototype by using other stemmers or syntactical parsers. However, using other types of relations such as hyponymy might impact its performance.

5   Conclusion

In this paper, we have proposed a new method of measuring semantic similarity between sentences from the LMF standard (ISO-24613). The originality of this approach lies in the use of a unique, finely-structured source which is rich in lexical and conceptual knowledge. The measure of semantic sentence similarity is started with a pre-processing phase and is based on lexical information on the similarity between words in the pairs of sentences, semantic information on   synonymy relations between the words composing the sentences and similarity between semantic arguments (semantic class+ thematic role) of the words of the sentence. Finally, we have presented and discussed a series of experiments to demonstrate the effectiveness of our method on a large set of Arabic sentences. The results indicate that the prototype realized has the capability to detect precisely the semantic similarity between Arabic sentences. Besides, the proposed approach has proven to be reliable through the applications carried out on the Arabic language whose choice is explained by two main motives. The first one is the great lack of research on measuring sentence similarity for the Arabic language and the second is the availability within our research team of an LMF standardized Arabic dictionary as well as NLP tools. Currently, we enriched a varied sentence pair to compare with human ratings. As for the future perspectives to our work, we will improve the algorithm to disambiguate word sense using the surrounding words to give a little contextual information. In addition, we will ameliorate other types of relationships such as hyponymy. Finally, we will apply machine learning to determine the good coefficients (λ, β, γ) for computing semantic similarity.

References

1. Mandreoli, F., Martoglia, R., Tiberio, P.: A syntactic approach for searching similarities within sentences. In: The Proceedings of the Eleventh International Conference on Information and Knowledge Management, McLean, Virginia, USA (2002)
2. Hatzivassiloglou, V., Klavans, J.L., Eskin, E.: Detecting Text Similarity over Short Passages: Exploring Linguistic Feature Combinations via Machine Learning, pp. 203–212 (1999)
3. Mihalcea, R., Corley, C., Strapparava, C.: Corpus-based and knowledge-based measures of text semantic similarity. In: The Proceedings of the 21st National Conference on Artificial Intelligence, Boston, Massachusetts, vol. 1 (2006)
4. Yuhua, L., McLean, D., Bandar, Z.A., O'Shea, J.D., Crockett, K.: Sentence similarity based on semantic nets and corpus statistics. IEEE Transactions on Knowledge and Data Engineering 18, 1138–1150 (2006)
5. Xiaoying, L., Yiming, Z., Ruoshi, Z.: Sentence Similarity based on Dynamic Time Warping. In: The International Conference on Semantic Computing, ICSC 2007, pp. 250–256 (2007)
6. Islam, A., Inkpen, D.: Semantic text similarity using corpus-based word similarity and string similarity. The ACM Transactions Knowledge Discovery Data 2, 1–25 (2008)
7. Miller, G.A., Beckwith, R., Fellbaum, C., Gross, D., Miller, K.J.: Introduction to WordNet: an on-line lexical database. The International Journal of Lexicography 3, 235–244 (1990)
8. Francopoulo, G., George, M.: Language Resource Management.2008. Lexical Markup Framework (LMF). Technical report, ISO/TC 37/SC 4 N453, N330 Rev.16 (2008)
9. Khemakhem, A., Gargouri, B., Ben Hamadou, A.: LMF standardized dictionary for Arabic Language. In: The International Conference on Computing and Information Technology, ICCIT 2012 (2012)
10. Jaccard, P.: Etude comparative de la distribution florale dans une portion des Alpes et des Jura. Bulletin del la Société Vaudoise des Sciences Naturelles 37, 547–579 (1901)
11. Mustaph, I., A. O. t. A., Language.:Al-Wassit Arabic dictionary. Dar Ihia' al Tourath al-Arabi
12. Bustani, B.: Arabic Dictionary (Muhit al-Muhit). International Book Centre, Incorporated (1983)
13. Manzur, I.: Lisan al Arab.Al-dar al-Misriya li-l-talif wa-l-taryamar (1290)
14. Sawalha, M., Atwell, E.: Comparative evaluation of Arabic language morphological analysers and stemmers. In: The Proceedings of 22nd International Conference on Computational Linguistics, COLING 2008, Manchester, UK, pp. 107–110 (2008)
15. Roth, R., Rambow, O., Habash, N., Diab, M.T., Rudin, C.: Arabic Morphological Tagging, Diacritization, and Lemmatization Using Lexeme Models and Feature Ranking. In: The Annual Meeting of the Association for Computational Linguistics, ACL 2008, Columbus, Ohio, pp. 117–120 (2008)
16. Green, S., Manning, C.D.: Better Arabic Parsing: Baselines, Evaluations, and Analysis. In: The Proceedings of 23nd International Conference on Computational Linguistics, COLING 2010, Beijing, China, pp. 394–402 (2010)

Impact of the Sakoe-Chiba Band
on the DTW Time-Series Distance Measure
for *k*NN Classification

Zoltan Geler[1], Vladimir Kurbalija[2], Miloš Radovanović[2], and Mirjana Ivanović[2]

[1] Faculty of Philosophy, University of Novi Sad
Dr Zorana Đinđića 2, 21000 Novi Sad, Serbia
gellerz@gmail.com
[2] Department of Mathematics and Informatics, Faculty of Sciences, University of Novi Sad
Trg D. Obradovića 4, 21000 Novi Sad, Serbia
{kurba,radacha,mira}@dmi.uns.ac.rs

Abstract. For classification of time series, the simple 1-nearest neighbor (1NN) classifier in combination with an elastic distance measure such as Dynamic Time Warping (DTW) distance is considered superior in terms of classification accuracy to many other more elaborate methods, including *k*-nearest neighbor (*k*NN) with neighborhood size $k > 1$. In this paper we revisit this apparently peculiar relationship and investigate the differences between 1NN and *k*NN classifiers in the context of time-series data and constrained DTW distance. By varying neighborhood size k, constraint width r, and evaluating 1NN and *k*NN with and without distance-based weighting in different schemes of cross-validation, we show that the first nearest neighbor indeed has special significance in labeled time-series data, but also that weighting can drastically improve the accuracy of *k*NN. This improvement is manifested by better accuracy of weighted *k*NN than 1NN for small values of k (3–4), better accuracy of weighted *k*NN than unweighted *k*NN in general, and reduced need to use large values of constraint r with weighted *k*NN.

Keywords: Time series, Dynamic Time Warping, global constraints, classification, *k*-nearest neighbor.

1   Introduction

A time series represents a series of numerical data points in successive order, usually with uniform intervals between them. This form of data can appear in almost every aspect of human activity including: representing social, economic and natural phenomena, medical observations, results of scientific and engineering experiments, etc. Time-series mining is the subfield of artificial intelligence where different data mining methods are applied on time-series data in order to understand the phenomenon which generated those time series. These methods include classification, clustering, anomaly detection, prediction, and indexing.

R. Buchmann et al. (Eds.): KSEM 2014, LNAI 8793, pp. 105–114, 2014.

The choice of appropriate distance/similarity measure is a crucial aspect of time-series mining since all mentioned methods explicitly or implicitly use distance measures. These measures should be carefully defined in order to reflect the essential similarities between time series which are commonly based on shapes and trends. Research in this field yielded several distance measures – from Euclidean distance [1] as the most simple and intuitive to the more sophisticated distance measures such as Dynamic Time Warping (DTW) [2], Longest Common Subsequence (LCS) [3], Edit Distance with Real Penalty (ERP) [4] and Edit Distance on Real sequence (EDR) [5].

Unfortunately, the quality of distance measures is usually hard to evaluate since the notion of similarity is a very subjective and data-dependent issue. The most common approach to the assessment of distance measures in the literature [6,7,8] is through evaluation of classification accuracies of distance-based classifiers. The quality of the nearest-neighbor based techniques strongly depends on the quality of the used distance measures, which makes the NN classifier very suitable for distance-measure assessment. Furthermore, the simple 1NN classifier is selected in several works [7, 9], as one of the most accurate classifiers for time-series data, demonstrating comparable and even superior performance than many more complex classification approaches, including the k-nearest neighbor classifier with $k > 1$.

The main goal or this paper is to provide a more detailed investigation of differences between 1NN and kNN classifiers in the context of time-series data and DTW distance. We will show that the accuracy of kNN can be improved and made superior to 1NN when the importance of the first neighbor is taken into account. The rest of the paper is organized as follows: next section gives some basic facts and an overview of the recent work in this area. Section 3 presents the detailed results of our experiments which are conducted on 46 datasets available from [10]. The final section contains conclusions drawn from the experiments, as well as possibilities for future work.

2   Background and Related Work

The advantages of Euclidean distance (easily implementable, fast to compute and represents a distance metric) have made it, over time, probably one of the most commonly used similarity measure for time series [11,12,13,14]. However, due to the linear aligning of the points of the time series it is sensitive to distortions and shifting along the time axis [15, 16]. To address this shortcoming, many different elastic similarity measures were proposed. Among them, some of the most widely used and studied are Dynamic Time Warping (DTW) and Longest Common Subsequence (LCS), and their extensions, Edit Distance with Real Penalty (ERP) and Edit Distance on Real sequence (EDR).

Implementations of these elastic similarity measures are based on dynamic programming: in order to determine the similarity between two time series we need to compare each point of one time series with each point of the other one. This can lead to pathological non-linear aligning of the points (where a relatively small part of one time series maps onto a large section of the other time series) and slow down the computations. One way to avoid these adverse effects is to constrain the warping path using the Sakoe-Chiba band [17].

It is reported that the elastic measures can have better classification accuracy than Euclidean distance and that constraining the warping window can further improve the accuracy of these measures [7, 9]. In [18] and [19] we have shown that when the constraint parameter is tight enough (less than 15%-10% of the length of the time series), constrained versions of the elastic measures (DTW, LCS, ERP and EDR) become qualitatively different from their unconstrained counterparts (in the sense of producing significantly different 1-nearest neighbor graphs). In [9] and [15], based on experiments using a limited number of datasets it is reported that narrow constraints (less than 10% of the length of time series) are necessary for accurate DTW and that a warping window which is too large may actually deteriorate classification accuracy.

All mentioned experiments for distance-measure assessment were conducted with 1NN classifier as it was shown that it gives among the best results (compared to many not only distance-based classifiers) with time-series data [7, 9]. This fact strongly indicates that the first neighbor has particularly important meaning in the time-series datasets. In [20], the reasons and origins of this special behavior of the first neighbor are investigated, and related with the observed diversity of class labels in k-neighborhoods. In this paper, we will compare the accuracies of 1NN and kNN classifiers when using the DTW time-series distance measure in order to understand the special meaning of the first neighbor. Furthermore, we will attempt to improve the accuracy of kNN by favoring the first (few) neighbors.

3   Experimental Results

Through extensive experiments in this section we will investigate the suggestions and findings regarding the influence of the Sakoe-Chiba band on the Dynamic Time Warping similarity measure, 1NN and kNN classifiers discussed above. We will observe the following widths of the warping window: 100% (the unconstrained similarity measure), 90%, 80%, 70%, 60%, 50%, 45%, 40%, 35%, 30%, and all values from 25% to 0% in steps of 1%. These values were chosen based on reports that the measures with larger constraints behave similarly to the unconstrained ones, while the smaller constraints show more apparent discrepancies [7, 9, 15, 18, 19].

We are going to report the minimal value of the warping window that maximizes the classification accuracy of the k-nearest neighbor classifier for a large number of datasets. This classifier is chosen taking into account that among many classification methods (decision trees, neural networks, Bayesian networks, support vector machines, etc.) simple nearest-neighbor methods often give the best results when working with time series [7, 9]. In addition to that, the quality of distance/similarity measure directly influences the accuracy of the NN classifier, which makes it appropriate for distance/similarity measure assessment.

To obtain a better insight into the impact of constraining the warping window our experiments encompass five different evaluation methods of classification accuracy: leave-one-out (LOO), stratified 9-fold cross-validation (SCV1x9), 5 times repeated stratified 2-fold cross-validation (SCV5x2), 10 times repeated stratified 10-fold cross validation (SCV10x10) and 10 times repeated stratified holdout method (SHO10x) using two-thirds of available time series for training and one third for testing. The datasets are randomly shuffled in each run. Furthermore, we observe the unweighted and the weighted kNN classifier with the values of parameter k in range from 1 to 30. Weights are calculated by the formula $1/d(q,c)^2$ where $d(q,c)$ denotes the distance between the time series q and c [21].

The experiments were conducted on 46 datasets from [10], which includes the majority of all publicly available, labeled time-series datasets in the world (Table 1). In addition to that, this collection of datasets is most commonly used for validation of different time-series mining concepts. They originate from a plethora of different domains, including medicine, robotics, astronomy, biology, face recognition, handwriting recognition, etc. The length of time series varies from 24 to 1882 depending of the dataset. The number of time series per dataset varies from 56 to 9236 and the number of classes varies from 2 to 50.

Table 1. Properties of the data sets

Data set	Size	Length	Classes	Data set	Size	Length	Classes
50words	905	270	50	mallat	2400	1024	8
adiac	781	176	37	medicalimages	1141	99	10
beef	60	470	5	motes	1272	84	2
car	120	577	4	noninvasivefatalecg_thorax1	3765	750	42
cbf	930	128	3	noninvasivefatalecg_thorax2	3765	750	42
chlorineconcentration	4307	166	3	oliveoil	60	570	4
cinc_ecg_torso	1420	1639	4	osuleaf	442	427	6
coffee	56	286	2	plane	210	144	7
cricket_x	780	300	12	sonyaiborobotsurface	621	70	2
cricket_y	780	300	12	sonyaiborobotsurfaceii	980	65	2
cricket_z	780	300	12	starlightcurves	9236	1024	3
diatomsizereduction	322	345	4	swedishleaf	1125	128	15
ecg200	200	96	2	symbols	1020	398	6
ecgfivedays	884	136	2	synthetic_control	600	60	6
faceall	2250	131	14	trace	200	275	4
facefour	112	350	4	twoleadecg	1162	82	2
fish	350	463	7	twopatterns	5000	128	4
gun_point	200	150	2	uwavegesturelibrary_x	4478	315	8
haptics	463	1092	5	uwavegesturelibrary_y	4478	315	8
inlineskate	650	1882	7	uwavegesturelibrary_z	4478	315	8
italypowerdemand	1096	24	2	wafer	7164	152	2
lighting2	121	637	2	wordssynonyms	905	270	25
lighting7	143	319	7	yoga	3300	426	2

The Unweighted kNN Classifier. In Fig. 1 we can clearly notice that the relationship between the parameter k and the average smallest error rate is almost linear – the growth of parameter k leads to the decline of classification accuracy. The highest average classification accuracy (88.772%) was achieved with the 1NN classifier and the LOO evaluation method and the lowest one (74.536%) with the 30NN classifier and the SCV5x2 evaluation method (Table 2).

In case of the unweighted kNN classifier the average width of the smallest warping window which gives the lowest error rate for DTW varies in the range from 3.783 to 10.087. We can see that the increase of the parameter k implies the growth of the average warping window widths (Fig. 2): we need wider and wider windows to get the best accuracy. The smallest average warping window (3.783) was obtained using the LOO evaluation method and the 1NN classifier and the largest one (10.087) with the SHO10x evaluation method and the 24NN classifier (Table 3).

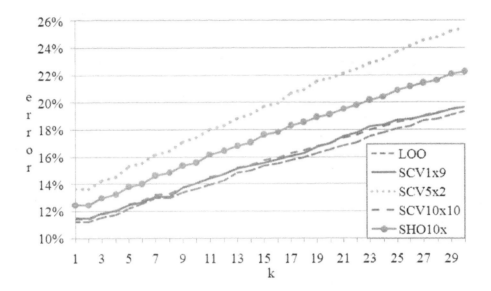

Fig. 1. Average lowest error rates for DTW with unweighted kNN

Table 2. Minimum and maximum of the average lowest error rates for DTW with unweighted kNN

	MIN error	k	MAX error	k	MAX-MIN
LOO	**11.228%**	1	19.317%	30	8.089
SCV1x9	11.494%	1	19.636%	30	8.142
SCV5x2	13.628%	1	**25.464%**	**30**	**11.836**
SCV10x10	11.410%	1	19.701%	30	8.291
SHO10x	12.471%	1	22.223%	30	9.752

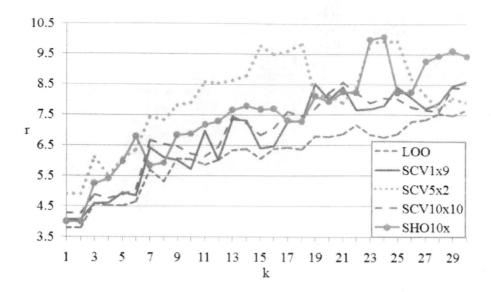

Fig. 2. Average smallest warping window widths for DTW with unweighted kNN

Table 3. Minimum and maximum of the average smallest warping window widths for DTW with unweighted kNN

	MIN		MAX		MAX-MIN
	r	*k*	*r*	*k*	
LOO	3.783	1	7.652	30	3.870
SCV1x9	4.065	1	8.587	30	4.522
SCV5x2	4.913	1	9.935	24	5.022
SCV10x10	4.261	1	8.565	21	4.304
SHO10x	4.000	1	**10.087**	**24**	**6.087**

The Weighted kNN Classifier. Looking at the chart in Fig. 3 we can see that in the case of DTW the use of weights changes the influence of the parameter k on the accuracy of classification: instead of 1NN the smallest average error rates were achieved with 3NN (or 4NN in the case of SCV5x2 and SHO10x). After a brief decline and reaching the minimum value, the error rates begin to grow again, similarly as in the case of the unweighted kNN classifier but visibly slower. The attained maximum values of the classification errors are more than 1.5 times less than without weights (Table 4). The highest average classification accuracy was achieved by LOO and the lowest one by SCV5x2.

Fig. 4 shows that the introduction of weights into the kNN classifier noticeably alleviates the growth of the average warping window widths. In this case the largest average warping window (6.848) was achieved by the combination of the 8NN classifier and the SCV5x2 evaluation method (Table 5). The smallest average warping window (3.783) was obtained using the 1NN classifier and the LOO evaluation method. The differences between the minimum and maximum average r values are about two times smaller than in the case of the unweighted kNN classifier.

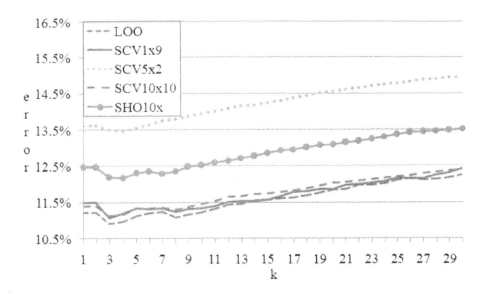

Fig. 3. Average lowest error rates for DTW with weighted kNN

Table 4. Minimum and maximum of the average lowest error rates for DTW with weighted kNN

	MIN error	k	MAX error	k	MAX-MIN
LOO	**10.923%**	**3**	12.256%	30	1.333
SCV1x9	11.072%	3	12.426%	30	1.354
SCV5x2	13.468%	4	**14.970%**	**30**	**1.502**
SCV10x10	11.134%	3	12.412%	30	1.278
SHO10x	12.177%	4	13.527%	30	1.350

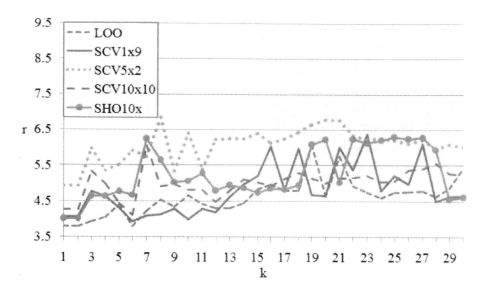

Fig. 4. Average smallest warping window widths for DTW with weighted kNN

Table 5. Minimum and maximum of the average smallest warping window widths for DTW with weighted kNN

	MIN		MAX		MAX-MIN
	r	k	r	k	
LOO	**3.783**	**1**	6.087	19	2.304
SCV1x9	3.935	6	6.370	23	**2.435**
SCV5x2	4.913	1	**6.848**	**8**	1.935
SCV10x10	4.109	6	6.043	7	1.935
SHO10x	4.000	1	6.304	25	2.304

4   Conclusions and Future Work

The results of experiments clearly confirmed the special importance of the first neighbor in time-series data. As seen in Fig. 1, the error rate of the unweighted kNN classifier almost linearly grows as the number of neighbors k grows. The kNN classifier actually gives the best results for the value $k = 1$ when considering k neighbors without a weighting scheme. On the other hand, when the weighting scheme is introduced (Fig. 3) the situation is changed to some extent. The best results are obtained for the value $k = 3$. Furthermore, the weighting scheme which favors the first neighbor significantly improved the accuracy for all values ok k.

When observing the value of constraint (Fig. 2 and 4) the introduction of the weighting scheme has an important impact. For unweighted kNN, the value of the constraint grows as k grows. On the other hand, with the weighting scheme the value of the constraint remains approximately the same for all values of k. In addition, the difference between minimum and maximum values of constraints is about two times smaller with the weighting scheme.

All these observations indicate that favoring the first neighbor with a weighting scheme improves the quality and stability of kNN. The first neighbor has a special meaning in time-series data and taking this fact into consideration can significantly improve the quality of kNN for all values of k, by making it even more accurate than 1NN for some small values of k.

In future work, it would be interesting to investigate the influence of weighting on other popular time-series distance measures like Euclidian distance, LCS, EDR, ERP, etc. In addition, the behavior of other weighting schemes [21, 22, 23, 24] we believe also warrants further investigation.

Acknowledgments. The authors would like to thank Eamonn Keogh for collecting and making available the UCR time series data sets, as well as everyone who contributed data to the collection, without whom the presented work would not have been possible. V. Kurbalija, M. Radovanović and M. Ivanović thank the Serbian Ministry of Education, Science and Technological Development for support through Project no. OI174023, "Intelligent Techniques and their Integration into Wide-Spectrum Decision Support."

References

1. Faloutsos, C., Ranganathan, M., Manolopoulos, Y.: Fast subsequence matching in time-series databases. ACM SIGMOD Rec. 23, 419–429 (1994)
2. Berndt, D., Clifford, J.: Using dynamic time warping to find patterns in time series. In: KDD Workshop, Seattle, WA, pp. 359–370 (1994)
3. Vlachos, M., Kollios, G., Gunopulos, D.: Discovering similar multidimensional trajectories. In: Proceedings of the 18th International Conference on Data Engineering (ICDE), pp. 673–684. IEEE Comput. Soc. (2002)
4. Chen, L., Ng, R.: On The Marriage of Lp-norms and Edit Distance. In: Proceedings of the 30th International Conference on Very Large Data Bases (VLDB), pp. 792–803. VLDB Endowment (2004)
5. Chen, L., Özsu, M.T., Oria, V.: Robust and fast similarity search for moving object trajectories. In: Proceedings of the ACM SIGMOD International Conference on Management of Data, pp. 491–502. ACM, New York (2005)
6. Keogh, E., Kasetty, S.: On the need for time series data mining benchmarks: A survey and empirical demonstration. In: Proceedings of the 8th ACM SIGKDD International Conference on Knowledge Discovery and Data Mining, pp. 102–111. ACM, New York (2002)
7. Ding, H., Trajcevski, G., Scheuermann, P., Wang, X., Keogh, E.: Querying and mining of time series data: Experimental comparison of representations and distance measures. Proc. VLDB Endow. 1, 1542–1552 (2008)
8. Esling, P., Agon, C.: Time-series Data Mining. ACM Comput. Surv. 45, 12:1–12:34 (2012)
9. Xi, X., Keogh, E., Shelton, C., Wei, L., Ratanamahatana, C.A.: Fast time series classification using numerosity reduction. In: Proceedings of the 23rd International Conference on Machine Learning (ICML), pp. 1033–1040. ACM, New York (2006)

10. Keogh, E., Zhu, Q., Hu, B., Y., H., Xi, X., Wei, L., Ratanamahatana, C.A.: The UCR Time Series Classification/Clustering Homepage, http://www.cs.ucr.edu/~eamonn/time_series_data/,citeulike-article-id:2139261

11. Agrawal, R., Faloutsos, C., Swami, A.: Efficient similarity search in sequence databases. In: Lomet, D.B. (ed.) FODO 1993. LNCS, vol. 730, pp. 69–84. Springer, Heidelberg (1993)

12. Chan, K.-P., Fu, A.W.-C.: Efficient time series matching by wavelets. In: Proceedings of the 15th International Conference on Data Engineering (ICDE), pp. 126–133. IEEE Comput. Soc. (1999)

13. Keogh, E., Chakrabarti, K., Pazzani, M., Mehrotra, S.: Dimensionality reduction for fast similarity search in large time series databases. Knowl. Inf. Syst. 3, 263–286 (2001)

14. Keogh, E., Chakrabarti, K., Pazzani, M., Mehrotra, S.: Locally adaptive dimensionality reduction for indexing large time series databases. In: Proceedings of the ACM SIGMOD International Conference on Management of Data, pp. 151–162. ACM, New York (2001)

15. Ratanamahatana, C.A., Keogh, E.: Three myths about dynamic time warping data mining. In: Proceedings of the 5th SIAM International Conference on Data Mining (SDM), pp. 506–510 (2005)

16. Keogh, E.: Exact indexing of dynamic time warping. In: Proceedings of the 28th International Conference on Very Large Data Bases (VLDB), pp. 406–417. VLDB Endowment (2002)

17. Sakoe, H., Chiba, S.: Dynamic programming algorithm optimization for spoken word recognition. IEEE T. Acoust. Speech Signal Process. 26, 43–49 (1978)

18. Kurbalija, V., Radovanović, M., Geler, Z., Ivanović, M.: The influence of global constraints on DTW and LCS similarity measures for time-series databases. In: Dicheva, D., Markov, Z., Stefanova, E. (eds.) Third International Conference on Software, Services and Semantic Technologies S3T 2011. AISC, vol. 101, pp. 67–74. Springer, Heidelberg (2011)

19. Kurbalija, V., Radovanović, M., Geler, Z., Ivanović, M.: The influence of global constraints on similarity measures for time-series databases. Knowl-Based Syst. 56, 49–67 (2014)

20. Radovanović, M., Nanopoulos, A., Ivanović, M.: Time-series classification in many intrinsic dimensions. In: Proceedings of the 10th SIAM International Conference on Data Mining (SDM), Columbus, Ohio, USA, pp. 677–688 (2010)

21. Dudani, S.A.: The distance-weighted k-nearest-neighbor rule. IEEE T. Syst. Man Cy. 6, 325–327 (1976)

22. Macleod, J.E.S., Luk, A., Titterington, D.M.: A re-examination of the distance-weighted k-nearest neighbor classification rule. IEEE T. Syst. Man Cy. 17, 689–696 (1987)

23. Gou, J., Xiong, T., Kuang, Y.: A novel weighted voting for k-nearest neighbor rule. J. Comput. 6, 833–840 (2011)

24. Gou, J., Du, L., Zhang, Y., Xiong, T.: A new distance-weighted k-nearest neighbor classifier. J. Inf. Comput. Sci. 9, 1429–1436 (2012)

Weighting Exponent Selection of Fuzzy C-Means via Jacobian Matrix

Liping Jing, Dong Deng, and Jian Yu

Beijing Key Lab. of Traffic Data Analysis and Mining,
Beijing Jiaotong University, Beijing, China
lpjing@bjtu.edu.cn

Abstract. FCM is a popular clustering algorithm and applied in various areas. However, there are still some problems to be solved including the selection of weighting exponent m and convergence analysis. In this paper, we present an efficient method to identify the proper range of m and convergence rate by a new Jacobian matrix of FCM. A series of experimental results on both synthetical data and real-world data validate the proposed theoretical results.

Keywords: FCM, Convergence, Parameter Selection, Jacobian Matrix.

1 Introduction

Clustering analysis is an unsupervised learning method, which aims to partition data into groups such that the objects in each group share some similarity. The partitions should be such that patterns are homogeneous within the groups and heterogeneous between the groups. Clustering analysis has been used as a knowledge discovery tool or a preprocessing step in various areas [10, 11]. The existing methods can be roughly divided into two categories: hard clustering and fuzzy clustering. Among them, fuzzy clustering extends the notion of partition to associate each point with every cluster using a membership function whose values span from zero to one [2, 5] . Fuzzy C-means (FCM) is one of the most important fuzzy clustering algorithms. It has been widely used in various areas such as pattern recognition, text mining, image processing, bioinformatics, social computing, and etc.

Even though FCM is popular, there are still some problems to be solved. One is the selection of weighting exponent m. When m approaches one, the FCM algorithm is close to hard C-means algorithm, while FCM will output a mass center of the data set when m approaches infinity. Thus, the weighting exponent plays an important role in FCM and it is necessary to determine proper value for m. In the literature, the heuristic strategies are usually used to set m. Pal and Bezdek [14] showed that it is probably in the interval [1.5, 2.5], and similar suggestions are given in [4, 6]. Ozkan and Turksen [13] demonstrated that the upper and lower bounds of fuzziness values are 2.6 and 1.4. Most researchers have empirically proposed $m = 2$ in practice. The above recommendations are based on empirical studies. Yu et al. [16] firstly gave the theoretical rules to select m

R. Buchmann et al. (Eds.): KSEM 2014, LNAI 8793, pp. 115–126, 2014.

via Hessian matrix of the FCM objective function. However, it is time-consuming to compute the Hessian matrix.

The other issue is the convergence analysis of FCM clustering process. The convergence of FCM has been studied in literature [2,3,7,8] which proved that FCM can converge to a local minimum. The question is how fast FCM can obtain its optimal solution. Groll and Jakel [7] showed that FCM converges linearly near a nonsingular local minimum, which is obtained by using the Taylor expansion of the objective function and its corresponding Hessian matrix. Selim and Ismail [15] used the Hessian matrix to show the convergence property of FCM. Recently, Yu et al. [16] adopted Hessian matrix as a criterion to decide the local optimality of a point for the FCM algorithm. However, the elements of Hessian matrix have different expressions which increases the difficulty of studying its properties.

In order to handle these issues, in this work, we adopt a Jacobian matrix of the FCM algorithm to identify the proper range of weighting exponent m and provide the convergence rate of FCM at the convergence point. To our best knowledge, this is the first work to demonstrate the relation between the convergence rate at the optimal point and the clustering performance of FCM.

The rest of the paper is organized as follows. Section 2 gives the FCM algorithm and the related theoretical result on m selection. In Section 3, we consider the Jacobian matrix of the FCM and its theoretical analysis. In Section 4, we conduct a series of experiments to validate our theoretical results. A brief conclusion is given in Section 5.

2   The FCM Algorithm

FCM is known to produce reasonable partitionings of the original data in many areas since Dunn [5] and Bezdek [2] proposed it. The aim of FCM is to find a fuzzy partition of a data matrix $X = \{x_1, x_2, \ldots, x_n\}$ consisting of n data points $x_k \in R^s$. The fuzzy partition of X forms c clusters $(2 \leq c < n)$ with centers $V = \{v_1, v_2, \ldots, v_c\}$ and $v_i \in R^s$. The corresponding fuzzy membership are $(c \times n)$-matrices U from the set

$$M = \left\{ U = [u_{ik}]_{c \times n} \middle| \forall i, \forall k, u_{ik} \in [0,1], \sum_{i=1}^{c} u_{ik} = 1, \sum_{k=1}^{n} u_{ik} > 0 \right\}. \tag{1}$$

Then the objective function of FCM is defined as

$$J_m(u, v; x) = \sum_{j=k}^{n} \sum_{i=1}^{c} u_{ik}^m \|x_k - v_i\|^2 \tag{2}$$

where $\| \cdot \|$ denotes the Euclidean distance between the point x_k and the ith cluster center v_i. The parameter m $(1 < m < \infty)$ is the weighting exponent or fuzzifier.

The minimization problem in (2) can be implemented by the alternating optimization procedure (AO) [2]. Starting with an initial selection of cluster centroids, AO iteratively updates the fuzzy membership U by fixing cluster centers V, and updates V by fixing U with the following equations until the stop condition is satisfied.

$$u_{ik} = \frac{d_{ik}^{\frac{1}{1-m}}}{\sum_{i=1}^{c} d_{ik}^{\frac{1}{1-m}}}, \tag{3}$$

where

$$d_{ik} = (x_k - v_i)^T (x_k - v_i),$$

and

$$v_i = \frac{\sum_{k=1}^{n} u_{ik}^m x_k}{\sum_{k=1}^{n} u_{ik}^m}. \tag{4}$$

A detailed description of the classical FCM algorithm and its variants is offered in [9]. To date, there are several key issues to be solved including the analysis of convergence, setting the weighting exponent value, the effect of centers initialization, the influence of the distance measure, and etc. In this paper, we will limit to highlight the former two issues.

From the solutions (3) and (4) for U and V, it can be seen that choosing a suitable weighting exponent is very important to obtain final reasonable clustering results. In the literature, m is usually chosen based on empirical studies. Recently, Yu et al. [16] firstly gave the following two theoretical rules to select m via Hessian matrix of the objective function (2).

Rule 1: $\quad m \leq \frac{\min\{s, n-1\}}{\min\{s, n-1\}-1}$, if $\min\{s, n-1\} \geq 3$,

Rule 2: $\quad m \leq \frac{1}{1-2\lambda_{\max}(F_{U^*})}$, if $\lambda_{\max}(F_{U^*}) < 0.5$.

where $F_{U^*} = (f_{kr}^{U^*})_{n \times n}$, $f_{kr}^{U^*} = \frac{1}{n} \frac{(x_k - \bar{x})^T}{\|x_k - \bar{x}\|} \frac{(x_r - \bar{x})^T}{\|x_r - \bar{x}\|}$ with $\bar{x} = \frac{1}{n} \sum_{k=1}^{n} x_k$, and $\lambda_{\max}(F_{U^*})$ is the maximum eigenvalue of the matrix F_{U^*}. However, the Hessian matrix is too complex to theoretically judge whether or not the FCM algorithm converges to a local minimum although it can lead to a theoretical range for weighting exponent. In next section, we will use a new Jacobian matrix of the FCM alternating optimization algorithm to find the proper range of m and the convergence rate at the convergence point.

3   Jacobian Matrix Analysis of the FCM Algorithm

In theory, the Hessian matrix of objective function is able to judge whether the algorithm can converge to an optimal point. In literatures, the Hessian matrix has been used to show the algorithm convergence properties such as EM algorithm [12] and FCM algorithm [15, 16]. However, the elements of Hessian matrix have different expressions, which increases the difficulty of studying its properties. In this section, therefore, we study the Jacobian matrices with respect to fuzzy membership U and cluster centroids V, and give the theoretical analysis of the FCM algorithm.

In order to get a simpler criterion, we choose the symmetrical variable space for u

$$\Psi = \left\{ u = [u_{ik}]_{(c-1)\times n} \,\Big|\, 1 \le i \le (c-1), 1 \le k \le n, u_{ik} \in [0,1], \right.$$

$$\left. \sum_{i=1}^{c-1} u_{ik} \le 1, \sum_{k=1}^{n} u_{ik} > 0 \right\}. \tag{5}$$

and $u_{ck} = 1 - \sum_{i=1}^{(c-1)} u_{ik}$. Then two variable mapping functions w.r.t u and v are defined as follows.

$$u_{ik}^{(t+1)} = \theta_{ik}(u^{(t)}) = \frac{\left(d_{ik}^{(t)}\right)^{\frac{1}{1-m}}}{\sum_{i=1}^{c} \left(d_{ik}^{(t)}\right)^{\frac{1}{1-m}}}, \tag{6}$$

where $d_{ik}^{(t)} = (x_k - v_i^{(t)})^T (x_k - v_i^{(t)})$, and $v_i^{(t)}$ is the centroid of the ith cluster in the tth iteration.

$$v_i^{(t+1)} = \mu_i(v^{(t)}) = \frac{\sum_{k=1}^{n} \left(u_{ik}^{(t+1)}\right)^m x_k}{\sum_{k=1}^{n} \left(u_{ik}^{(t+1)}\right)^m}, \tag{7}$$

where $u_{ik}^{(t+1)}$ is the fuzzy membership of the kth point to the ith cluster in the $(t+1)$th iteration. In the iterative procedure of FCM, if (u, v) converges to some point (u^*, v^*), then u^* and v^* must satisfy

$$u^* = \theta(u^*), \tag{8}$$

and

$$v^* = \mu(v^*). \tag{9}$$

Lemma 1. *Based on the mapping function (8), the convergence rate of FCM at its local optima is*

$$\frac{\partial \theta_{ik}}{\partial u_{jr}} = \frac{u_{ik} u_{jk}}{(1-m) d_{jk}} \left\{ \frac{2m u_{jr}^{m-1}}{\sum_{k=1}^{n} u_{jk}^{m}} (x_r - v_j)^T (x_k - v_j) \right\} \tag{10}$$

$$+ \frac{u_{ik} u_{ck}}{(1-m) d_{ck}} \left\{ -\frac{2m u_{cr}^{m-1}}{\sum_{k=1}^{n} u_{ck}^{m}} (x_r - v_c)^T (x_k - v_c) \right\}$$

$$- \frac{2m \delta_{ij} u_{ir}^{m-1} u_{ik}}{(1-m) d_{ik} \sum_{k=1}^{n} u_{ik}^{m}} (x_r - v_i)^T (x_k - v_i)$$

where $\delta_{ij} = 1$ if $i = j$, otherwise $\delta_{ij} = 0$.

Proof. From (4), we know

$$\frac{\partial v_i}{\partial u_{jr}} = \delta_{ij} \frac{m u_{ir}^{m-1} (x_r - v_i)}{\sum_{k=1}^{n} u_{ik}^{m}},$$

Then, the element at the $(i-1) \times n + k$ row and the $(j-1) \times n + r$ column of the Jacobian matrix $\frac{\partial \theta(u)}{\partial u}$ for the variable mapping $\theta(u)$ in (6) can be expressed as

$$
\frac{\partial \theta_{ik}}{\partial u_{jr}} = \frac{-(d_{ik})^{\frac{1}{1-m}}(d_{jk})^{\frac{1}{1-m}-1}}{(1-m)\left(\sum_{i=1}^{c}(d_{ik})^{\frac{1}{1-m}}\right)^2}\left\{ -\frac{2mu_{jr}^{m-1}}{\sum_{k=1}^{n}u_{jk}^m}(x_r - v_j)^T(x_k - v_j)\right\} \quad (11)
$$

$$
+ \frac{-(d_{ik})^{\frac{1}{1-m}}(d_{ck})^{\frac{1}{1-m}-1}}{(1-m)\left(\sum_{i=1}^{c}(d_{ik})^{\frac{1}{1-m}}\right)^2}\left\{ -\frac{2mu_{cr}^{m-1}}{\sum_{k=1}^{n}u_{ck}^m}(x_r - v_c)^T(x_k - v_c)\right\}
$$

$$
- \frac{2m\delta_{ij}u_{ir}^{m-1}(d_{ik})^{\frac{1}{1-m}-1}}{(1-m)\sum_{i=1}^{c}(d_{ik})^{\frac{1}{1-m}}\sum_{k=1}^{n}u_{ik}^m}(x_r - v_i)^T(x_k - v_i)
$$

According to (3), (10) can be obtained from (11). $\qquad\square$

Lemma 2. *Based on the mapping function (9), the convergence rate of FCM at its local optima is*

$$
\frac{\partial \mu_i}{\partial v_j} = \frac{-2\delta_{ij}m\sum_{k=1}^{n}u_{ik}^m\frac{(x_k-v_i)(x_k-v_i)^T}{d_{ik}} + 2m\sum_{k=1}^{n}u_{ik}^m u_{jk}\frac{(x_k-v_i)(x_k-v_j)^T}{d_{jk}}}{(1-m)\sum_{k=1}^{n}u_{ik}^m}
$$
$$
\tag{12}
$$

where $\delta_{ij} = 1$ if $i = j$, otherwise $\delta_{ij} = 0$.

Proof. From (3), $\frac{\partial u_{ik}}{\partial v_j}$ can be got via

$$
\frac{\partial u_{ik}}{\partial v_j} = -2\delta_{ij}\frac{u_{ik}}{(1-m)d_{ik}}(x_k - v_i) + 2\frac{u_{ik}u_{jk}}{(1-m)d_{jk}}(x_k - v_j) \quad (13)
$$

Then, the element at the submatrix at the ith row and the jth column of the Jacobian matrix $\frac{\partial \mu(v)}{\partial v}$ for the variable mapping $\mu(v)$ can be expressed as

$$
\frac{\partial \mu_i}{\partial v_j} = \frac{m\sum_{k=1}^{n}u_{ik}^{m-1}x_k\left(\frac{\partial u_{ik}}{\partial v_j}\right)^T}{\sum_{k=1}^{n}u_{ik}^m} - \frac{m\left(\sum_{k=1}^{n}u_{ik}^m x_k\right)\sum_{k=1}^{n}u_{ik}^{m-1}\left(\frac{\partial u_{ik}}{\partial v_j}\right)^T}{\left(\sum_{k=1}^{n}u_{ik}^m\right)^2}
$$

According to (4), $\frac{\partial \mu_i}{\partial v_j}$ can be further represented by

$$
\frac{\partial \mu_i}{\partial v_j} = \frac{m\sum_{k=1}^{n}u_{ik}^{m-1}x_k\left(\frac{\partial u_{ik}}{\partial v_j}\right)^T}{\sum_{k=1}^{n}u_{ik}^m} - \frac{mv_i\sum_{k=1}^{n}u_{ik}^{m-1}\left(\frac{\partial u_{ik}}{\partial v_j}\right)^T}{\sum_{k=1}^{n}u_{ik}^m} \quad (14)
$$
$$
= \frac{m\sum_{k=1}^{n}u_{ik}^{m-1}(x_k - v_i)\left(\frac{\partial u_{ik}}{\partial v_j}\right)^T}{\sum_{k=1}^{n}u_{ik}^m}
$$

By substituting (13) in (14), we can have (12). $\qquad\square$

According to [2, 3], the mass center of the data set, i.e., $v_i|_{\forall i} = \bar{x}$ is a fixed point of FCM for any $m > 1$. In other words, the mass center $v_i|_{\forall i} = \bar{x}$ is the fixed point of $\mu(v)$, and $u_{ik}|_{\forall i, \forall k} = c^{-1}$ is the fixed point of $\theta(u)$. Then, we can get

$$\frac{\partial \theta_{ik}}{\partial u_{jr}}\Big|_{\forall i,k,u_{ik}=c^{-1}} = -\frac{2m\delta_{ij}}{1-m}\frac{(x_r - \bar{x})^T(x_k - \bar{x})}{n\|x_k - \bar{x}\|^2}. \tag{15}$$

and

$$\frac{\partial \mu_i}{\partial v_j}\Big|_{\forall j, v_j=\bar{x}} = \frac{2m}{m-1}(\delta_{ij} - c^{-1})\sum_{k=1}^{n}\frac{(x_k - \bar{x})(x_k - \bar{x})^T}{n\|x_k - \bar{x}\|^2}. \tag{16}$$

Naturally the weighting exponent m contributes to the generation of the uncertainty in predictions. If the mass center $v_i|_{\forall i} = \bar{x}$ is the solution of FCM, the membership value of each data point to all clusters is c^{-1} which clearly does not depend on the level of fuzziness. Thus, the jacobian matrix at mass center point should be not less than 1 (otherwise, the mass center will be a local minimal point which is not expected in real application). By (15) and (16), we know that

$$\lambda_{\max}\left[\frac{\partial \mu(v)}{\partial v}\Big|_{\forall i, v_i=\bar{x}}\right] = \frac{2m}{m-1}\lambda_{\max}(F) \geq 1$$

where $F = \sum_{k=1}^{n}\frac{(x_k-\bar{x})(x_k-\bar{x})^T}{n\|x_k-\bar{x}\|^2} \in R^{s\times s}$ (It can be seen that F is a symmetric matrix.), and

$$\lambda_{\max}\left[\frac{\partial \theta(u)}{\partial u}\Big|_{\forall i,k,u_{ik}=c^{-1}}\right] = \frac{2m}{m-1}\lambda_{\max}(\eta) \geq 1$$

where $\eta = [\eta_{kr}]_{n\times n} = [b_r^T \times a_k]_{n\times n}$ with $a_k = \frac{x_k-\bar{x}}{n\|x_k-\bar{x}\|^2}$ and $b_r = x_r - \bar{x}$ (It can be seen that η is a symmetric matrix.). It is easy to know $\lambda_{\max}(F) = \lambda_{\max}(\eta)$, we can get $m \leq \frac{1}{1-2\lambda_{\max}(F)}$ or $m \leq \frac{1}{1-2\lambda_{\max}(\eta)}$. As expected, the theoretical region of weighting exponent is same with the region obtained by Yu [16].

4   Experimental Results

4.1   Clustering Evaluation Methods

The clustering quality is typically assessed using different types of validity measure. When the true cluster labels are unknown, the internal validity measures, including Compactness (CP), Davies-Bouldin (DB) and etc., are used to evaluate the goodness of a data partition using only quantities and features inherited from the data set. including internal and external validity. Given a dataset whose correct clusters are known, we can assess how accurately a clustering method partitions the data via the external validity measures including Clustering Accuracy (CA), Adjusted Rand Index (ARI) and etc.. In this paper, we used CP, DB, CA and ARI to evaluate the clustering results.

CP measures the average distance between every pair of data points which belong to the same cluster. More precisely, it is defined as

$$CP = \frac{1}{n} \sum_{k=1}^{c} n_k \left(\frac{\sum_{x_i, x_j \in C_k} d(x_i, x_j)}{n_k(n_k - 1)/2} \right)$$

where c is the number of clusters, n_k is the number of data points in the kth cluster, N is the total number of data points, and $d(x_i, x_j)$ is the distance between data point x_i and x_j. In FCM, $d(x_i, x_j)$ denotes the Euclidean distance between points. Ideally, the members of each cluster should be as close to each other as possible, thus, lower value of CP means better cluster performance.

DB index makes use of the similarity R_{ij} between cluster C_i and C_j which is defined on the cluster dispersion and dissimilarity between two cluster and formulated as

$$R_{ij} = \frac{s_i + s_j}{d_{ij}}$$

where $s_i = \frac{1}{n_i} \sum_{x \in C_i} d(x, v_i)$, $d_{i,j} = d(v_i, v_j)$, v_i denotes the center of cluster C_i, n_i is the number of data points in C_i. Following that, the DB index is defined as

$$DB = \frac{1}{c} \sum_{k=1}^{c} \max_{\forall j, j \neq k} R_{kj}$$

DB measures the average of similarity between each cluster and its most similar one. The lower DB index indicates better goodness of a data partition.

CA measures the number of correctly classified data points of a clustering solution compared with known class labels. Before computing CA, each cluster is relabeled with the majority cluster label, which most of data points in that cluster come from. Let m_k is the number of points with the majority cluster label in cluster k, the CA can be defined as

$$CA = \sum_{k=1}^{c} m_k / n.$$

The CA ranges from 0 to 1. Larger value indicates better clustering performance.

ARI takes into account the number of object pairs that exist in the same and different clusters, which is defined as

$$ARI = \frac{n_{11} - \frac{(n_{11} + n_{10})(n_{11} + n_{01})}{n_{00}}}{\frac{(n_{11} + n_{10})(n_{11} + n_{01})}{2} - \frac{(n_{11} + n_{10})(n_{11} + n_{01})}{n_{00}}}$$

where n_{10} is the number of object pairs belonging to the same cluster but are in the different categories (Note, cluster is obtained by FCM, category is predefined by the ground truth label), n_{01} indicates the number of object pairs belonging to the same category but in the different clusters, n_{11} indicates the number of object pairs that are in the same cluster and the same category, n_{00} indicates the number of object pairs that are placed in the different clusters and the different categories. The ARI has a value between 0 and 1, with the more the value approximates to 1 the higher the agreement is.

4.2 Synthetic Data

For a preliminary comparison study, we test the proposed method on simulated data composed by a variable number of independent, uncorrelated and
randomly distributed Gaussian clusters. The Gaussian model is often regarded
as a benchmark in literature for studying the clustering performance including
the convergence rate. Simulations allow controlling the parameters influencing
cluster recovery performance such as the feature space dimensionality s, the true
number of clusters s, and the separation among clusters α (in standard deviation
units).

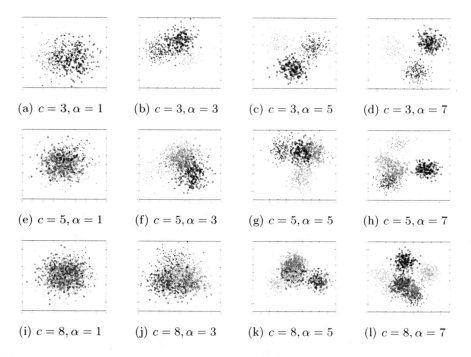

(a) $c = 3, \alpha = 1$ (b) $c = 3, \alpha = 3$ (c) $c = 3, \alpha = 5$ (d) $c = 3, \alpha = 7$

(e) $c = 5, \alpha = 1$ (f) $c = 5, \alpha = 3$ (g) $c = 5, \alpha = 5$ (h) $c = 5, \alpha = 7$

(i) $c = 8, \alpha = 1$ (j) $c = 8, \alpha = 3$ (k) $c = 8, \alpha = 5$ (l) $c = 8, \alpha = 7$

Fig. 1. Demonstration of simulated data sets based on their first two principal components

In the experiments, c cluster centers are randomly drawn according to a mul-
tivariate normal distribution in s dimensions:

$$N\left(0, \frac{\alpha^2}{2s}\mathbf{I}_{s \times s}\right)$$

Using $\frac{\alpha^2}{2s}$ as scaling factor of the variance, the expectation value of the Euclidean distance between two centers is equal to α^2, independently of s. In order
to control the minimum clusters separation, any two cluster centers are closer

Table 1. FCM results on synthetic data with varying the clusters overlap degree α and the number of clusters c

c	α	UpM	m	λ_{max}	$\left.\frac{\partial\theta(u)}{\partial u}\right|_{u^*}$	CP	DB	ARI	CA
	1	1.3778	1.20	0.9671	4.0251	2.7719	0.1915	0.5983	
	2	1.4528	1.10	0.8047	3.9657	2.5173	0.4405	0.7750	
	3	1.7202	1.15	0.5638	3.7788	1.9961	0.6852	0.8817	
3	4	1.8166	1.10	0.4159	3.5687	1.6844	0.8444	0.9450	
	5	1.8940	1.10	0.2123	3.4910	1.4893	0.9554	0.9850	
	6	2.2158	1.10	0.1327	3.1940	1.2869	0.9900	0.9967	
	7	2.5051	1.10	0.0947	3.1532	1.1701	0.9950	0.9983	

c	α	UpM	m	λ_{max}	$\left.\frac{\partial\theta(u)}{\partial u}\right|_{u^*}$	CP	DB	ARI	CA
	1	1.3089	1.25	0.9957	3.9024	2.5402	0.0610	0.3670	
	2	1.4017	1.30	0.9885	3.8665	2.6029	0.2577	0.5480	
	3	1.5389	1.15	0.8575	3.6804	2.1985	0.4535	0.7300	
5	4	1.6837	1.05	0.8046	3.4782	1.9920	0.6468	0.8370	
	5	1.8419	1.40	0.7505	3.3336	1.9895	0.6487	0.8800	
	6	1.9882	1.10	0.1044	2.7554	1.0766	0.9875	0.9950	
	7	2.1261	1.05	0.0814	2.6863	1.0659	0.9900	0.9960	

c	α	UpM	m	λ_{max}	$\left.\frac{\partial\theta(u)}{\partial u}\right|_{u^*}$	CP	DB	ARI	CA
	1	1.2928	1.05	0.9682	3.7504	2.2583	0.0221	0.2281	
	2	1.3465	1.25	0.9651	3.6717	2.2270	0.1968	0.4481	
	3	1.4585	1.25	0.9495	3.5706	2.1423	0.3709	0.6406	
8	4	1.4379	1.25	0.8841	3.4808	2.1113	0.4748	0.7181	
	5	1.5281	1.10	0.4803	2.9897	1.3823	0.7964	0.8850	
	6	1.7079	1.50	0.4043	2.8264	1.3379	0.9001	0.9525	
	7	1.7718	1.45	0.3687	2.7685	1.3226	0.9315	0.9694	

than $\alpha/2$. For each cluster, we generated 200 Gaussian distributed samples with unit variance. With this strategy, FCM can provide optimal clustering results.

We generated 27 data sets by varying two types of spatial parameters: decreasing the degree of overlap between the clusters ($\alpha \in \{1, 2, 3, 4, 5, 6, 7, 8, 9\}$), and increasing the number of clusters ($c = \{3, 5, 8\}$) so that data are more structured and then more complex. To provide an insight into the spatial distribution of the simulated clusters, the principal component analysis (PCA) is performed to reduce the dimensionality. Fig. 1 shows the clusters distribution of the 12 data sets with $s = 10$ in different clusters as the parameter α spans from 1 to 7. It can be seen that the clusters are separated with the increasing of α. For $\alpha = 1$, the clusters overlap almost completely which results in few chances of recovering the clustering structure. For $\alpha = 7$, the clusters touch each other only by a small amount on the borders which leads to easy identifying the data clusters.

The data-driven optimal upper bound of the weighting exponent m of each simulated data set can be calculated as shown in the UpM column of Table 1. It can be seen that the upper bound of m tends to become smaller with the increasing number of clusters c or with the decreasing of α. The reason is that both of them increase the pattern complexity . Especially, as the decreasing of α, the overlapping of Gaussians in the original mixture becomes larger, i.e., the identification of hidden cluster structure becomes more difficult.

For each data set, different m values are tested from 1 to UpM with step 0.05 as shown in Fig.2. From this figure, it is interesting that the best clustering accuracy is obtained at the point where the spectral radius of Jacobian matrix is smallest, which means that the final optimal is a point with fast convergence rate. In this case, we can use the spectral radius to select the proper value for the weighting exponent m.

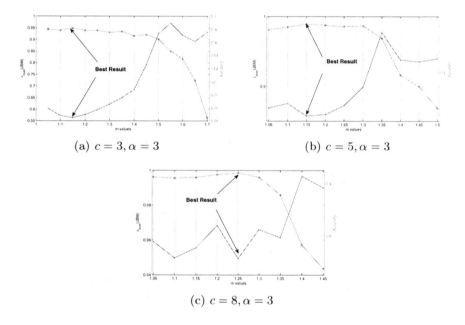

(a) $c = 3, \alpha = 3$ (b) $c = 5, \alpha = 3$

(c) $c = 8, \alpha = 3$

Fig. 2. Demonstration of the spectral radius of the Jacobian matrix ($\lambda_{max}(JBM)$) and the clustering accuracy by varying the weighting exponent (m) values on simulated data sets

Table 1 listed four evaluation metrics CP, DB, ARI and CA. Meanwhile, the corresponding Jacobian matrix $\frac{\partial \theta(u)}{\partial u}$ and its spectral radius at the local optimal point u^* are computed as shown in the fifth column of Table 1. As the increasing of α, i.e., as the overlapping of the Gaussians in the original mixture becomes smaller, the clustering performance becomes better (i.e., CA and ARI becomes larger, while CP and DB becomes smaller), which indicates that it becomes easier for FCM to identify the hidden cluster patterns. The $\lambda_{\max}\left[\frac{\partial\theta(u)}{\partial u}\Big|_{u^*}\right]$ decreases rapidly with the increasing of α, i.e., it is faster for FCM to converge to the corresponding local minimum when the overlapping of the Gaussians becoming smaller. In this case, we confirm that $\frac{\partial\theta(u)}{\partial u}$ can be used to estimate the convergence rate of the FCM algorithm.

4.3 Real-World Data

Five real world data sets from the UCI Repository of Machine Learning Databases [1] are used to verify the theoretical conclusions in last section. Table 2 gives the detailed information about the data sets. Each data set is standardized via zero-mean normalization method as follows. If x_i stands for the ith point in the data set, x_{ij} indicates the jth feature of x_i, then

$$X_{ij} = \frac{x_{ij} - \sum_{k=1}^{n} x_{kj}/n}{\sqrt{\sum_{k=1}^{n}(x_{kj} - \sum_{l=1}^{n} x_{lj}/n)^2/(n-1)}}.$$

Table 2. Real World Data Sets

Dataset	n	s	c	UpM	m	CP	DB	ARI	CA
Sonar	208	60	2	1.6388	1.6	9.7927	2.4583	0.5243	0.6154
Glass	214	9	6	3.1726	1.35	2.2885	1.1492	0.6834	0.5701
PimaIndiansDiabetes	768	8	2	2.0475	1.95	3.4035	2.1115	0.5918	0.7148
Vowel	990	10	11	1.7787	1.15	3.1803	1.4419	0.8458	0.2677
Waveform	5000	21	3	2.8935	2.85	5.3962	1.5882	0.6820	0.6360

Since the true label of the real-world data is known, the external evaluation metrics can be computed. By computing $\lambda_{\max}(F)$ for each data set, the theoretical upper bound of weighting exponent m can be given as shown in the last two columns of Table 2 . It can be seen that the result is same with the result listed by Yu [16], which further indicates that Jacobian matrix is a useful way to analyze the property of FCM. For each data set, different m values are tested in the range of (1,UpM], and the results with the smallest spectral radius of the Jacobian Matrix at the convergence point are recorded in Table 2.

5 Conclusions

In order to analyze the property of FCM, we introduce Jacobian matrix of the mapping function in alternative iteration. By analyzing the spectral radius of the Jacobian matrix at the mass center, a region of weighting exponent m is theoretically given. The theoretical results are proven by a series of experiments. In this paper, we provide a new method to judge the convergence rate of algorithm via Jacobian matrix. In the future, more algorithm will be analyzed in this framework.

Acknowledgement. This work was supported in part by the National Natural Science Foundation of China under Grant 61375062, Grant 61370129, and Grant 61033013, the Ph.D Programs Foundation of Ministry of Education of China under Grant 20120009110006, the Opening Project of State Key Laboratory of Digital Publishing Technology, and the Fundamental Research Funds for the Central Universities under Grant 2014JBM029 and Grant 2014JBZ005.

References

1. Bache, K., Lichman, M.: UCI machine learning repository (2013)
2. Bezdek, J.: Patern recognition with fuzzy objective function algorithms. Plenum Press (1981)
3. Bezdek, J., Hathaway, R., Howard, R., Wilson, C., Windham, M.: Local convergence analysis of a grouped variable version of coordinated descent. J. Class. Theory Appl. 54, 471–477 (1987)
4. Cannon, R., Dave, J., Bezdek, J.: Efficient implementation of the fuzzy c-means clustering algorithms. IEEE Trans. on Pattern Anal. Machine Intell. 8, 248–255 (1986)
5. Dunn, J.: A fuzzy relative of the isodata process and its use in detecting compact, well-separated clusters. J. Cybern. 3, 32–57 (1973)
6. Fadili, M., Ruan, S., Bloyet, D., Mayoyer, B.: On the number of clusters and the fuzziness index for unsupervised fca application to bold fmri time series. Med. Image Anal. 5, 55–67 (2001)
7. Groll, L., Jakel, J.: A new convergence proof of fuzzy c-means. IEEE Trans. on Fuzzy Systems 13, 717–720 (2005)
8. Hoppner, F., Klawonn, F.: A contribution to convergence theory of fuzzy c-means and derivatives. IEEE Trans. on Fuzzy Systems 11, 682–694 (2003)
9. Hoppner, F., Klawonn, F., Kruse, R., Runkler, T.: Fuzzy cluster analysis: methods for classification, data analysis and image recognition. John Wiley & Sons (1999)
10. Jain, A., Dubes, R.: Algorithms for clustering data. Prentice-Hall, Englewood Cliffs (1988)
11. Kaufman, L., Rousseeuw, P.: Finding groups in data: an introduction to cluster analysis. Wiley-Blackwell, New York (2005)
12. Ma, J., Xu, L., Jordan, M.: Asymptotic convergence rate of the em algorithm for gaussian mixtures. Neural Computation 12, 2881–2907 (2000)
13. Ozkan, I., Turksen, I.: Upper and lower values for the level of fuzziness in fcm. Information Sciences 177, 5143–5152 (2007)
14. Pal, N., Bezdek, J.: On cluster validity for the fuzzy c-mean model. IEEE Trans. on Fuzzy Systems 3, 370–379 (1995)
15. Selim, S., Ismail, M.: On the local optimality of the fuzzy isodata clustering algorithms. IEEE Trans. on Pattern Anal. Machine Intell. 8, 284–288 (1986)
16. Yu, Y., Cheng, Q., Huang, H.: Analysis of the weighting exponent in the fcm. IEEE Trans. on Syst., Man, Cybern, Part B. 34, 634–639 (2005)

Dividing Traffic Sub-areas
Based on a Parallel K-Means Algorithm

Binfeng Wang[1], Li Tao[1], Chao Gao[1], Dawen Xia[1,2], Zhuobo Rong[1],
Wu Wang[1], and Zili Zhang[1,3,*]

[1] School of Computer and Information Science
Southwest University, Chongqing 400715, China
[2] School of Information Engineering
Guizhou Minzu University, Guiyang 550025, China
[3] School of Information Technology
Deakin University, VIC 3217, Australia
zhangzl@swu.edu.cn

Abstract. In order to alleviate the traffic congestion and reduce the
complexity of traffic control and management, it is necessary to exploit
traffic sub-areas division which should be effective in planing traffic. Some
researchers applied the K-Means algorithm to divide traffic sub-areas
on the taxi trajectories. However, the traditional K-Means algorithms
faced difficulties in processing large-scale Global Position System(GPS)
trajectories of taxicabs with the restrictions of memory, I/O, computing performance. This paper proposes a Parallel Traffic Sub-Areas Di-
vision(PTSD) method which consists of two stages, on the basis of the
Parallel K-Means(PKM) algorithm. During the first stage, we develop a
process to cluster traffic sub-areas based on the PKM algorithm. Then,
the second stage, we identify boundary of traffic sub-areas on the base
of cluster result. According to this method, we divide traffic sub-areas
of Beijing on the real-word (GPS) trajectories of taxicabs. The experiment and discussion show that the method is effective in dividing traffic
sub-areas.

Keywords: Traffic Sub-Areas, GPS Trajectories, K-Means, MapReduce.

1 Introduction

With the rapid development of urbanization and the explosive growth of vehicles,
traffic congestion becomes a critical problem in metropolis [1]. Most of these
cities spend much money in planning urban traffic for alleviating the traffic
congestion. The strategy of dividing traffic sub-areas[2] is adopted to plan traffic
effectively, control the traffic flows and alleviate traffic congestion. The division
of traffic sub-areas is to divide the whole traffic area into many sub-areas as a
multi-area hierarchical control system, based on a certain extent of similarity and
correlation. Nowadays, traffic sub-areas division has become the significant part

---

* Corresponding author.

R. Buchmann et al. (Eds.): KSEM 2014, LNAI 8793, pp. 127–137, 2014.

of traffic planning. Beyond that, traffic sub-areas division is a powerful tool to analyze complex traffic network. A complete urban traffic system is so large and complex that it has difficulties in analyzing the traffic problem. We divide the urban traffic system into different traffic sub-areas and then study each traffic sub-areas in order to reduce the complexity of analysis, in which case we can improve the availability, reliability, instantaneity of the traffic system.

The K-Means algorithm is a well-known partition algorithm [3], and is commonly used in analyzing the data of GPS trajectories for mining internal relationships among these data. Lv et al. [4] adopted K-Means algorithm in dividing traffic sub-areas. The experiment showed that the result of dividing conformed to the existing traffic sub-areas to some extent. However, with the explosive growth of GPS trajectories of taxicabs, the traditional K-Means algorithm exists some bottlenecks in processing massive data, such as high memory cost, out of memory, high I/O overload, low computing performance, poor scalability and reliability and so on. So the K-Means algorithm of the single machine environment is not suitable for processing massive GPS data of taxicabs.

To meet the requirement of processing large-scale data, some researchers proposed few parallel K-Means algorithms [5-7], but these algorithms still have difficulties in processing massive data(e.g., high I/O overload, low computing performance). In 2006, Apache Software Foundation proposed the Hadoop framework, including Hadoop Distributed File System (HDFS [8]) and Hadoop MapReduce [9], etc. As a typical method, the MapReduce framework provides effective techniques for dividing traffic sub-areas by processing massive data of GPS trajectories. Nguyen et al. implemented parallel two-phase K-Means algorithm (Par2PK-means) [10], devoting to overcome the limitations of high consumption and it was divided into two phases which include Mapper and Reducer. Nonetheless, the algorithm ignored the combine stage. Moreover, Zhou et al. [11] designed a parallel K-Means algorithm based on the MapReduce framework, and implement automatic classification of large-scale document. However, he failed to consider fully the choice of initial cluster centers. At the same time, these methods mentioned above don't be applied in traffic sub-areas division based on large-scale GPS trajectories of taxicabs.

In this paper, we present a effective method for dividing traffic sub-areas. The contributions of this work are summarized as follows:

- We implement the Parallel K-Means(PKM) algorithm in the MapReduce environment, devoting to solve effectively the existing problems of processing large-scale data of GPS trajectories. According to the experiments, we find that the method have a better speedup and a higher efficiency.
- We apply the method based on the Parallel Traffic Sub-Areas Division(PTSD) to divide traffic sub-areas by using the GPS trajectories of taxicabs in Beijing. Especially, this work can provide helpful suggestion for constructing the traffic system reasonably.

The remainder of this paper is organized as follows. First, the method of dividing traffic sub-areas is showed in the Section 2. Then, the process of dividing traffic sub-areas is described in detail in Section 3. Next, the discussion about the method is presented in Section 4. Finally, the paper is concluded in Section 5.

2   The Method of PTSD

In this section, the implemented method of dividing traffic sub-area is depicted in detail.

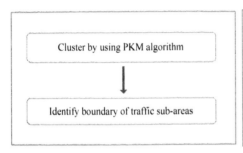

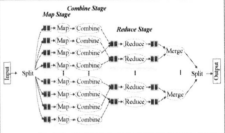

Fig. 1. The procedure of dividing traffic sub-areas

Fig. 2. The execution procedure of PKM algorithm

The process of dividing traffic sub-areas is divided into two steps, as shown in Fig. 1.

(1)We apply the PKM algorithm to cluster on the basis of GPS trajectories of taxicab in Beijing.

(2)We divide the traffic sub-areas based on the cluster result.

2.1   PKM Algorithm

The computational work of PKM algorithm is mainly composed of the following two aspects.

(1) The algorithm assign each data vector to the closest cluster.

(2) The algorithm compute new cluster center.

In this work, we implement the aforementioned steps based on Hadoop using MapReduce. In the beginning, we select randomly K vectors as the initial cluster center which are stored in the HDFS as global variables. Then, we update the cluster centers after iterating. The execution procedure of the PKM algorithm is depicted in Fig. 2, and the iteration of the PKM algorithm includes Map, Combine and Reduce stages.

2.1.1   The Map Stage

The Map stage is composed of two steps. First, it calculates distance between each data object and each cluster center. Then, it assigns data object to the closest cluster according to the shortest distance, and then generate the output of a $< key, value >$ pairs.

Algorithm 1. Map(key, $value$)

Input:

key: the offset of data, $value$: the corresponding data, $pre\_centers$: the cluster centers.

Output:

$< key_1, value_1 >$, key_1: the index of the closest center, $value_1$: the data.

1: Get centers from the $pre\_centers$;

2: Store vectors from value to an array;

3: min_Dis = Double. MAX_VALUE;

4: $index = 0$;

5: **for** i = 0 to $centers$.length **do**

6:    $distance$ = GetEnumDistance (values[i], centers[i]);

7:      **if** $distance < min\_Dis$ **then**

8:         $min\_Dis = distance$;

9:         $index$ = i;

10: **end for**

11: Set $index$ as key_1;

12: Set $value_1$ as a string consist of the sum of different dimensions and num;

13: **return** $< key_1, value_1 >$ pairs;

The execution of map function is shown in Algorithm 1.

Algorithm 2. Combine(key_1, $iteration$)

Input:

key_1: the identifier of clusters, $iteration$: the vectors assigned to the same cluster.

Output:

$< key_2, value_2 >$, key_2: the identifier of clusters, $value_2$: the sum of the vector in the same cluster and the number of number.

1: Record the sum of vector of each dimension in the same cluster to an array;

2: Record the sum of vector number in the same cluster to the variable($sumCount$);

3: **for** i = 0 to $iteration$.length **do**

4:    The number of vectors plus one;

5:       **for** j=0 to centers [0]. length **do**

6:            Sum up the values of different dimensions of the vector to the array;

7:       **end for**

8: **end for**

9: Set index as key_2;

10: Set $value_2$ as a string consists of vector number and sum of vector;

11: **return** $< key_2, value_2 >$ pairs;

2.1.2   The Combine Stage

The Combine stage mainly includes two steps. Above all, it extracts all vectors from the value which is the output of Map function. Then, it sums up all the vector extracted, and recorde the number of vectors in the collection.

The execution of Combine function is shown in Algorithm 2.

2.1.3   The Reduce Stage

The Reduce stage consists of two steps. First of all, it extracts vectors from the value which is the output of Combine function. Then, it sums up all the vector and the number of vectors in the same cluster as well as the combine stage. Finally, it divides the sum of vector by the number of vector.

The execution of reduce function is shown in Algorithm 3.

Algorithm 3. Reduce(key_2, *iteration*)

Input:

key_2: the identifier of clusters, *iteration*: the list of the vector assigned to the same cluster.

Output:

$< key_3, value_3 >$, key_3: the identifier of clusters, $value_3$: the new cluster center.

1: Record the sum of the vector of each dimension in the same cluster to an array;

2: Record the sum of the vector number in the same cluster to the variable($sumCount$);

3: for i = 0 to *iteration.length* **do**

4:   The number of vectors plus one;

5:   **for** j=0 to centers [0]. length **do**

6:   Summing up the values of different dimensions of the vector to the array;

7:   **end for**

8: end for

9: Divide the value of the different dimension array by the sum of the vector number in the same cluster, and then get the new center;

10: Set *index* as key_3;

11: Set $value_3$ as a string consist of the new cluster center;

12: return $< key_3, value_3 >$ pairs;

2.2   Boundary Identification for Traffic Sub-areas

Since distinguishing the boundary among the different areas by the coordinates based on the ARCGIS platform is very difficult, we need to divide the boundaries of each area based on the result of clustering. The method of dividing boundary consists of three steps, as shown in Fig. 3.

Step 1. We establish the coordinate system, and take (0, 0) as the origin of coordinates, and then divide equally coordinate system into n parts ,as shown in Fig. 3(a).

Step 2. We match each center of the cluster to the origin of coordinates, and the other point is mapped to the coordinate system of the same cluster. And the farthest point of each part is selected (e.g., the P in Fig. 3(b)).

Step 3. We connect these selected points, and then gain a border area ,as shown in Fig. 3(c).

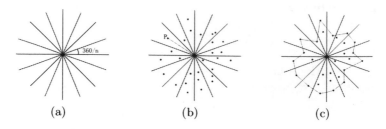

Fig. 3. The methods of division: (a) The coordinates of N equal, (b)The selection of boundary point, (c)The connection of boundary point

3   Traffic Sub-areas Division

In this section, we divide the traffic sub-areas of Beijing based on GPS trajectories of taxicabs by using the method of PTSD.

3.1   Data

In this work, we use the GPS trajectories of taxicab in Beijing from datatang[1], which includes 12000 GPS trajectories of taxicab generated in Nov. 2012. After processing, the data scale is compressed from 50GB to 15.1GB. And the data is shown as ASCII text with a comma separator (e.g., data item and order: car single, events, running status, GPS time, GPS longitude, GPS latitude, GPS speed, direction of GPS, GPS status, and the corresponding records: 123456, 0, 0, 20110414160613, 116.4078674, 40.2220650, 21274, 1 <0x0d0x0a>). In this work, we extract the latitude and longitude attribute of getting on and off taxi from the datasets mentioned above, for dividing traffic sub-areas.

3.2   The Results of Cluster

The traffic sub-areas are the collection of nodes with correlation, which can provide support for the management and planning of urban traffic system.

We apply the PKM algorithm to cluster the GPS trajectories of taxicab, and divide the overall area of Beijing into one hundred traffic sub-areas. Meanwhile, we present the results of clustering on the ARCGIS platform based on the road network of Beijing. The results of clustering are shown in Fig. 4.

In the Fig. 4, each point is the coordinate point of latitude and longitude where passengers get on or off taxicab, and each area of different colors represents a cluster. Of course, the different clusters stand for different traffic sub-areas of Beijing. In the lower right corner of the figure, the green area is a larger version of the corresponding area on the left.

---

[1] http://www.datatang.com

Fig. 4. The results of clustering

3.3   The Results of Division

We divide the boundary of each traffic sub-areas using the method mentioned in section 2.2, and the results of partition are shown in Fig. 5.

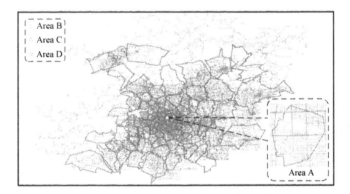

Fig. 5. The results of dividing

Compared with the real map, Area B, C and D are the typical areas which have obvious characteristics of partition area, and the stream of people and automobile in these areas is high. The area B is a traffic area includes the national stadium, the area C includes the Beijing exhibition center, and the area D includes the Village and the workers' stadium. A traffic sub-areas is shown particularly in the lower right corner of the figure, which includes Beijing North Railway Station, Beijing University of Aeronautics and Astronautics, Tsinghua University, Beijing University of Posts and Telecommunications, Renmin University of China, Zhongguancun, Beijing's Olympic Sports Center Gymnasium and so on. According to the analysis of the famous places, we find that the traffic sub-areas have some similarities in the traffic and business. For example, in some ways, population of these universities have a consistent travel condition. Moreover, these area have some similar traffic condition in some extent.

Based on the analysis above, we find that the results of traffic sub-areas division are consistent with the traffic condition of Beijing. It indicates that traffic sub-area division is effective and can provide support for planning traffic.

4   Discussion

In this section, we discuss the reasons for employing the solution to dividing traffic sub-areas of Beijing. It is equivalent to evaluating the performance of the solution through several experiments.

In experiment, the Hadoop cluster consists of one master machine and five slave machines. Meanwhile, all experiments are performed on Intel Xeon (R) E7-4820 2.00GHz CPU (4-core), 4.0GB Memory, and 150GB Hard Disk.

First, we use three real datasets (Iris, Haberman's Survival and Ecoil)[2] which are shown in table 1, in order to evaluate the accuracy of the PKM algorithm. Then, according to the properties of the Iris datasets, we process the real Iris datasets into 80MB, 160MB, 320MB, 640MB, and 1280MB in the other experiments, for evaluating the performance of the PKM algorithm on feasibility, speedup[12], scalability[12] and reliability.

Table 1. The detail datasets of evaluating the accuracy

Datasets	The number of objects	The number of attributes
Iris	150	4
Haberman's Survival	306	3
Ecoil	336	8

4.1   The Accuracy of PKM Algorithm

The accuracy of algorithm is denoted as $R = A/C$, where A is the number of correct cluster objects, and C is the total number of cluster objects. The accuracy comparison of K-Means algorithm in the single machine and the PKM algorithm based on MapReduce is shown in Fig. 6.

Figure 6 shows that the accuracy of the PKM algorithm is mostly equal to the K-Means algorithm in the single machine environment. The experimental results indicate that the PKM algorithm is effective.

4.2   The Feasibility of PKM Algorithm

Firstly, the K-Means algorithm is executed in the single machine environment using the five different datasets which are 80MB, 160MB, 320MB, 640MB and 1280MB. Then, PKM algorithm also run on the five different datasets mentioned

---

[2] http://archive.ics.uci.edu/ml/datasets.html

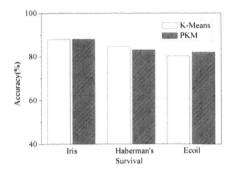
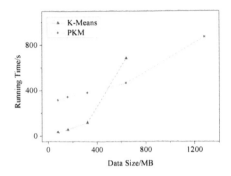

Fig. 6. The accuracy of the K-Means algorithm and the PKM algorithm

Fig. 7. The Feasibility of PKM algorithm

above. Finally, we compare the feasibility of the two different algorithms. The experimental result is shown in Fig. 7.

According to this experiment, we find the K-Means algorithm based on the single machine environment has some restrictions in processing large-scale data, such as high cost of memory and other resources, the degradation of performance. When the K-Means algorithm processes 640MB datasets, the report "out of memory" appear in the console window, which means the algorithm can not work noamally. (So, Fig. 7 does not show the running time of the K-Means algorithm when the datasets is over 640MB datasets). However, the K-Means algorithm based on MapReduce environment can preform tasks of computing effectively in facing large-scale data, which indicates that the PKM algorithm is feasible.

At the same time, the running time of the K-Means algorithm in the single machine environment is shorter than the PKM algorithm in processing small data. In the MapReduce environment, the communication and interaction of each node consume a certain amount of time (e. g. , the start of Job task and Task task, communication between NameNode and DataNode), leading the running time longer than actual computation time. With the gradual growth of datasets scale, the efficiency of PKM algorithm improve obviously, and the superiority of processing efficiency is obvious. The analysis shows that the PKM algorithm is feasible.

4.3   The Speedup of PKM Algorithm

With the number of DataNodes increasing gradually, we evaluate the speedup of the PKM algorithm on datasets with different sizes. The number of DataNode vary from 1 to 5. And the size of the datasets increases from 80MB to 1280MB. The experimental results are shown in Fig. 8.

Figure 8 shows that the running time of the PKM algorithm is proportionally decreasing with the number of Datanodes increasing. In addition, the PKM algorithm can efficiently complete the computing tasks in large-scale datasets. The results indicate that the PKM algorithm has a good speedup.

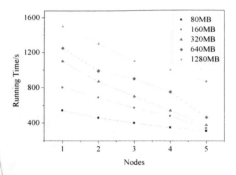 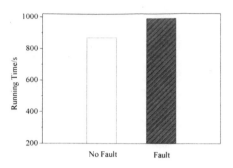

Fig. 8. The speedup of PKM algorithm   **Fig. 9.** The reliability of PKM algorithm

4.4 The Reliability of PKM Algorithm

To evaluate the reliability of PKM algorithm, we shut down a node to check whether the PKM can normally run and obtain the same results. The experimental results are shown in Fig. 9.

In the single machine environment, the K-Means algorithm can't be executed successfully when the machine is faulty. So the reliability of the K-Means algorithm in the single machine is very poor.

In the MapReduce environment, although the execution time of PKM algorithm is longer than normal condition with a DataNode stopping, the PKM algorithm can execute normally and output the same results. It shows that the cluster of Hadoop has a good reliability and fault tolerance. When a node cannot execute tasks, the JobTracker will automatically assign failed tasks of the fault node to other free nodes in the cluster. The analysis mentioned above shows that the PKM algorithm has a good reliability.

According to the discussion above, it is obvious that the method can find out accurately and rapidly the result of cluster. Moreover, the mothod can easily cope with the large-scale GPS trajectories.

5   Conclusion

In this paper, we propose the method of dividing traffic sub-areas which consists of two stages. Then, according to the method, we divide the traffic sub-areas of Beijing. Finally, we discuss the performance of the PKM algorithm. The experiment and discussion show that the method is effective in dividing traffic sub-areas.

Acknowledgment. This work was partially supported by the National Science and Technology Support Program of China (No. 2012BAD35B08), and the Fundamental Research Funds for the Central Universities (Nos. XDJK2012C028, XDJK2014C142).

References

1. Palma, D.A., Lindsey, R.: Traffic congestion pricing methodologies and technologies. Transportation Research Part C: Emerging Technologies 19(6), 1377–1399 (2011)
2. Wåhlberg, A.E., Dorn, L., Kline, T.: The effect of social desirability on self reported and recorded road traffic accidents. Transportation Research Part F: Traffic Psychology and Behaviour 13(2), 106–114 (2010)
3. Han, J., Kamber, M., Pei, J.: Data mining: concepts and techniques. Morgan Kaufmann (2006)
4. Lv, Y.Q., Qin, Y., Jia, L.M., et al.: Dynamic Traffic Zone Partition Based on Cluster Analysis of Taxi GPS Data. Logistics Technology 29(9), 86–88 (2010)
5. Pham, D.T., Dimov, S.S., Nguyen, C.D.: A two-phase k-means algorithm for large datasets. Journal of Mechanical Engineering Science 218(10), 1269–1273 (2004)
6. Kantabutra, S., Couch, A.L.: Parallel K-means clustering algorithm on NOWs. NECTEC Technical Journal 1(6), 243–247 (2000)
7. Kraj, P., Sharma, A., Garge, N., et al.: ParaKMeans: Implementation of a parallelized K-means algorithm suitable for general laboratory use. BMC Bioinformatics 9(1), 200 (2008)
8. Shvachko, K., Kuang, H., Radia, S., Chansler, R.: The Hadoop Distributed File System. In: 26th IEEE Symposium on Mass Storage Systems and Technologies, pp. 1–10. Incline Village (2010)
9. Dean, J., Ghemawat, S.: MapReduce: simplified data processing on large clusters. Communications of the ACM 51(1), 107–113 (2008)
10. Nguyen, C.D., Nguyen, D.T., Pham, V.-H.: Parallel two-phase K-means. In: Murgante, B., Misra, S., Carlini, M., Torre, C.M., Nguyen, H.-Q., Taniar, D., Apduhan, B.O., Gervasi, O. (eds.) ICCSA 2013, Part V. LNCS, vol. 7975, pp. 224–231. Springer, Heidelberg (2013)
11. Zhou, P., Lei, J., Ye, W.: Large-Scale Data Sets Clustering Based on MapReduce and Hadoop. Journal of Computational Information Systems 7(16), 5956–5963 (2011)
12. Xu, X., Jager, J., Kriegel, H.P.: A Fast Parallel Clustering Algorithm for Large Spatial Databases. Data Mining and Knowledge Discovery 3, 263–290 (1999)

On the Social Network Based Semantic Annotation of Conceptual Models

Hans-Georg Fill

University of Vienna, RG Knowledge Engineering, 1090 Vienna, Austria
hans-georg.fill@univie.ac.at
http://homepage.dke.univie.ac.at/fill

Abstract. The semantic annotation of conceptual visual models permits to leverage the contained semantic information to a machine processable level. At the same time the intuitive and non-technical nature of graphical models can be maintained. Although this leads to direct benefits in terms of potential analysis functionalities, the addition of annotations for large model corpora requires quite some effort by the modelers. In order to contribute to the efficiency when adding semantic annotations, we propose an approach that links information on semantic annotations to information available from social network applications. In this way the addition of semantic annotations can be facilitated in terms of communication support, economic benefits, and technical opportunities. For the design of the approach we use a specifically developed modeling method, which allows to reason about the requirements at a high abstraction level and permits to add more formal specifications, e.g. to define behavior. A first prototype of the modeling method has been implemented using the freely available ADOxx meta modeling platform based on the SeMFIS toolkit.

Keywords: Semantic annotation, Conceptual Model, Social Network Application, Meta Modeling, ADOxx.

1 Introduction

In the past, several areas in information systems research and practice have made use of conceptual visual models. Prominent examples can be found in the field of requirements analysis and engineering [30], business process management and business-IT alignment [25] as well as in software engineering and IT architecture management [26]. Among their many applications they are used to support the communication between developers and users [30], to help analysts understand and evaluate a domain by simulating certain behavior [19], for documenting requirements and for the configuration and the engineering of IT systems, e.g. to realize model-driven-architectures. A core aspect of these types of models is that they are directed towards supporting human communication and understanding and do not aim for an entirely machine processable representation of a domain [27]. They are thus based on a formal syntax that is complemented by

R. Buchmann et al. (Eds.): KSEM 2014, LNAI 8793, pp. 138–149, 2014.

formal semantic defintions where needed. In this way the models can be used by domain experts in an intuitive way and then gradually enriched with formal semantics depending on the required machine processing capabilities.

One particular type of such enrichments are *semantic annotations* that offer a way to make unstructured natural language information contained in models processable [15]. The main idea hereby is to add mappings to formal semantic schemata without changing the original structure and purpose of conceptual models but to provide semantic processing functionalities as needed. Especially in business process management where conceptual models are used extensively in many companies, several approaches have been described for applying and using semantic annotations. Besides tasks such as semantic similarity matching of models [6], the user-centric visualization of process views [15], or the automation of processes using web services [18], semantic annotations have recently also been used to conduct complex process analysis, e.g. for risk management [10].

Despite the potential advantages the effort for adding this specific semantic information is still very large. It is therefore of particular interest to enhance this process and thus increase its efficiency. Although several techniques have been developed for suggesting annotations based on natural language processing or context-based similarity matching, the user still needs to select the correct annotation based on his or her knowledge. Another direction to enhance the process is by supporting *human annotators* in a way that they can perform the annotations more efficiently. Besides organizational measures, also the use of crowd-sourcing mechanisms that distribute the annotation tasks to a multitude of people may offer potential benefits for this purpose. However, to realize such approaches and ideally combine them with the automated techniques for bringing up suggestions for possible annotations, the appropriate technical foundations have to be established and made available to the community.

Following ideas presented in [16], we describe an approach that integrates the information provided by social network applications with semantic annotations. In particular we build upon the SeMFIS approach for conducting loosely-coupled semantic annotations of arbitrary conceptual models using separate annotation models [11]. By integrating information on actors, resources, and groups from social network applications, additional functionalities for supporting the definition of semantic annotations can be provided. This concerns in particular functionalities for supporting the collaboration on annotation tasks and for the personalization of the user experience. In addition, the integration of social network applications provides technical opportunities through re-using services such as the workflows for user authentication. In order to analyze the requirements and dependencies of such an approach, we extend the SeMFIS modeling method and present a first implementation on the ADOxx meta modeling platform.

The remainder of the paper is structured as follows: In section 2 we will briefly discuss some foundations for our approach. Subsequently, in section 3 the approach itself is presented which is then discussed in section 4. The paper ends with a conclusion and an outlook in section 5.

2   Foundations

In order to clarify some of the terms we are going to use within the scope of this paper in regard to conceptual models, their semantic annotations, and social network functionalities, we will briefly outline these three areas.

2.1   Conceptual Visual Models

For characterizing the core constituents of conceptual models we refer to a framework that has been been described by Karagiannis and Kuehn [23]. In this framework the top level concept are modeling methods which consist of a modeling technique and mechanisms and algorithms. The modeling technique is further specialized by a modeling language and an according modeling procedure. The modeling language is then defined by its syntax, semantics, and visual notation and the modeling procedure by its steps and results. In contrast to other approaches - as e.g. described in [17] - the notation of the modeling language defines the graphical representation of the syntax and may thus also be modified independently of the syntax [9].

The semantics of the modeling language is assigned to the syntax by mapping the elements of the syntax to a semantic schema. This schema may be natural language or may be some formal semantics, e.g. when using a formal schema such as an ontology. The mechanisms and algorithms of the modeling method can be specialized by generic, specific, and hybrid mechanisms and algorithms. Thereby, generic mechanisms and algorithms can be applied to any modeling language, specific mechanisms and algorithms only to a particular subset of modeling languages and hybrid mechanisms and algorithms can be parameterized to be applicable for several modeling languages.

The syntax of a modeling language can be further specialized by an abstract syntax, which may be represented by a meta model, and the concrete syntax, which is represented by the resulting model instance. The meta model is thus a model that defines the language for expressing a model. The model is then the concrete realization using this language.

2.2   Semantic Annotation of Conceptual Models

Based on these definitions we can now detail the notion of semantic annotations. The grammars of conceptual models traditionally give their users a large degree of freedom when it comes to describing their content. Therefore, the labels and attached comments of the elements and relations are often expressed in natural language. It may however be beneficial to be able to process the information contained in these descriptions at a later stage. To make the information contained in these descriptions machine processable one approach is to add mappings to formal semantic schemata [8]. Thereby, the semantic information can be raised to a machine processable level without requiring a modification of the original modeling language. This has two particular advantages. The first is that the consistency to other models that are based on the original specification of the modeling language is ensured. It also means that any algorithms that were based on the original state of the language do not need to be adapted. The second advantage is that these annotations and according processing capabilities can also be added ex-post, i.e. after the conception of the modeling language and after the creation of according model instances. It is thus not necessary for the user to deal with formal semantic definitions at first hand and only add them as needed.

For the representation of the formal semantic schema it is common to use ontologies which define the vocabulary used to describe and represent an area of knowledge. They are usually expressed in a logic-based language that permits to make fine, accurate, consistent, sound, and meaningful distinctions among the elements of the ontology [29]. Furthermore, ontology languages may permit to conduct reasoning tasks to ensure the consistency of the ontology and automate the classification of new terms [20]. Today it can be chosen from a wide variety of ontology languages including W3C standards such as OWL or RDF.

Several benefits have been described by using semantic annotations. During the stage of model creation the ontologies can be used to check whether the attributes of the model elements are valid [1] or to provide auto-completion functionalities for increasing the efficiency of modelling [2]. For the analysis of models semantic annotations have been described to measure the similarity between process models [6] and to validate models against formal specifications [22]. The benefits for the execution of processes using semantic annotations have been described for the case of semantic web services by [5].

2.3   Social Network Based Applications

In the last years several *social network sites* have emerged. They are commonly defined as "web-based services that allow individuals to (1) construct a public or semi-public profile within a bounded system, (2) articulate a list of other users with whom they share a connection, and (3) view and traverse their list of connections and those made by others within the system" [3][p. 211]. Due to their enormous participation rates they have achieved so far, they have been an attractive subject for researchers as well as the advertising and media industry.

More recently, the opportunities of *social network based applications* have been investigated [28]. These are software applications that access the data of social network sites either by using proprietary protocols such as the Facebook Graph API [7] or open source third-party APIs. The data provided by the social network sites can then be used to retrieve information about a user's personal profile as well as information about the profile of his or her fellows and other resources.

Depending on the social network site and also which information the user is willing to share, this includes for example information about the user's age, spoken languages, home town, current workplace or his or her interests in literature, philosophy or music. In addition, several of the social network sites allow users to form special interest groups or to group together by expressing that they like a particular resource on the social network such as a fan page. This data can then be analyzed and used for scientific analyses [24], e.g. for determining subgroups with similar properties [21].

3   Social Network Based Semantic Annotation of Conceptual Models

With the foundations presented in the previous chapter, we can present our approach as a synthesis of a. conceptual models, b. their semantic annotations with ontologies, and c. social network applications. The core idea as shown in figure 1 is to use the information provided through social network applications in conducting semantic annotations. In order to add semantic annotations to conceptual models, additional knowledge needs to be provided by human actors. These human actors in turn - exemplarily numbered from 1 to 7 in the figure - have connections to other actors and other social network ressources, e.g. in the way of subscriptions to information feeds in the network. These connections are made explicit by the data structures provided by the social network applications. This information therefore be used to support the tasks in conducting the annotations.

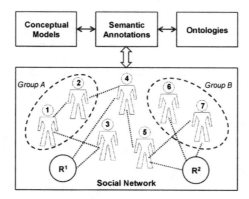

Fig. 1. Major Components of the Proposed Approach

For detailing the relationships between conceptual models, semantic annotations, ontologies and social network data we will use in the following a semi-formal, model-based representation [4,23]. This will not only permit us to discuss how the three parts interact in detail and which information is accessed by each part. It will provide a foundation for subsequently specifying how to process the contained information in a formal way.

3.1   Model-Based Representation

To represent the information about the linkage of conceptual models, semantic annotations, ontologies, and social network applications we designed a specific modeling language as an extension of the SeMFIS modeling language [11] - an excerpt of its meta model is shown in figure 2. To illustrate the application to particular conceptual models, this modeling language contains a simplified version of the BPMS modeling language for business processes [13] – shown in the upper left of figure 2. It provides elements for representing the *information flow* and the *control flow* of a business process as well as the according *sequence flow* relations.

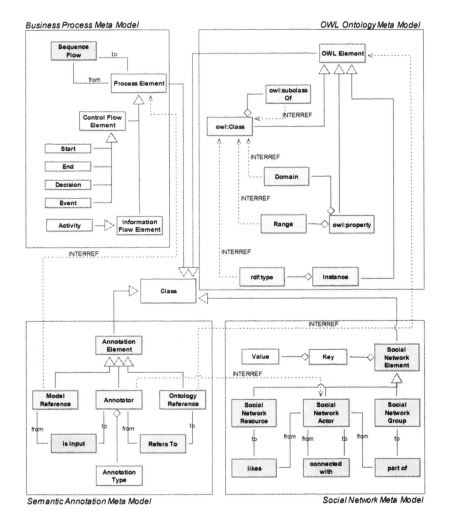

Fig. 2. Excerpt of the Alignment between a Business Process, an OWL Ontology, a Semantic Annotation, and a Social Network Meta Model

Furthermore, to represent ontologies that can be later used for the semantic annotations of the conceptual models, an OWL ontology meta model is added for representing ontologies in the web ontology format – see the upper right corner in figure 2. The meta model does not provide any formal semantic definition for the OWL ontologies. It is assumed that a consistent and sound representation of an OWL ontology is available, e.g. by using an import from an ontology management toolkit such as Protégé [12]. For the representation of the annotations, the meta model provides an *annotation element*. This is specialized in the form of a *model reference*, an *ontology reference*, and an *annotator* element. Thereby, the annotation can be specified independently both of the used conceptual modeling language and the used ontology language. The annotation can be further detailed by specifying an *annotation type*, e.g. to express is-broader-than or is-narrower-than relationships.

For integrating the information from social network applications, the meta model is complemented by a *social network element* and its specializations. These are used to represent the core concepts of social network websites. Thereby the following relations can be represented: Between *social network actors* by using the *connected* relation, between *social network actors* and *social network resources* by using the *likes* relation, and between *social network actors* and *social network groups* by using the *part of* relation. Additionally, each *social network element* can have *key/value* pairs assigned to it. These can be used to integrate the range of attributes that are available for detailing the entities in the social network, e.g. the personal data of the actors or the description of a group. To integrate the semantic annotations and the social network entities, a reference relation (*INTERREF*) is shown between the *social network element* and the *annotator* element. This permits to express that an annotation is conducted by an actor in the social network.

The meta model has been implemented as an extension of the SeMFIS toolkit on ADOxx - see figure 3 [11,13]. Based on this implementation concrete model instances of the meta model elements can be created and used as input for the development of algorithms for processing this data. Thus, semantic model annotations respectively the information provided by social network applications, can be represented and linked using the references as defined by the meta model. In addition to the early analysis of these relationships, the meta model serves as a starting point for further implementations.

3.2   Addition of Behavior

With the information structures provided by the semi-formal meta models, more formal specifications can be realized. For example, the behavior can be specified using a rule-based approach as shown in the following. Based on the linkage of semantic annotations of model contents with the actors in the social network, it could be of benefit to identify other actors in the network who engage in similar annotation tasks. Thus, several additional scenarios can be realized. E.g., it could be searched for actors who possess similar knowledge based on their annotations to incite communication between actors with similar interests; or, existing semantic annotations could be proposed for review and evaluation to other actors for quality or correctness assessments. Another scenario would be to

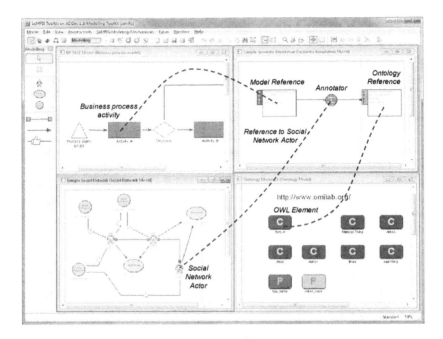

Fig. 3. Implementation of the Meta Models on the ADOxx Meta Modeling Platform

propose the membership in a group of the social network for actors with similar interests.

In the following we outline a formal definition of rules based on the structures in the meta model - a more thorough formal foundation may be achieved with the FDMM formalism but is omitted here due to limited space [14]. At first, we define the antecedent that has to be met. The rules thus requires the existence of two social network actors a, b who engage in semantic annotation tasks on process elements p_1, p_2. The annotations are specified by model references m_1, m_2, ontology references r_1, r_2 and annotator elements t_1, t_2. It is assumed by the rule that one common ontology element o from an OWL ontology is used, as shown in equation 1.

$$\exists a, b, p_1, p_2, o, m_1, m_2, r_1, r_2, t_1, t_2$$
$$SocialNetworkActor\,(a) \wedge SocialNetworkActor\,(b) \wedge$$
$$ProcessElement\,(p_1) \wedge ProcessElement\,(p_2) \wedge$$
$$OWLElement\,(o) \wedge ModelReference\,(m_1) \wedge ModelReference\,(m_2) \wedge$$
$$OntologyReference\,(r_1) \wedge OntologyReference\,(r_2) \wedge$$
$$Annotator\,(t_1) \wedge Annotator\,(t_2) \tag{1}$$

The linkages between the elements can then be defined by *interref*, *isInput*, and *refersTo* predicates as shown in 2.

$$interref\,(m_1, p_1) \wedge interref\,(m_2, p_2) \wedge interref\,(r_1, o) \wedge$$
$$interref\,(r_2, o) \wedge isInput\,(m_1, t_1) \wedge isInput\,(m_2, t_2) \wedge$$
$$refersTo\,(t_1, r_1) \wedge refersTo\,(t_2, r_2) \wedge interref\,(t_1, a) \wedge interref\,(t_2, b) \qquad (2)$$

Finally, we can refer to the connections between the two social actors via the *connectedWith* predicate and define different variants for consequents of the rule based on the antecedents defined in 1 and 2. In the first variant in equation 3 it is assumed that the actors were not connected before and are now being connected based on their annotation activity. In equation 4 it is shown how an existing connection between the two actors leads to their inclusion in the socical network group *g* for bringing together actors with similar interests based on their common usage of the ontology concept *o* for their annotations.

$$(1) \wedge (2) \rightarrow connectedWith\,(a, b) \qquad (3)$$
$$(1) \wedge (2) \wedge connectedWith\,(a, b) \rightarrow \exists g, SocialNetworkGroup\,(g),$$
$$partOf\,(a, g), partOf\,(b, g) \qquad (4)$$

However, from a more practical perspective it would be desirable not to force actors in the social network into groups automatically or connect them with other actors. Rather, a less deterministic approach may be more appropriate that gradually increases the likelihood of the consequents to become effective. We show this formally in the following via the function *recommendConnection* that maps to a real number. The consequent of the rule is thus modified as shown in 6, where a modification factor β is added each time the rule fires. In this concrete case this means that whenever two actors refer to the same ontology concept in their annotations, the likelihood of connecting them is increased. Based on the attainment of a threshold value θ, as in equation 7, the actual connection is established in 8.

$$recommendConnection : (a \times b) \rightarrow [0 \dots 1] \qquad (5)$$
$$(1) \wedge (2) \rightarrow (recommendConnection(a, b) + \beta) \qquad (6)$$
$$relevantConnection\,(a, b) = \begin{cases} true \text{ if } recommendConnection(a, b) \geq \theta \\ false \text{ if } recommendConnection(a, b) < \theta \end{cases}$$
$$\qquad (7)$$
$$(1) \wedge (2) \wedge relevantConnection\,(a, b) \rightarrow connectedWith\,(a, b) \qquad (8)$$

4   Discussion

With the above descriptions we can now discuss the benefits and limitations of the chosen approach. In particular we will refer to three aspects that have been proposed in [16] for using social network information in the collaborative formalization of semantics. These are: *communication support, economic benefits,* and *technical opportunities.*

Regarding communication support, the major advantage of the approach is that it enables the *collaboration on annotation tasks* while at the same time preserving existing conceptual model and ontology handling interfaces. A user who does not want to engage in the annotations but is just interested in modeling tasks is not affected by the new annotation functionalities. By integrating the functionalities offered by the social networking websites, new possibilities for interaction emerge: Not only can it be shared with connections in the social network, which annotations an actor has conducted by making this information public on the social network. This opens the potential for integrating a multitude of users in the sense of crowd-sourcing for the annotation tasks, which leads to economic benefits in terms of *reduced effort* for annotations by a single user. In contrast to other annotation solutions it can be easily communicated and made visible to other users which annotation tasks have already been conducted and where additional information is required. Besides the information about the semantic annotation also the content of the models and the ontologies can be shared. At the same time of course privacy aspects may constitute a limitation that needs to be specifically considered.

The second characteristic of the proposed approach is that it allows to *personalize the user experience* in regard to semantic annotations. By accessing the social network information together with the information about the annotations, new ways for suggesting annotations to users can be designed. This has already been shown by the descriptions in section 3.2. It could be further extended by specifying a number of additional rules, e.g. to suggest connections based on commonly used social network resources, commonly used model references, ontology references or more detailed subsets of these elements such as elements of the information and control flow of business processes. Furthermore, additional information available in the social network that can be stored using the key/value pairs can be accessed. This concerns for example information about the work experience of actors or common professional interests.

From a technical side, many social network applications provide single-sign-on functionalities that can be easily re-used for implementing semantic annotation applications. Due to the central handling of the *user authentication*, all implementation effort that usually has to be devoted to creating separate authentication services can be omitted. This concerns also the workflows for password changes, retrieving forgotten passwords or changes in the user profile. Therefore, additional economic benefits may be reaped through less implementation effort. Nevertheless it may be additionally necessary to restrict the access to the information to certain groups of the social network. However, these group policies may be handled using the social network's functionalities, e.g. based on the administrative rights for the groups.

5 Conclusion and Outlook

With the presented approach we could show how information from social network applications can be integrated with the semantic annotation of conceptual models. Furthermore, based on the semi-formal definition of the according meta models, more formal specifications for the behavior could be added using a rule-based approach. The next steps will include the further implementation of the approach by adapting and extending existing tools for conceptual modeling and ontology handling. Based on such an implementation it can then be investigated which of the many options for combining social network information, conceptual models, ontologies and the annotations best meets the user requirements for conducting efficient annotations. For the future it is planned to integrate more information from social network applications in the domain of conceptual modeling, e.g. by analyzing statements from social network actors made in natural language and transferring them to the domain of conceptual models.

References

1. Becker, J., Delfmann, P., Herwig, S., Lis, L., Stein, A.: Towards Increased Comparability of Conceptual Models - Enforcing Naming Conventions through Domain Thesauri and Linguistic Grammars. In: ECIS 2009, Verona, Italy (2009)
2. Betz, S., Klink, S., Koschmider, A., Oberweis, A.: Automatic user support for business process modeling. In: Workshop Semantics for Business Process Management, Budva, Montenegro (2006)
3. Boyd, D.M., Ellison, N.B.: Social Network Sites: Definition, History, and Scholarship. Journal of Computer-Mediated Communication 13, 210–230 (2008)
4. Demirkan, H., Kauffman, R.J., Vayghan, J.A., Fill, H.-G., Karagiannis, D., Maglio, P.P.: Service-oriented technology and management: Perspectives on research and practice for the coming decade. Electronic Commerce Research and Applications 7(4), 356–376 (2008)
5. Drumm, C., Lemcke, J., Namiri, K.: Integrating semantic web services and business process management: A real use case. In: Semantics for Business Process Management 2006, Budva, Montenegro (2006)
6. Ehrig, M., Koschmider, A., Oberweis, A.: Measuring Similarity between Semantic Business Process Models. In: Roddick, J.F., Hinze, A. (eds.) APCCM 2007, vol. 67, pp. 71–80. ACM (2007)
7. Facebook: Facebook Developers - Graph API (2011), http://developers.facebook.com/docs/reference/api/
8. Fill, H.G.: Design of Semantic Information Systems using a Model-based Approach. In: AAAI Spring Symposium. AAAI, Stanford University (2009)
9. Fill, H.G.: Visualisation for Semantic Information Systems. Gabler (2009)
10. Fill, H.G.: An Approach for Analyzing the Effects of Risks on Business Processes Using Semantic Annotations. In: Proceedings of ECIS 2012. AIS (2012)
11. Fill, H.G.: SeMFIS: A Tool for Managing Semantic Conceptual Models. In: Kern, H., Tolvanen, J.P., Bottoni, P. (eds.) GMLD Workshop (2012)
12. Fill, H.G., Burzynski, P.: Integrating Ontology Models and Conceptual Models using a Meta Modeling Approach. In: 11th International Protégé Conference, Amsterdam (2009)

13. Fill, H.G., Karagiannis, D.: On the Conceptualisation of Modelling Methods Using the ADOxx Meta Modelling Platform. Enterprise Modelling and Information Systems Architectures 8(1), 4–25 (2013)

14. Fill, H.-G., Redmond, T., Karagiannis, D.: FDMM: A Formalism for Describing ADOxx Meta Models and Models. In: Proceedings of ICEIS 2012, Wroclaw, Poland, vol. 3, pp. 133–144 (2012)

15. Fill, H.-G., Reischl, I.: An Approach for Managing Clinical Trial Applications Using Semantic Information Models. In: Rinderle-Ma, S., Sadiq, S., Leymann, F. (eds.) BPM 2009. Lecture Notes in Business Information Processing, vol. 43, pp. 581–592. Springer, Heidelberg (2010)

16. Fill, H.G., Tudorache, T.: On the Collaborative Formalization of Agile Semantics Using Social Network Applications. In: AAAI Spring Symposium 2011 - AI for Business Agility. AAAI (2011)

17. Harel, D., Rumpe, B.: Modeling Languages: Syntax, Semantics and All That Stuff - Part I: The Basic Stuff. Tech. Rep. MCS00-16, The Weizmann Institute of Science (August 22, 2000)

18. Hepp, M., Leymann, F., Domingue, J., Wahler, A., Fensel, D.: Semantic business process management: a vision towards using semantic Web services for business process management. In: ICEBE Conference 2005, pp. 535–540 (2005)

19. Herbst, J., Junginger, S., Kuehn, H.: Simulation in Financial Services with the Business Process Management System ADONIS. In: 9th European Simulation Symposium (ESS 1997). Society for Computer Simulation (1997)

20. Horrocks, I., Patel-Schneider, P., Van Harmelen, F.: From SHIQ and RDF to OWL: The Making of a Web Ontology Language. Web Semantics: Science, Services and Agents on the World Wide Web 1(1), 7–26 (2003)

21. Hsieh, M.H., Magee, C.L.: A new method for finding hierarchical subgroups from networks. Social Networks 32, 234–244 (2010)

22. Ibanez, M.J., Alvarez, P., Bhiri, S., Ezpeleta, J.: Unary RDF-Annotated Petri Nets: A Formalism for the Modeling and Validation of Business Processes with Semantic Information. In: SBPM 2009 Workshop. ACM (2009)

23. Karagiannis, D., Kühn, H.: Metamodelling platforms. In: Bauknecht, K., Min Tjoa, A., Quirchmayr, G. (eds.) EC-Web 2002. LNCS, vol. 2455, p. 182. Springer, Heidelberg (2002)

24. Lewis, K., Kaufman, J., Gonzalez, M., Wimmer, A., Christakis, N.: Tastes, ties, and time: A new social network dataset using facebook.com. Social Networks 30, 330–342 (2008)

25. List, B., Korherr, B.: An evaluation of conceptual business process modelling languages. In: SAC 2006. ACM, Dijon (2006)

26. Moser, C., Bayer, F.: IT Architecture Management: A Framework for IT Services. In: Desel, J., Frank, U. (eds.) EMISA Workshop, Klagenfurt, Austria. LNI (2005)

27. Mylopoulos, J.: Conceptual Modeling and Telos. In: Loucopoulos, P., Zicari, R. (eds.) Conceptual Modelling, Databases and CASE: An Integrated View of Information Systems Development, pp. 49–68. Wiley (1992)

28. Nazir, A., Raza, S., Chuah, C.N.: Unveiling facebook: A measurement study of social network based applications. In: IMC 2008. ACM, Vouliagmeni (2008)

29. Obrst, L.: Ontologies for semantically interoperable systems. In: Proceedings of the 12th International Conference on Information and Knowledge Management. ACM Press, New Orleans (2003)

30. Wand, Y., Weber, R.: Research Commentary: Information Systems and Conceptual Modeling - A Research Agenda. Information Systems Research 13(4), 363–376 (2002)

Development of an Evaluation Approach for Customer Service Interaction Models

Hisashi Masuda[1], Wilfrid Utz[2], and Yoshinori Hara[3]

[1] Japan Advanced Institute of Science and Technology, Ishikawa, Japan
masuda@jaist.ac.jp
[2] University of Vienna, Research Group Knowledge Engineering, Vienna, Austria
wilfrid.utz@dke.univie.ac.at
[3] Graduate School of Management, Kyoto University, Kyoto, Japan
hara@gsm.kyoto-u.ac.jp

Abstract. In today's service economy, the evaluation of interaction between service providers and different consumer segments/target groups is an important topic. In context-sensitive settings, the service provider adapts the interaction with the consumer by selecting a fitting design pattern, also including mechanisms for validation and evaluation. In current service evaluation models, the extraction of dynamic characteristics of consumers poses a challenge, as simplification of methods for externalization of patterns is needed to enable an improved understanding of consumer types. The paper at hand aims for a contribution on visualization of the service interaction for each consumer type/pattern supported. A prototypical visualization approach has been implemented using meta-modelling concepts and technologies as an realization environment, validated in case studies from the food service industry.

Keywords: Service evaluation, Customer interaction modeling, Modelling, Meta-modelling.

1 Introduction

1.1 Background

Continuous improvement through evaluation loops and aligning operations to feedback gathered as a means to increased productivity, quality, performance, is a well-established and researched discipline, supported by various lifecycle-driven approaches (see for example Deming's PDCA - Plan - Do - Check – Act [16], or CMMI Capability Maturity Model Integration by Carnegie Mellon University [17]). The common baseline of all these approaches is to successively monitor, evaluate and improve the quality of service/product provision through pre-structured phases feeding back to the design, operation and also evaluation phases of the system.

The servicing economy faces similar issues, but has to consider that the process to deliver a specific service to the consumer is "value co-creating". This means, that in addition to realizing provider internal mechanism, also the consumer perspective has

R. Buchmann et al. (Eds.): KSEM 2014, LNAI 8793, pp. 150–161, 2014.
© Springer International Publishing Switzerland 2014

to be considered. Different consumer expectation and experience levels – summarized as the service literacy of the consumer – need to be analyzed and dynamically combined with the provider's view. Dimensions with respect to service literacy relate to the consumer maturity, the interaction situation/objective as well as the frequency of service usage.

A comprehensive evaluation approach can be constructed through a combination/integration of these dynamic aspects on consumer side with the providers' environment. Within this paper we focus on the development of a visualization approach for such a system. Contextualized service interaction processes are the baseline for the visualization approach developed; these processes are analyzed with the aim to enable a continuous improvement cycle of consumer satisfaction. As a realization technique, meta-modeling concepts and technologies are introduced to support multiple representations of views within the visualization approach.

1.2   Research Question and Objective

The work presented in this paper is motivated by the research question how to combine design level representation on provider and consumer side for service interaction and evaluation. Different representation formats for evaluation/analysis and visualization are considered, realizing a comprehensive, view-based environment for continuous service (consumer satisfaction) improvement. The evaluation model as presented in [5] is extended with an adaptive approach on consumer processes and visualization support. This extension results in a proposal how the combination between the service processes (from different viewpoints and including additional, contextual consumer characteristics) and service evaluation can be established.

1.3   Approach

The system under study is presented in Figure 1 as the "service enterprise", defining the system boundaries whereas a distinction is made between the consumer and provider aspects. Both aspects result in views, constructed using model-based realizations. Each view and the related models are briefly introduced below:

- *Service Enterprise View:* as an overview representation of the service enterprise from an organizational perspective, setting the boundaries of the system under study in the model-based environment.
- *Consumer-Oriented Service Process View:* as a representation of the consumer interaction processes, developed following a bottom-up, data-driven approach accomplished by ethnographical acquisition techniques. This view and its models represent the foreground for the work presented in this paper.
- *Provider-Oriented Service Process View:* as a representation of the service provider level (how a service is provided, what interactions are foreseen, etc.), typically constructed following a top-down approach based on the service enterprise's strategy. This view and its models are underpinned by pre-existing background.

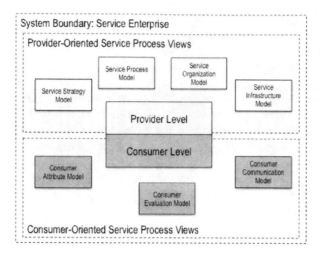

Fig. 1. Combined/Integrated Method: Service Process Views and Service Evaluation

The integration method considers the views and models of above in a comprehensive, cross-referenced and linked manner as a common knowledge base for improvement and management. This means that changing service structures on provider side would directly influence the consumer interaction models and vice-versa, resulting in an adaptive method for service evaluation.

The paper at hand is structured in the following chapters: section 2 positions the research work in the context of previous work. Section 3 introduces the structure/purpose of the views and models, evaluated using the prototypical realization and case studies in section 4. The implications of the results found are presented in section 5 before concluding in section 6.

2   Related Works

The research presented in this paper aims to contribute to service evaluation in general, focusing concretely on how the evaluation of services can be transformed from a static to a dynamic approach that considers flexible situation in service process design. The baseline for customer satisfaction research is the disconfirmation model as presented in [6] and SERVQUAL [1] in relation to service quality. The dynamic aspect as introduced above relate to research for time frequent analysis as presented in [7,8], in addition to research for dynamic aspect of customer satisfaction in the KANO model [4].

With respect to service process design, the representation of dynamic structures in models poses a challenge, since tasks/activities are not predefined, but are concretized during runtime. Different approaches exist such as rule-based mechanisms as described in [11].

Meta-modelling concepts and technologies represent the third pillar of related work to the research performed. These concepts enable an agile development on one hand,

and provide means for flexible integration/hybrid combination of methodological building blocks as introduced on the other hand. The generic meta-modelling framework as defined in [3] is applied as a guiding principle in the implementation of the prototype.

3   Method and Concept: Evaluation of Consumer Service Interaction Models

The research presented follows an experimental approach: a prototypical implementation of a modelling method is realized on the meta-modelling platform ADOxx [12] and validated in the context of case studies derived within the Japanese Creative Service (JCS) project [9]. The foundation of the approach relates to the research work of the Open Models Initiative Laboratory (OMILab) [13], applying the baseline of method and modelling language integration as described in [3, 14] and using the Generic Modelling Method Specification Framework shown in Figure 2 to define the models and views of above.

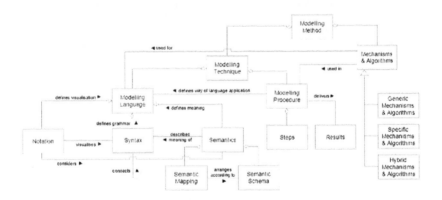

Fig. 2. Modelling Method Specification Framework [14]

Related work on modelling method integration by Kuehn [15] supports the integration method proposed in this paper.

In the following the envisioned service process review platform is introduced from a procedural perspective, substantiated with model/view level definition.

3.1   Service Process Review Platform

The service provider defines the steps/tasks and activities as nodes using an arbitrary process-modeling notation. The outcomes of this phase are models on an abstract level that define how the service is offered to the consumer from the provider's perspective. The abstract representation is further refined by consumer views to integrate aspects such as experience during interaction, frequency-based modification, and expectation, resulting in a concrete representation of the service process.

These updates are considered as instances of the abstract level and present feedback to understand the consumer heterogeneity and dynamic aspect of their characteristics. The review platform itself integrates with the evaluation view and provides statistical assessments/visualization for assessment of service value dimensions.

3.2   Service Interaction Views

The definition of the service process presents a complex challenge due to the high variability and complexity to define them including different views (customer vs. provider perspective) within the same knowledge base.

In order to understand the characteristics of the service process from the provider perspective, 2 alternative representations have been developed and presented in [10]. A distinction is made between context-free and context-dependent service processes. The first model structure does not consider adaptation or heterogeneity of consumers whereas the second structure enables the definition of models that are adaptive to consumer groups and their characteristics.

Context-Free Service Model
The context-free service model enables the definition of static service processes. The model provides concepts to define nodes; their order/control flow and input to service evaluation using fixed nodes and pre-set references to the evaluation view. Service evaluation is enabled using static evaluation questionnaires with the dimensions of satisfaction, expectation, and gap of expectation.

Context-Dependent Service Model
The context-dependent service model uses a hybrid structure for service processes. The structure enables the pre-design of default models including adaptation patterns that react to the actual instantiation information for different types of consumers.

The interaction models are developed and presented from a consumer's viewpoint with a focus on the dialectic provider – the customer interaction in a highly contextual setting. To evaluate the customer's variety and adaptation, we construct an integrated approach from three perspectives (see Figure 3): a) the customer attribute, b) service evaluation and c) customer communication perspective. An understanding of the customer's variety is realized by linking the customer evaluation and their context information accordingly.

Customer Attribute Model
The Customer Attribute Model represents the characteristics of the customers. Customer's criteria change depending on their circumstances, context and experience; e.g. the consumer is a beginner, repeater or expert user of the service.

Evaluation Visualization Model
The Evaluation Visualization Model defines the evaluation of the overall service process. The individual service evaluation are connected with concrete service processes taking into account the customer attribute model, for instance, a beginner and experts of one service evaluates the service with score "5" in five-scale customer satisfaction survey. We represent the difference of the two evaluations based on the service process.

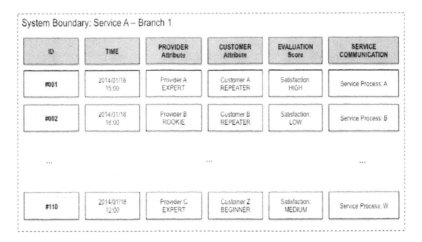

Fig. 3. Overview: Service interaction model for reviews

Customer Communication Model

The Customer Communication Model describes the actual interaction between the service provider and consumer, narrowing down the context-free or dependent service model introduced in [10]. We are able to represent the process/communication flow on the time axis. The nodes and control flow in this model explain the behavior or conversation required for consuming the service.

4 Implementation Results: Evaluation of Consumer Service Interaction Models

Within this section we introduce the realized tool support for the design of the models defined. In a first step the results of the implementation are shown on a generic level, further assessed in distinct application cases for a context-free as well as a context-dependent model.

4.1 Overview: Customer Service Interaction Modelling Environment

The evaluation of the theoretical approach has been performed through a prototypical implementation of the integrated modelling method, applying meta-modeling techniques for realization. The implementation provides tool support for the design of customer service interaction. The implementation considers – on design level – process representation for the provider and consumer view, questionnaire design through decomposition of the service process and visualization support for analytical evaluation of gather customer data. The focus within this paper lies upon the data/information acquisition process using questionnaire models and related visualization techniques.

4.1.1   Customer Attribute Model

The customer attribute model represents the characteristics of consumers. Using the meta-modeling approach (Figure 4) the characteristics are adaptable and configurable according to requirements. Figure 5 shows two variables explaining consumer information: experiences and the consumer risk level, e.g. risk appetite or risk averse.

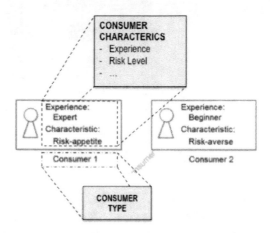

Fig. 4. Definition of the Customer Attribute Model

4.1.2   Service Evaluation Model

Figure 5 shows the results of implementation with respect to visualization of analytical assessments. Different visualization types have been implemented (line charts, spider diagrams, and heat map representation). All visualizations build upon a common input dataset derived from questionnaire data.

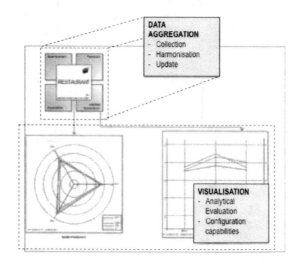

Fig. 5. Definition of the Analytical Visualization Approach

4.1.3 Service Communication Model

The prototypical implementation supports the decomposition of service processes into different levels of abstraction and granularity. On an initial level the overall service process is presented in phases, further detailed on decomposition levels as concrete steps and tasks of the interaction process. For the context-free model a direct mapping is used; for the context-dependent an adaptive mapping approach is applied.

Figure 6 shows an example representation within the prototype including the mentioned decomposition capabilities. The service process is translated to a set of questions focusing on quantitative data for a) customer satisfaction, b) expectation and c) modified expectation. In the case presented above a 5-point evaluation scale is applied. Additionally, as qualitative data, open questions and comments for each node are foreseen.

For the example in Figure 6, the decomposed model is build following a predesignate method, foreseen for the context-free model introduced in [10]. Adaptive sub-process binding, e.g. for the context-dependent model, could be realized through rule-based or semantic reasoning approaches.

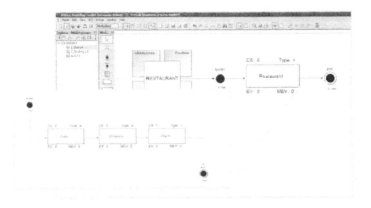

Fig. 6. Decomposition of Service Processes

4.2 Case Study: Service Interaction Models for Food Services

In this section, the case studies to validate the approach are introduced. Two cases are constructed to present the static, context-free structure on one hand and the adaptive, context-dependent structure on the other. With respect to the validation of the context-free structure, the "Fast Food" case has been developed, for the context-dependent, the "Edomae Sushi Restaurant" case is introduced based on the outcomes of the Japanese Creative Services (JCS) project [9]. For the sushi case, we focus on "Okonomi" style, where the consumer of the service is the decision maker during the service and a pre-design of the model is therefore only partially possible.

4.2.1   Case Study: Fast Food Restaurant

The "Fast Food" case (as shown in a mockup in Figure 7) presents the design of a service interaction process in a "one-size-fits-all" manner, targeting a scaling model on a global basis for any type of consumer. Independent of the consumer's experience/service literacy, the service is offered and consumed. The interaction process is static and does not adapt to contextual characteristics.

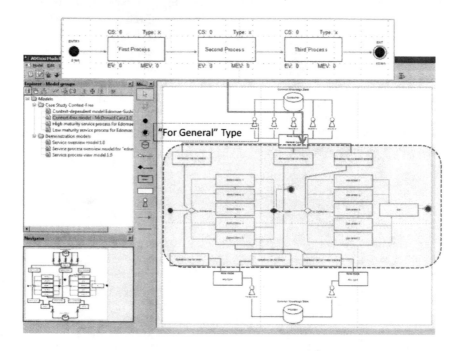

Fig. 7. Context-Free Service Model: Fast Food Case

4.2.2   Case Study: Edomae-sushi Restaurant

In contrast to the initial case, the "Edomae Sushi Restaurant" presents highly contextual and adaptive requirements to the model structure.

In Japan, high-end sushi service restaurant (as shown in a mockup in Figure 8) require a certain knowledge level of the service by the consumer [2].

The sushi chef understands the consumer knowledge in a first impression during the ordering process for drinks. The chef modifies the service process based on consumer interaction and decisions taken throughout the interaction. A clear understanding and analysis on the communication and interaction patterns is integrated in this view to provide evidence for pattern selection and application on type and instance level (see Figure 9 for potential patterns to select from on communication level).

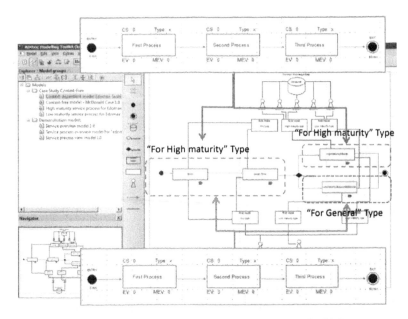

Fig. 8. Context-Dependent Service Model: Edomae Sushi Case

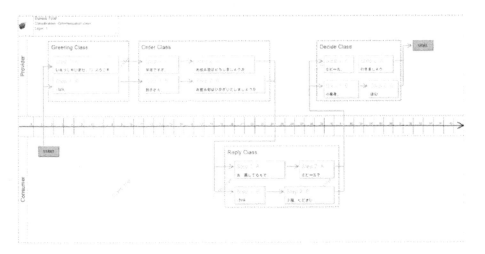

Fig. 9. Communication/Interaction Model: Edomae Sushi Case

For this flexible service process, the service provider decides on a pattern-basis on steps to perform as the service process considering consumer reaction. The evaluation of the service quality is performed along the line of this evolving service process in the form of dynamically constructed questions.

5   Implication

A main implication and outcome of the research performed relates to the visualization of contextual consumer/provider interaction patterns. Following this approach we are able to construct the consumer context on a theoretical level. The visualization approach enables us to understand the interaction process (including the consumer characteristics and their communication processes) as an adaptive to changing consumer groups and their demands. Means to create understanding on the consumer and their behavior to flexibly react/adapt to changes in current economy settings are regarded an important challenge.

6   Conclusion

The dynamic approach for service interaction process design, evaluation and visualization in an integrated way allows us to extend existing continuous improvement and feedback loop mechanisms by including the consumer perspective in the assessment of service value and quality. The proposed visualization approach contributes to allow an easy understanding of the dynamics of customers' variety in the service economy sector. As further steps of our research we aim to develop mechanisms for dynamic web/mobile questionnaire creation, utilizing the adaptive service process modeling approach for collecting data of the customer service interactions. Empirical research will be conducted based on proposed model to get further insights in the consumer categories and used/applied patterns, also investigating in different cultural settings/communication models within increasing global economics.

Acknowledgment. Empirical research of Japanese Sushi restaurant was performed by Dr. Yutaka Yamauchi as a baseline for our development presented in this paper. We appreciate comments and input from his research.

References

1. Parasuraman, A., Zeithmal, V.A., Berry, L.L.: SERVQUAL; A Multiple-Item Scale for Measuring & Consumer Perceptions of Service Quality. Journal of Retailing 64(1) (1998)
2. Hara, Y., Yamauch, Y., Yamakawa, Y., Fujisawa, J., Ohshima, H., Tanaka, K.: How Japanese traditional "Omonpakari" services are delivered - a multidisciplinary approach. In: SRII 2012, pp. 906–913 (2012)
3. Karagiannis, D., Grossmann, W., Hoefferer, P.: Open model initiative - a feasibility study (April 4, 2010, 2008), http://cms.dke.univie.ac.at/uploads/media/OpenModelsFeasibilityStudSEPT2008.pdf
4. Nilsson-Witell, L., Fundin, A.: Dynamics of service attributes: a test of Kano's theory of attractive quality. International Journal of Service Industry Management 16(2), 152–168 (2005)
5. Masuda, H., Hara, Y.: A Dynamic Evaluation Model based on Customer Expectation and Satisfaction. In: SRII 2011, pp. 401–408 (2011)

6. Oliver, R.L.: The Expectancy Disconfirmation Model of Satisfaction. In: Oliver, R.L. (ed.) Satisfaction: A Behavioral Perspective on the Consumer, pp. 98–131. McGraw Hill, New York (1997)
7. Bolton, R.N.: A Dynamic Model of the Duration of the Customer's Relationship with a Continuous Service Provider: The Role of Satisfaction. Marketing Science 17(1), 45–65 (1998)
8. Boulding, W., Kalar, A., Staelin, R., Zeithmal, V.A.: A Dynamic Process Model of Service Quality: From Expectations to Behavioral Intentions. Journal of Marketing Research, 7–27 (February 1993)
9. Japanese Creative Service (JCS) Project,
10. Masuda, H., Utz, W., Hara, Y.: Context-Free and Context-Dependent Service Models Based on "Role Model" Concept for Utilizing Cultural Aspects. In: Wang, M. (ed.) KSEM 2013. LNCS, vol. 8041, pp. 591–601. Springer, Heidelberg (2013)
11. Leutgeb, A., Utz, W., Woitsch, R., Fill, H.-G.: Adaptive Processes in E-Government - A Field Report about Semantic-Based Approaches from the EU-Project "FIT". In: ICEIS (3), pp. 264–269 (2007)
12. Metamodelling Platform ADOxx, http://www.adoxx.org
13. Open Models Initiative Laboratory, http://www.omilab.org
14. Karagiannis, D., Kühn, H.: Metamodelling platforms. In: Bauknecht, K., Min Tjoa, A., Quirchmayr, G. (eds.) EC-Web 2002. LNCS, vol. 2455, p. 182. Springer, Heidelberg (2002)
15. Kühn, H.: Methodenintegrationim Business Engineering, Dissertation Thesis, University of Vienna (2004)
16. Deming, W.E.: The New Economics. MIT Press (1993)
17. Carnegie Mellon University Software Engineering InstituteCMMI v1.3, http://whatis.cmmiinstitute.com/

Developing Conceptual Modeling Tools Using a DSL

Niksa Visic and Dimitris Karagiannis

Faculty of Computer Science, University of Vienna, Austria
{niksa.visic,dimitris.karagiannis}@univie.ac.at

Abstract. There are multiple ways one can pursue when developing conceptual modeling tools. The most common are the ones where modeling tool engineers implement by using multiple graphical editors and various programming languages to realize the requirements of a modeling method. In this case, implementing artifacts such as abstract and concrete syntax or algorithms is linked to a specific technological platform. This motivated us to develop a DSL for this area, which entailed designing, specifying and implementing it. In this paper we propose a domain-specific language (MM-DSL) based on a metamodeling approach, which gives us the ability to be technology independent. With MM-DSL a specification for a conceptual modeling tool is programmed on an abstract level. The code can be compiled and executed on different metamodeling platforms. We tested the MM-DSL 1.0 using a prototype-based evaluation.

Keywords: MM-DSL, metamodeling platform, conceptual modeling tool.

1   Introduction

During the last couple of decades of applied research in metamodeling approaches and technologies many meta-metamodels ($meta^2$models) have emerged, some of them very complex, the others very simple and easy to use, but typically created for the same purpose: to capture the requirements imposed on them by the modeling domain. Initially, most of the $meta^2$models have been designed to be instantiated in a single domain. For example, EMF Ecore is mostly used in the development of Eclipse applications, GME primarily in the area of electrical engineering, ConceptBase [1] for conceptual modeling and metamodeling in software engineering, and ADONIS was designed for describing and simulating business processes. However, over time it has been discovered that they are also applicable in similar domains. A good example is the extension of the ADONIS $meta^2$model, which later became the ADOxx $meta^2$model. The ADOxx $meta^2$model and the ADOxx metamodeling platform have until today been used for the realization of a myriad of domain-specific modeling methods. A couple of dozen of implemented modeling tools can be found on the OMiLAB web site [2]. Another very successful example is the GOPPRR $meta^2$model employed by the MetaEdit+ language workbench, which has been used in many industrial projects during the last decade [3].

Meta²models are typically joined together with a metamodeling tool (Ecore comes with EMF and Eclipse IDE, ADOxx comes with its own metamodeling platform,

R. Buchmann et al. (Eds.): KSEM 2014, LNAI 8793, pp. 162–173, 2014.

MetaEdit+, ConceptBase, and GME as well), which makes them available for use out-of-the-box. A brief comparison of meta^{2}models and their accompanying tools is compiled in [4].

As we can see, metamodeling approaches have been applied in many real world scenarios, like functional and non-functional requirements definition in software engineering, social modeling for requirements engineering [5], as well as modeling tool development, where metamodeling platforms have become a very popular go-to software. The answer behind the metamodeling platform attractiveness lies within the fact that one gets the meta^{2}model and the tools that work with it for free. What is left for developers to do is: (1) instantiate the meta^{2}model and (2) apply the platform's functionality to bring a modeling method to life. This process is explained in more detail in [6] and [7].

A notable difference between MM-DSL and metamodeling platforms is in their way of tackling with the development of conceptual modeling tools. MM-DSL provides domain-specific functionality which is not linked to a specific technological platform. Developing by using only a metamodeling platform is very technology-specific. One has the entire platform's functionality out-of-the-box and the possibility to reuse and extend it. However, this comes with a drawback of locking in the future development only to one platform. There is, in most cases, no way of reusing any developed artifacts (e.g., code, files, etc.) on another metamodeling platform. This is one of the primary issues MM-DSL tries to address.

In the next section (section 2) related work and its influence on MM-DSL is discussed. Section 3 focuses on the design and specification of MM-DSL. In section 4 we evaluate our DSL by implementing a pseudo modeling method. This paper concludes with advantages MM-DSL brings to the modeling tool development process and further evaluation plans (section 5).

2   Related Work

There are dozens of DSLs designed to describe a metamodel of a modeling language, the most significant being KM3 [8], HUTN [9], and Emfatic [10]. KM3 is designed particularly for specifying the abstract syntax. HUTN (Human Usable Textual Notation) is an OMG standard for specifying a default textual notation for each metamodel, developed with a purpose of solving the problem of the XMI/XML (XML Metadata Interchange) format, which is intended to be processed by the machine, therefore it is neither succinct, nor easily readable or writable. Emfatic is a language used to represent EMF Ecore models in a textual form. There exist even more DSLs that specialize in model processing, most of them connected with the Eclipse community and EMF. Epsilon [11], for example, is a family of languages and tools for code generation, model-to-model transformation, model validation, comparison, migration and refactoring that works out-of-the box with EMF and other types of models. FunnyQT [12] is a querying and transformation DSL embedded in JVM-based Lisp dialect Clojure. FunnyQT primarily targets the modeling frameworks JGraLab and EMF. GReTL [13] is an extensible, operational, graph-based transformation language, which allows specifications of transformations in plain Java using the GReTL API.

Among all of these different languages and frameworks, there are two similar to our approach: Graphiti [14] and XMF (Xmodeler) [15]. Graphiti is an Eclipse-based graphics framework that allows the development of graphical editors from domain models typically specified with EMF. It is a significant productivity improvement when compared to Eclipse GMF. XMF is a meta-programming language with an OCL-like syntax that allows construction of an arbitrary number of classification levels. Xmodeler is a metamodeling platform developed around XMF. Both of these technologies try to improve the productivity of modeling tool development, and both of them suffer from the same issues: platform independency and complexity. Graphiti currently only supports Eclipse. XMF only supports Xmodeler and it is relatively hard to learn for domain experts with limited programming knowledge. The mentioned technologies are specialized in realizing modeling languages as modeling tools, but none of them fully covers the realization of modeling methods (e.g., possibility to implement all the essential parts: abstract syntax, concrete syntax, and algorithms that can be executed on models).

In addition to the related research, the work at hand is mostly influenced by the following concepts and technologies: Language Oriented Programming [16], Language Driven Design [17], Model Driven Architecture [18], Software Factories [19], Language Workbenches [20], and Metamodeling Platforms [21]. All of them are used to ease the development of various kinds of software (e.g., systems, applications, tools) by providing the means to define custom programming or modeling artifacts and code generation facilities.

3   MM-DSL

The $\underline{M}$odeling $\underline{M}$ethod $\underline{D}$omain-$\underline{S}$pecific $\underline{L}$anguage (MM-DSL) has been designed to simplify the realization of modeling methods. It is a platform independent language for describing modeling methods – their abstract syntax, concrete syntax, and algorithms. MM-DSL is supported by an integrated development environment (MM-DSL IDE) which provides several helpful features: (1) compile time error checking, (2) code suggestions and auto completion, (2) code templates which can be extended, (3) and code highlighting. The MM-DSL IDE is built upon Eclipse technology and can be easily modified and extended with new features.

3.1   Clarifying Design Decisions

The requirements for MM-DSL have been gathered using a top-down and bottom-up approach. In top-down approach several modeling methods have been analyzed. The inspected modeling methods are compiled in [22]. The purpose was to determine the most commonly used concepts and to establish their appropriate abstractions (e.g., "*class*" is an appropriate abstraction for concept "*actor*"). The bottom-up approach gave insight on how are metamodeling technologies applied for realization of modeling methods. Several meta^{2}models have been carefully examined to determine their building blocks. Abstractions of concepts extracted with top-down approach have been matched with meta^{2}model concepts. In this process the backbone for abstract and concrete syntax of the MM-DSL has been formed, as well as mechanisms that allow compilation to different metamodeling technologies. The MM-DSL abstract syntax is conforming to the already established modeling method generic framework [21]. The chosen concrete syntax (e.g., language keywords) was adopted from the modeling community our research group belongs to (see the ADOxx meta^{2}model depicted in Figure 1).

The similarity between meta^{2}models is what MM-DSL compilers take advantage of. By leveraging this property new compilers do not need to be implemented from scratch, but configured from the same template. Figure 1 shows similarities between meta^{2}models. The concept of an "*attribute*" is highlighted with a rectangle, the concept of a "*class*" with an ellipsis, and the concept of a "*model type*" with a round-rectangle. These highlighted concepts are all singletons (consist only of one element). However, there exist concepts that are represented with multiple elements, such as "*relation*". Relation in ADOxx is represented through *Relation Class* and its two *End Points*. In GOPPRR it is represented with elements: *Relationship*, *Binding* and *Connection*. In Ecore and GME relation is a single element: *EReference*, respectively *Reference*. From this observation, it is clear that most of the concepts described with different meta^{2}models are semantically identical (or at least very similar).

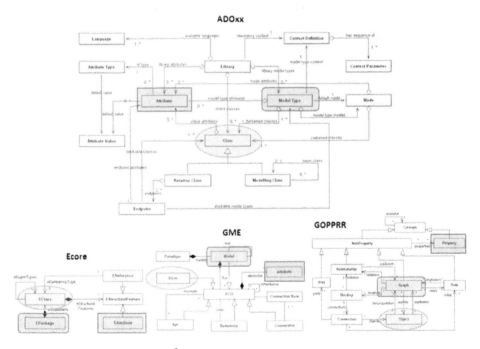

Fig. 1. Meta^{2}model Comparison (adapted from [4])

Additionally, the overall design of the language was driven with the idea of minimizing the lines of code, in other words: to reduce the programming effort and raise productivity. This is the reason we decided to introduce concepts such as inheritance (transmission of characteristics from parent object to child object) and referencing (reusing previously defined objects by passing their identifier to other objects). How these concepts work in practice is demonstrated in section 4 (see Figure 4).

3.2   Specification

In this section, we will briefly elaborate the syntax and semantics of the most relevant program statements that can be used in MM-DSL. Keep in mind that the current specification is a work in progress and is susceptible to changes.

Table 1. Excerpt from the MM-DSL's Grammar Specification in EBNF

Statement Name	Statement Specification in EBNF
Root	root ::= methodname embedcode* method
Method Name	methodname ::= 'method' name
Embed	embedcode ::= 'embed' name '<' name-embedplatformtype (':' name-embedcodetype)? '>' 'start' embeddedcodegoeshere 'end'
Method	method ::= enumeration* symbolstyle* symbolclass* symbolrelation* metamodel algorithm* event*
Enumeration	enumeration ::= 'enum' name '{' enumvalues+ '}'
Metamodel	metamodel ::= class+ relation* attribute* modeltype+
Class	class ::= 'class' name ('extends' name-class)? ('symbol' symbolclass)? '{' (attribute | insertembedcode)* '}'
Relation	relation ::= 'relation' name ('extends' name-relation)? ('symbol' name-symbolrelation)? 'from' name-class 'to' name-class '{' (attribute | insertembedcode)* '}'
Attribute	attribute ::= 'attribute' name ':' type ('access' ':' acesstype)?
Access	acesstype ::= 'write' | 'read' | 'internal'
Type	type ::= simpletype | enumtype
Simple Type	simpletype ::= 'string' | 'int' | ...
Enumeration Type	enumtype ::= 'enum' name
Model Type	modeltype ::= 'modeltype' name '{' 'classes' name-class+ 'relations' ('none' | name-relation+) 'modes' ('none' | name-mode+) '}'
Mode	mode ::= 'mode' name 'include' 'classes' name-class+ 'relations' ('none' | name-relation+)
Class Symbol	symbolclass ::= 'classgraph' name ('style' name-symbolstyle)? '{' (svgcommand | insertembedcode)* '}'
Relation Symbol	symbolrelation ::= 'relationgraph' name ('style' name-symbolstyle)? '{' 'from' (svgcommand | insertembedcode)* 'middle' (svgcommand | insertembedcode)* 'to' (svgcommand | insertembedcode)* '}'
SVG Command	svgcommand ::= (rectangle | circle | ellipse | line | polyline | poligon | path | text) symbolstyle
Symbol Style	symbolstyle ::= 'style' name '{' 'fill' ':' ('none' | fillcolor) 'stroke' ':' strokecolor 'stroke-width' ':' strokewidth ('font-family' ':' fontfamily)? ('font-size' ':' fontsize)? '}'
Algorithm	algorithm ::= 'algorithm' name '{' (algorithmoperation | insertembedcode)* '}'
Event	event ::= 'event' name-event '.' 'execute' '.' name-algorithm

Every program starts with a keyword *method* (also known as the **method name statement**), followed by the user defined name of a modeling method (e.g., `method CarParkModelingLanguage`). The statements that can be used afterwards are specified by the language's grammar (Table 2 contains the compressed overview in EBNF). The most relevant statements, their brief description and a simple usage example are given:

— **enumeration statement**, which starts with a keyword *enum*, and allows the definition of user defined data types:

```
enum EnumParkType { "car" "truck" "motorcycle" "bicycle" }
```

— **class statement**, which starts with a keyword *class*, supports inheritance through optional keyword *extends*, association with graphical symbols through optional keyword *symbol*, and can contain attributes:

```
class Car extends Vehicle symbol CarGraph {}
```

— **relation statement**, which starts with a keyword *relation*, supports inheritance, association with graphical symbols, and can contain attributes:

```
relation isParked symbol IsParkedGraph from Vehicle to Park {}
```

— **attribute statement**, which starts with a keyword *attribute*, supports the ability to define attribute name, data type, and access type (write, read, or internal) that the user has to the attribute value:

```
attribute lenght:int access:write
```

— **model type statement**, which starts with a keyword *modeltype*, contains references to classes, relations, and allows the definition of various modes (or views) on the contained objects:

```
modeltype CarPark {
  classes Car Truck Motorcycle Bicycle ParkingLot ParkingGarage
  relations isParked
  modes none }
```

— **class graph statement**, which starts with a keyword *classgraph*, supports referencing to styles, and contains the sentences describing the primitive graphical objects, like circle, or rectangle:

```
classgraph CarGraph style Blue { rectangle x=-10 y=-10 w=20 h=20 }
```

— **relation graph statement**, which starts with a keyword *relationgraph*, contains the same possibilities as class graph statement with an addition of structuring sentences for primitive graphical objects in three sections – from, middle, and to:

```
relationgraph IsParkedGraph style Black {
  from rectangle x=-2 y=-2 w=4 h=4
  middle text "is parked in" x=0 y=0
  to polygon points=-2,2 2,0 -2,-2 }
```

— **scalable vector graphics (SVG) statements**, which is a group of statements for defining primitive graphical objects: rectangle, circle, ellipse, line, polyline, polygon, path, and text, with an option to include style definitions:

```
polygon points=-20,20 0,20 -20,0 style Red {fill:red stroke:black
stroke-width:1}
```

- **symbol style statement**, which starts with a keyword *style*, and describes the general style of a graphical object, including the stroke color and width, fill color, font size and font family:

```
style Blue { fill:blue stroke:black stroke-width:1 }
```

- **algorithm statement**, which starts with a keyword *algorithm*, and can contain multiple algorithm operation statements:

```
algorithm MyAlgorithm { model.create MyModel CarPark }
```

- **algorithm operation statements**, which is a group of statements supporting various operations on a modeling method data structure:

```
model.discard MyModel
```

- **event statement**, which starts with a keyword *event*, followed by the event type and event action, is a statement that triggers algorithms based on the events happening in the modeling tool:

```
event.AfterCreateModelingConnector.execute.MyAlgorithm
```

- **embed statement**, which starts with a keyword *embed*, is used in embedding of external code to the program, which can later be referenced by other statements using the insert sub-statement:

```
embed ADODrawRoundRectangle <ADOxx:GraphRep>
  start "embedded code goes here" end
```

- **insert statement**, which starts with a keyword *insert*, is used for referencing the parts of foreign code embedded using the embed statement:

```
insert ADODrawRoundRectangle
```

The statements in this list have been used while implementing a modeling tool for the Car Park Modeling Method (see section 4), with an exception of the algorithm specific statements (algorithm, algorithm operation, and event statements) and the embedding statements (embed, and insert statements). Algorithm specific statements allow us to describe generic functionality of a modeling tool (open, edit, close and delete models, modify attribute values, create modeling objects, etc.) in MM-DSL itself. Whether the defined algorithms can be realized depends on the execution platform. If the platform does not support it, the compiler will throw a warning and skip that part of the code. As with any other programming artifacts in MM-DSL, a programmer can extend the existing set of algorithm operations and events. Embedding is an important concept allowing a developer to insert a platform-specific code directly into the MM-DSL program. An example would be a code for a complex graphical representation, which one does not want to reimplement. Embedding of external code is possible through the special *embed* statement. Referencing of the embedded code in various parts of the program is done by using the *insert* statement. For more details and additional examples see the specification available in [2].

4   A Prototype-Based Evaluation

4.1   A Pseudo Modeling Method: Requirements

For the evaluation purpose an exemplary modeling method has been designed and implemented with MM-DSL. The pseudo modeling method, named *"Car Park Modeling Method"*, is a representative example, which comprises of the most important modeling method building blocks.

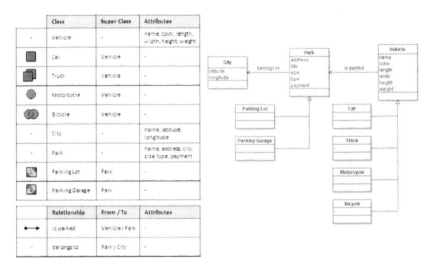

Fig. 2. The Car Park Modeling Method Requirements

The elements with the graphical representation are: *Car*, *Truck*, *Motorcycle*, *Bicycle*, *Parking Lot*, *Parking Garage*, and *is parked*. These are the ones we can model with. The rest of the elements are either abstract (e.g., *Vehicle* and *Park*) or do not have any kind of graphical representation (e.g., *City*, and *belongs to*). Together they form the metamodel of the Car Park Modeling Method. The UML Class Diagram notation has been used to specify the metamodel. The metamodel on the right side of Figure 3 shows the relations between various modeling method elements: *Vehicle* is associated with *Park* with a relationship *is parked*; *Park* is associated with *City* with a relationship *belongs to*; *Parking Lot* and *Parking Garage* are specializations of *Park*; *Car*, *Truck*, *Motorcycle* and *Bicycle* are specializations of *Vehicle*.

4.2   An Implementation Using the MM-DSL

In this section we will see how successfully MM-DSL can be applied in the development of modeling tools. The pseudo modeling method is used to illustrate the development process.

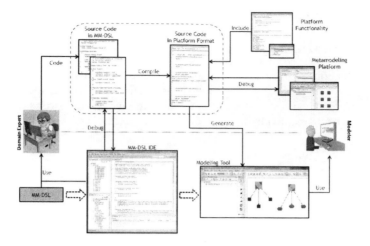

Fig. 3. Using MM-DSL for Modeling Tool Development

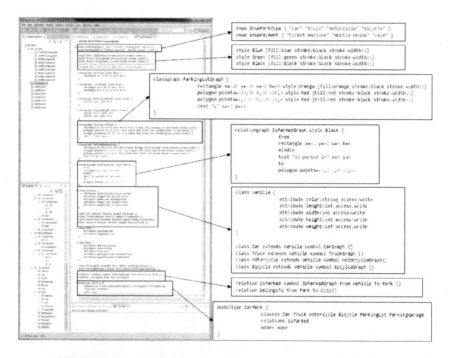

Fig. 4. The Car Park Modeling Method Implementation in MM-DSL

The modeling tool development using MM-DSL, depicted in Figure 3, is structured in the following steps: (1) code a modeling method, (2) compile the code to one or more execution environments (typically metamodeling platforms), (3) continue adding finishing touches using the functionality provided by execution environments, and (4) generate a modeling tool. Step (3) is optional. Continuing the development on the execution platform is only necessary if one requires features that are not covered with MM-DSL. In step (1) one works with MM-DSL IDE. The IDE discovers bugs and code errors before compilation. After the code has been compiled to the specific execution environment, future modifications inside the execution environment can no longer be debugged with the support of MM-DSL IDE. However, most of the metamodeling platforms provide the debugging support by themselves, which can be used in step (3).

There is no particular order in coding the modeling method with MM-DSL. One can start by defining a metamodel (abstract syntax), then structuring it in model types. Afterwards, graphical representation (concrete syntax) can be defined. Or vice versa, start with graphical representation, and then code the metamodel part. Nonetheless, some of the elements of a modeling method (e.g., classes, relations, attributes) need to be already present in the code before algorithms can be realized, because most of the algorithms require some sort of input. In our case, inputs can be any modeling elements and their values (attribute values, class or relation instances).

Looking at the code, it can be noted that enumerations have been defined first, followed by definition of styles. These artifacts need to be defined before they can be referenced. For example, *IsParkedGraph* is referencing *Black* style, and class *Park* contains attributes that are of type *EnumParkType* and *EnumPayment*. We can define styles inside class symbol and relation symbol statements as well. For example, style *Orange* is defined inside class style *ParkingLotGraph*. Symbols also need to be defined before they can be referenced by classes or relations. Concept of inheritance is used in definition of classes *Car*, *Truck*, *Motorcycle* and *Bicycle*. In the current version of MM-DSL inheritance only works for the structure. For example, attributes defined in the class *Vehicle* will be inherited by the class *Car*. However, if *Vehicle* had a symbol associated with it, it would not be inherited by the class *Car*. This was a design decision which helps in differentiating abstract and concrete classes (concrete classes always have a symbol assigned to them). There is one more relevant concept that needs addressing – model types. They are used to aggregate modeling elements into a diagram used inside a modeling tool. These elements are also visible inside a modeling tool tool-box (see Figure 5). For example, model type *CarPark* contains several classes, one relation, and only a default sub-view (indicated by modes none command). Modes or sub-views allow selecting which from the existing modeling elements in a diagram we want to show. There can be multiple modes in one model type. Resulting modeling tool is shown in Figure 5.

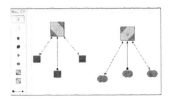

Fig. 5. The Resulting Modeling Tool

5   Conclusions and Further Evaluation Plans

There are three main concerns one needs to tackle with when implementing modeling tools for modeling methods: abstract syntax, concrete syntax, and algorithms. Here is a brief overview on how a metamodeling platform handles these issues, and how are those issues handled with MM-DSL:

— Before one can start with the development of a modeling tool, a metamodel of a modeling method needs to be designed. This is what represents the abstract syntax of a modeling method. How will this abstract syntax be implemented on a meta-modeling platform depends on the underlying meta^{2}model. Because MM-DSL is a language and can be freely extended, there is no dependency on a particular meta^{2}model.
— Secondly, graphical representation or concrete syntax needs to be specified. The realization of graphical representation depends on the functionality provided by the platform. Some of the platforms can only work with images (e.g. bitmaps), while the others have mechanisms for drawing vector objects. MM-DSL has SVG-like commands, which is a familiar (SVG is an open standard developed by W3C) and expressive approach for creating graphical objects.
— Thirdly, algorithms working on modeling elements need to be specified. Most of the metamodeling platforms use general purpose programming languages (e.g., Java, C++) to tackle with this issue. The minority has dedicated DSLs. MM-DSL includes concepts specialized for development of modeling algorithms, as well as commands for model manipulation.

The key advantages of using MM-DSL to code a modeling method and compile it into a modeling tool are: (1) independence from the execution platform, (2) modeling method concepts coded in MM-DSL can always be reused with or without modifications, and (3) reduced development effort. Because of the mentioned advantages, MM-DSL is best suited for (1) fast prototyping and (2) when support for multiple platforms is needed (write once, run everywhere principle).

Currently, MM-DSL usability and expressivity is being tested by implementing modeling tools for several modeling methods published in [22]. Further evaluation steps include a study where participants will have the opportunity to describe a modeling method with MM-DSL by using the provided language specification and modeling method requirements. No IDE will be provided, because we want to test understandability and learnability of the language. The IDE's usability will be tested in an exercise-like environment, where testers will have to extend the given MM-DSL code and compile it to a modeling tool.

References

1. Jarke, M., Gallersdörfer, R., Jeusfeld, M.A., Staudt, M., Eherer, S.: ConceptBase — A deductive object base for meta data management. Journal of Intelligent Information Systems 4, 167–192 (1995)
2. Visic, N.: MM-DSL: An EBNF Specification,
   http://www.omilab.org/web/guest/mm-dsl

3. Tolvanen, J.-P., Kelly, S.: MetaEdit+: defining and using integrated domain-specific mod-el-ing languages. In: Proceedings of the 24th ACM SIGPLAN Conference Companion on Object Oriented Programming Systems Languages and Applications, pp. 819–820. ACM Press, New York (2009)
4. Kern, H., Hummel, A., Kühne, S.: Towards a comparative analysis of meta-metamodels. In: Proceedings of the compilation of the co-located workshops on DSM 2011, TMC 2011, AGERE! 2011, AOOPES 2011, NEAT 2011, & VMIL 2011, pp. 7–12. ACM, New York (2011)
5. Yu, E.S.K.: Social modeling for requirements engineering. MIT Press, Cambridge (2011)
6. Karagiannis, D., Visic, N.: Very Lightweight Modeling Language (VLML): A Metamodel-based Implementation. In: Seyff, N., Koziolek, A. (eds.) Modelling and Quality in Requirements Engineering, Monsenstein und Vannerdat, Münster (2012)
7. Fill, H.-G.: On the Conceptualization of a Modeling Language for Semantic Model Annotations. In: Salinesi, C., Pastor, O. (eds.) CAiSE Workshops 2011. LNBIP, vol. 83, pp. 134–148. Springer, Heidelberg (2011)
8. Jouault, F., Bézivin, J.: KM3: A DSL for Metamodel Specification. In: Gorrieri, R., Wehrheim, H. (eds.) FMOODS 2006. LNCS, vol. 4037, pp. 171–185. Springer, Heidelberg (2006)
9. HUTN, http://www.omg.org/spec/HUTN/
10. Emfatic Language Reference, http://www.eclipse.org/epsilon/doc/articles/emfatic/
11. Epsilon, http://www.eclipse.org/epsilon/
12. Horn, T.: Model Querying with FunnyQT. In: Duddy, K., Kappel, G. (eds.) ICMB 2013. LNCS, vol. 7909, pp. 56–57. Springer, Heidelberg (2013)
13. Ebert, J., Horn, T.: GReTL: an extensible, operational, graph-based transformation language. Software & Systems Modeling, 1–21 (2012)
14. Graphiti, http://www.eclipse.org/graphiti/
15. Clark, T.: Xmodeler, http://www.eis.mdx.ac.uk/staffpages/tonyclark/Software/XModeler.html
16. Ward, M.P.: Language Oriented Programming. Software—Concepts and Tools 15, 147–161 (1995)
17. Clark, T., Sammut, P., Willans, J.: Applied metamodelling: a foundation for language driven development (2008), http://eprints.mdx.ac.uk/6060/
18. Frankel, D.S.: Model Driven Architecture: Applying MDA to Enterprise Computing. Wiley (2003)
19. Greenfield, J., Short, K.: Software factories: assembling applications with patterns, models, frameworks, and tools. Wiley Pub. (2004)
20. Fowler, M., Parsons, R.: Domain Specific Languages. Addison-Wesley Longman, Amsterdam (2010)
21. Karagiannis, D., Kühn, H.: Metamodelling platforms. In: Bauknecht, K., Tjoa, A.M., Quirchmayr, G. (eds.) EC-Web 2002. LNCS, vol. 2455, p. 182. Springer, Heidelberg (2002)
22. OMiLAB: The Modelling Methods Booklet, http://www.omilab.org/web/user/booklet

The Core Elements of Corporate Knowledge Management and Their Reflection in Research and Practice – The Case of Knowledge Management Systems

Nora Fteimi and Franz Lehner

Chair of Information Systems II, Passau, Germany
{nora.fteimi,franz.lehner}@uni-passau.de

Abstract. Knowledge management (KM) is nowadays a relevant topic of interest for numerous reference disciplines and involved parties in research and practice. Considering KM-literature, a fast-growing and heterogeneous collection of content exists, including various theories, topics, keywords and models which are discussed and used to handle KM-related problems and topics. The consequence is the absence of a common understanding and harmonization in KM. The paper presents a first approach towards a normative and scientifically evidenced corporate KM-framework that supports highlighting unexplored KM-topics and contributes to a common understanding of terminology, concepts and methods used in KM. The framework in this study focuses on KM-systems and is the result of the integration of most cited classification approaches. It serves as a starting point to consolidate the topics of the KM-discipline and helps in obtaining an overview of relevant topics to successfully address KM issues in research and praxis.

Keywords: Corporate knowledge management, Knowledge management framework, Design science research, Knowledge management systems.

1   Introduction

In recent years knowledge management (KM) became an increasingly important discipline [1]. The literature considering KM offers a fast-growing collection of insights consisting of various theories, concepts or topics. Depending on the relevant context and the underlying reference discipline this content might be quite heterogeneous or can be seen as inconsistent. Furthermore, different orientations are proposed for KM. In addition to the different reference disciplines and design possibilities, various concepts, attitudes and schools exist, which justify and explain KM. Moreover, it can be noticed that research and practice discuss and handle KM-related problems and topics differently. In conclusion, it seems that a common unified understanding of the discipline KM is lacking [2, 3].

The preceding overview illustrates how heterogenous KM is, both as a research discipline as well as from a practical perspective. Therefore, a main challenge in research will be to consolidate and order the various streams and trends of this discipline and to seek a common understanding of KM. Research will thereby be able to

R. Buchmann et al. (Eds.): KSEM 2014, LNAI 8793, pp. 174–185, 2014.

gain a base for the systematic comparison and classification of research results. Practitioners can rely on consistent methods and approaches when implementing KM-related activities and will be able to make the right decisions at the right time in the right place.

The paper presents a first approach towards a normative and scientifically evidenced corporate KM-framework that supports highlighting unexplored KM-topics and contributes to a common understanding of terminology, concepts, activities and methods used in KM.

The remainder of this paper is structured as follows: In Section 2 we shed some light on different streams of literature. Section 3 provides an overview of the research objectives and the research design followed by this paper. Subsequently we present a framework proposal for the corporate knowledge management beginning with Knowledge Management Systems (KMS). Finally we conclude with a summary of main results, limitations of the study and some directions and implications for future.

2   Literature Review

During past years, several attempts have been made, to consolidate and reorder the huge collection of findings in the KM-discipline e.g. by presenting a separate framework. These frameworks address and pick up special issues of KM and through the analysis of literature we could recognize four major streams of frameworks. Besides, the distinction according to these four streams of frameworks is one possible classification and does not claim the exclusive assignment of one referenced work to only one of the streams due to the multidimensionality and the different perspectives which could be adapted by the referenced papers.

Some of the studies conducted a meta-analysis of a huge number of KM-papers, whereas others outlined specific frameworks and models used in KM and classify them. Also, several studies examined and proposed frameworks for KMS or handled and discussed KM-related themes and frameworks in general. To date, it is widely accepted, that there is no generally and globally established consensus about these different KM-classifications [4]. This literature review builds on the above mentioned four streams and explains the main results according to them.

Studies of the **first steam** were carried out in form of a meta-analysis and investigate general issues and topics concerning KM such as most frequently used definitions of knowledge, often used research methodologies and most cited related work or productivity rankings of e.g. authors and countries. For instance, Nie et al. [5, 6] published a meta-analysis to answer questions about the importance of KM, actions and operations in the KM-field, the factors that enable the birth of KM, ways to implement and to support KM and applications of KM. Another example is an attempt of Heisig [2] to harmonize KM-frameworks. The author applied the method of content analysis to compare and analyze 160 frameworks with regard to the following categories: source (in the sense of title, author and year), origin according to country and region, type, knowledge definitions, frequently mentioned KM-activities and critical success factors.

Into the **second stream** fall studies, that examined and classified models, perspectives, schools of thought and approaches for KM. Some papers discussed KM models and approaches in terms of related processes and activities as in the study of Vorakul-pipat and Rezgui [1], who examined and evaluated different models, which fall into the category of knowledge category models. Other researchers proposed general own classification schemas or taxonomies to classify existing models and frameworks (cf. [4] or [7]). Lloria for example [4] reviewed seven different classifications of approaches, schools and models in this discipline. One of these approaches is that one of Nonaka and Takeuchi [8], who stated the following: whereas European countries are interested in the process of measuring knowledge, American ones manage knowledge and Japan is associated with the creating of knowledge. On the other side, Earl [9] specified three schools of KM: the technocratic school with its subdivisions into the system, the cartographic and the engineering school, the economic school and the behavioral school with its subdivisions into the organizational, the spatial and the strategic school. After reviewing these approaches they were integrated in an own classification proposal.

The **third stream** handled frameworks for KMS and contains, depending on the intended use, a variety of different classification proposals. Some approaches described KMS according to their support for the processes in the knowledge lifecycle or the SECI-model (e.g. [10–12]). Other approaches followed a strategy-oriented perspective and classified KMS according to their support for strategy [13] or according to different KM-perspectives like the transactional KM, the process-based KM or the analytical KM [14]. Besides, there exist some more technology-oriented [15, 16] or context-oriented approaches [17]. Nevertheless this diversity of different orientations may sometimes be helpful, depending on the considered context and application.

A representative of the **fourth stream** is the global KM-framework of Pawlowski and Bick [18]. The authors stated the absence of a clear task understanding of KM in praxis. In addition, it is often unclear, how benefits could be reached through the realization and implementation of KM. The global KM-framework is described in its core by processes differentiated according to knowledge processes, business processes and external processes. These processes are in relation with several other components like strategies, stakeholders, culture and instruments and results in outcomes like performance or valuable knowledge.

The collection of all these frameworks enables researchers and practitioners to get an overview of existing related work. Thus, the choice for a suitable approach or model is simplified and can be carried out faster.

3   Research Objectives of the Study and Research Design

The general purpose of our study is contributing to a common view of relevant research topics in KM and identifying gaps between research and practice including:

- Attainment of a common understanding of terminology, concepts and methods used in the field of KM
- Identification of white spots on the landscape of KMS (with no research activities or low number of studies)

In this paper, we present the results of a preliminary study starting with the reflection of KMS as a subdomain of KM and addressing the following research questions:

1. Does a core consensus or dissent exist about already available KMS-frameworks?
2. What are key areas addressed by academic research and which topics are seen as relevant by practice?

The study is following a design science oriented approach [19]. According to this approach, first the problem needs to be identified. Based on that, solution objectives should be identified, followed by the steps of design and development, demonstration, evaluation and communication of the results.

The first step of this approach is represented here by the demonstration of the overall importance and relevance of this work to the KM-field which is described in the introduction section. The steps 2-4 correspond to the presentation of a first draft of a framework for corporate KM focusing in particularly on KMS. The evaluation phase will take place in form of discussions within the KM-community in order to evaluate, validate and improve the framework suggested here.

First of all, a structured literature review was conducted aiming to examine and analyze already existing KM-frameworks in literature. This review follows a taxonomy presented by Cooper and adjusted by vom Brocke et al. [20, 21]. Besides papers in different high ranking scientific journals related to KM (e.g. "Knowledge Management Research and Practice" or "Journal of Knowledge Management"), also relevant conference papers were taken into consideration. Forward and backward search was performed too by running through the references of relevant papers and looking for further interesting papers respectively looking for papers citing these of our sample. This procedure helped us to find further papers of relevance which were included into our sample. The literature was searched based on relevant keywords such as "Knowledge Management" or "Knowledge Management Systems" in conjunction with "Meta-Analysis", "State of the Art", "Review" and "Framework". These keywords were used for the automated search of electronic databases e. g. AiSEL, Science Direct and journal websites. Initial hits were reduced by analyzing their titles in a first step. In a second synthesizing step, abstracts of the initial hits were analyzed in-depth by checking their relevance to the objectives of this paper. The resulting hit list consisted altogether of 24 relevant papers. A summary of the review results was already discussed in section 2.

4   First Results – Corporate KM-Framework

In this section we present a first approach towards a normative and scientifically evidenced KM-framework that contributes to a common understanding of terminology, activities, concepts and methods used in KM.

The divergence and heterogeneity of the existing classification approaches underlines the need for a common understanding but also the need to consolidate and harmonize the different frameworks.

This study can be seen as a first step towards this consolidation by suggesting and creating a normative framework. We understand this normative framework as a conceptual consolidation of already existing classification approaches that serves as a starting point for further analysis in our research. In the future, we will build on this framework, to empirically test its validity and improve it.

Table 1. Overview of reviewed publications according to the four categories of KM research

	Publication	Category			
		M	A/M	KMS	G
	Scholl et al. (2004) [22]	x			
	Serenko & Bontis (2004) [23]	x			
	Nie et al. (2007), Nie et al. (2009) [5, 6]	x			
	Heisig (2009) [2]	x	x		
	Serenko et al. (2009) [24]	x			
	Lee and Chen (2012) [25]	x			
Focus of the paper	**Alavi & Leidner (2001)** [10]			**x**	x
	Binney (2001) [14]			**x**	x
	Tyndale (2002) [16]			**x**	
	Liao (2003) [15]			**x**	
	Jennex (2006) [17]			**x**	
	Saito et al. (2007) [13]			**x**	
	Becerra-Fernandez & Sabherwal (2011) [11]			**x**	
	De Carvalho & Ferreira (2011) [12]			**x**	
	Holsapple & Joshi (1999) [7]		x		
	McAdam & McGreedy (1999) [26]		x		
	Kakabadse et al. (2003) [27]		x		
	Asl & Rahmanseresht (2007) [28]		x		x
	Vorakulpipat & Rezgui (2008) [1]		x		
	Lloria (2008) [4]		x		
	Jafari et al. (2009) [29]		x		x
	Moteleb and Woodman (2007) [30]				x
	Pawlowski (2012) [18]				x

M : Meta-Analysis; A/M : Approaches/Models; KMS : Knowledge Management Systems; G: General

The creation of the normative framework is based on the analysis and aggregation of already existing frameworks, which were identified in scientific journals ex ante (cf. section 2). The classifications (table 1 gives an overview of reviewed classifications with their categorization according to the four streams mentioned in section 2) were extracted and integrated into a mind map. The choice of this design format made it possible to visually illustrate the collection of classification approaches and thus to obtain an overall picture of the state of the art. Subsequently we started the aggregation process by looking for similar or related classifications, which could be integrated into a new classification schema. In this stage of progress, we had some discussions within our team, to ensure the validity of the aggregations. Finally we build new main categories, which include the integrated classifications.

Subject matter of this study is a normative approach starting with KMS. Despite of the fact that KMS are one of the basic and important fundaments of KM, no broad agreement about the term KMS exists [17]. A popular definition of KMS is related to Alavi and Leidner who proposed the following: *"KMS refer to a class of information systems applied to managing organizational knowledge. That is, they are IT-based systems developed to support and enhance the organizational process of knowledge creation, storage/retrieval, transfer, and application"* [10].

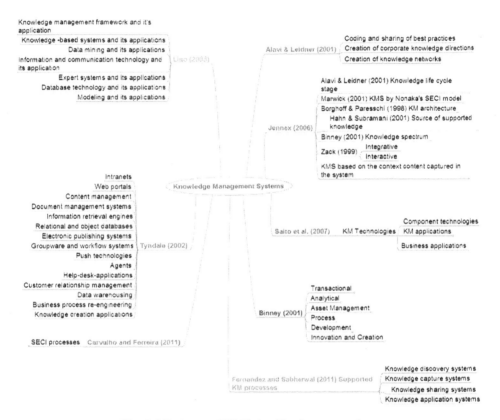

Fig. 1. Mind map of KMS classification approaches

Figure 1 illustrates the mind map of nine main KMS-classifications according to different authors, who propose or reflect our findings namely: Alavi and Leidner [10], Jennex [17], Saito et al. [13], Binney [14], Fernandez and Sabherwal [11], Liao [15], Tyndale [16], Carvalho and Ferreira [12].

The analysis of the mind map uncovers some noticeable aspects:

- KMS can be applied and adopted for a wide range of knowledge processes such as the generation, identification, structuring, storing and distribution of knowledge. Innovations in the field of information technologies offer new possibilities to support the organizational knowledge base as well as the tasks and processes of KM. This involves less the automated management of big data than the linkage between human and mechanical skills. This could be enhanced by the increasing integration and the combination of technologies, which could be used in an integrated or isolated manner.
- Some classification approaches follow a strategy oriented perspective and classify KMS according to their support of either codification or personalization strategy.

These classifications take into account the concept of the enterprise architecture with its distinction between several enterprise levels such as the strategic or the operational level.

- A final group of classification schemas characterize KMS differentiating between different task processing modes and the context of system use. This includes for example the distinction whether the task has been performed in an integrative or in an interactive manner but also the distinction between the task coordination in a distributive or collaborative working environment.

Based on the preceding analysis, the normative framework with its multidimensional categories and subcategories has been created. The design of this framework is based on the adaption and visualization of generally used classification schemas or classification schemas, which were discussed and mentioned frequently by more than one paper in literature. The framework presented here is an attempt to reflect and summarize already existing KMS classification schemas highlighting the consensus and dissents between them. The new integrated KM-framework will be created in the next steps and can be used to categorize research papers in the field of KM. As mentioned before this study is preliminary and describes only a part of the overall framework by focusing on KMS. Multidimensionality means here, that it is possible to assign elements within the framework to more than one single category. For example the element "social web" could represent a subcategory of a category "KM-processes", but it could also be assigned to a category named "Strategy oriented KMS" or to other suitable categories.

Figure 2 (cf. figure 2) shows the proposed framework structure for KMS. The initial framework consists of three main categories:

- **Category 1:** Process orientation
- **Category 2:** Strategy orientation
- **Category 3:** System type orientation

The decision for choosing these three categories reflects the three main classification approaches that can be found in the relevant literature.

The **"process"-category** subsumes approaches, which describe KMS according to their support of processes in the knowledge life cycle. Naming this category as the process oriented refers to the selected sources in the literature review. The authors of these sources investigate several KMS in terms of their support to the processes in the knowledge chain or activities as named in the well known life cycle model according to Probst et al. [31]. Advocates of this approaches are e. g. [10, 11] and [32] Based on the classification schemas we suggest to split this category into four subcategories according to the SECI Model [33]. The resulting subcategories (Socialization, Externalization, Combination and Internalization) could be refined by assigning suitable knowledge- or KM-processes such as sharing, transferring, distributing or storing of knowledge. These processes are tagged in the relevant literature by different keywords which are often used synonymously but describe the same process. To group and identify the process oriented keywords, a content analysis [34] on a sample of research articles of the Journal of Knowledge Management was done. The sample

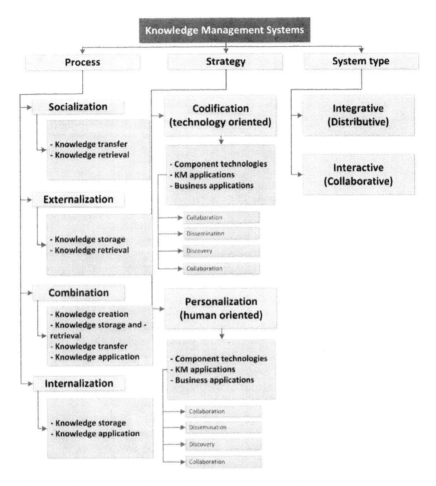

Fig. 2. Partial normative framework proposal for knowledge management systems

consists of 394 articles (time span 2005 to 2013) obtained from the abstract and citation database scopus (www.scopus.com). We decided to focus only on the abstracts and keywords during the content analysis because the summary contains the main aspects. Focusing only on parts of the papers like title, keywords and abstracts is a common procedure when doing a content analysis based upon literature. The abstracts and keywords of the sample were tagged and a word frequency count resulted in a list of 34 knowledge life cycle processes. In a synthesizing step these keywords were assigned to the four main categories by Alavi and Leidner [10]. The decision for choosing this categorization as a benchmark is motivated by the popularity and citation index of the author's publication (cf. table 2).

For example, all the processes of assimilating, recombining, generating and producing knowledge could be summarized as processes, in which new knowledge is created. On the other side, when sharing, diffusing or exchanging knowledge, a transfer of knowledge is taking place.

Table 2. Knowledge life cycle processes categorization (Basis sample (n= 394): Journal of Knowledge Management 2005-2013)

Knowledge creation	Knowledge storage / retrieval	Knowledge transfer	Knowledge Application
Assimilation	Access	Diffusion	Appropriation
Generation	Accumulation	Dissemination	Attrition
Production	Acquisition	Distribution	Conversion
Recombine	Capture	Exchange	Exploration
	Collection	Sharing	Exploitation
	Documentation		Integration
	Harvesting		Obtaining
	Maintaining		Recycling
	Preservation		Utilization
	Retention		Validation

Based on the categorization in table 2, the four main knowledge life cycle processes were assigned to the SECI-model (cf. figure 2). The stage of socialization could be characterized for example by the transfer and retrieval of knowledge, whereas the externalization phase could be described by the storage and retrieval of knowledge. In addition, whilst combining knowledge, new knowledge is created, transferred, stored and applied later on. In the internalization stage, most of the activities focus on the storage and application of knowledge.

Finally each one of these processes could be facilitated or supported by the use of several technologies. Social web based systems e.g. facilitate the processes of sharing, storage and transfer of knowledge, whereas databases fit best for storing and retrieving knowledge.

The strategy oriented categorization, as proposed by Saito et al. [13] classifies KMS according to their support of a certain strategy into a technology oriented or human-oriented approach. The first approach represents the codification of knowledge and puts the focus on technology support for KM-related tasks especially the creation and transfer of knowledge. The human-oriented approach focuses primarily on the personalization strategy and on creating and transferring knowledge between individuals. Saito et al. distinguished between component technologies, KM-applications and business applications and propose that each one of these technologies could be supported by special collaboration-, dissemination-, discovery- and repository technologies.

Last but not least the system type can be used as the third main category in the normative framework. According to Zack [35] some technologies support doing and processing tasks integrative, whilst others are suitable for those tasks that deliver the best outcomes when being executed interactively. Park and Jeong [36] present a corresponding approach distinguishing between distributive KMS and collaborative KMS according to integrative and interactive KMS. Integrative oriented KMS are for example data warehouse systems, data mining systems and databases, whereas groupware and instant messaging systems need to be used interactive in order to deliver the requested results.

5   Concluding Remarks and Limitations of the Study

In this paper we presented a first approach towards a corporate KM-framework by addressing two main research questions:

1. Does a core consensus or dissent exist about already available KMS-frameworks?
2. What are key areas addressed by academic research and which topics are seen as relevant ones for practice?

Focusing on KMS, we reviewed and analyzed the most cited classification approaches to identify similarities and differences in the proposed frameworks. Based on the analyses we combined three common classification approaches and built a new normative framework. The resulting schema sheds light on key areas of interest for research and practice. In the future we will build on these results to extend the framework by enabling elements and categories covering the KM discipline as a whole and not only KMS in particular.

Concluding, our study has some limitations that should be mentioned at this point. At the moment our framework is restricted to KMS and needs to be extended by integrating the missing topics but also the relationships and dependencies between the categories. This task will be done in the next steps as continuation of this study. Even though the first discussions and validations within our team have shown the applicability of the framework, we are aware, that our results still needs additional and scientific validation. The framework is normative and represents the result of a mainly conceptual work. Anyway, the idea is to set up this framework as a starting point of an iterative process until reaching the expected and desired final and common KM-framework. This includes qualitative research in form of an extensive content analysis, but also interviews within the community to test and evaluate the results.

We contribute to research by presenting a first step towards consolidation and obtaining a single common understanding of the KM-discipline. A unified view helps to reflect the research field with its core values, assumptions and attitudes, and supports a cumulative research process in this field. The systematic comparison and the proposed classification schema can be used as a starting point for further research in order to get an overview of the state of the art and to classify new research projects. With regard to the practical impact, businesses get help for introducing and implementing KM within the company.

References

1. Vorakulpipat, C., Rezgui, Y.: An evolutionary and interpretive perspective to knowledge management. J. Knowl. Manag. 12, 17–34 (2008)
2. Heisig, P.: Harmonisation of knowledge management – comparing 160 KM frameworks around the globe. J. Knowl. Manag. 13, 4–31 (2009)
3. Serenko, A., Bontis, N., Booker, L., Sadeddin, K., Hardie, T.: A scientometric analysis of knowledge management and intellectual capital academic literature (1994-2008). J. Knowl. Manag. 14, 3–23 (2010)
4. Lloria, M.B.: A review of the main approaches to knowledge management. Knowl. Manag. Res. Pract. 6, 77–89 (2008)

5. Nie, K., Ma, T., Nakamori, Y.: Building a Taxonomy for Understanding Knowledge Management. Electron. J. Knowl. Manag. 5, 453–466 (2007)
6. Nie, K., Ma, T., Nakamori, Y.: An Approach to Aid Understanding Emerging Research Fields — the Case of Knowledge Management. Syst. Res. Behav. Sci. 26, 629–644 (2009)
7. Holsapple, C.W., Joshi, K.D.: Description and Analysis of Existing Knowledge Management Frameworks. In: Proc. 32nd Hawaii Int. Conf. Syst., pp. 1–15 (1999)
8. Nonaka, I., Takeuchi, K.: The knowledge creating company: How Japanese companies create the dynamics of Innovation. Oxford University Press, Oxford (1995)
9. Earl, M.: Knowledge management strategies: toward a taxonomy. J. Manag. Inf. Syst. 18, 215–233 (2001)
10. Alavi, M., Leidner, D.E.: Knowledge Management and Knowledge Management Systems: conceptual Foundations and Research Issues. MIS Q. 25, 107–136 (2001)
11. Becerra-Fernandez, I., Sabherwal, R.: The role of information and communication technologies in knowledge management: A classification of knowledge management systems (2011)
12. De Carvalho, R.B., Ferreira, M.A.T.: Knowledge management software. In: Schwartz, D., Te'eni, D. (eds.) Enzyclopadie of Knowledge Management, pp. 738–749. IGI global, Hershey (2011)
13. Saito, A., Umemoto, K., Ikeda, M.: A strategy-based ontology of knowledge management technologies. J. Knowl. Manag. 11, 97–114 (2007)
14. Binney, D.: The knowledge management spectrum - understanding the KM landscape. J. Knowl. Manag. 5, 33–42 (2001)
15. Liao, S.: Knowledge management technologies and applications—literature review from 1995 to 2002. Expert Syst. Appl. 25, 155–164 (2003)
16. Tyndale, P.: A taxonomy of knowledge management software tools: origins and applications. Eval. Program Plann. 25, 183–190 (2002)
17. Jennex, M.E.: Classifying Knowledge Management Systems Based on Context Content. In: Proc. 39th Hawaii Int. Conf. Syst. Sci., pp. 1–8 (2006)
18. Pawlowski, J.M., Bick, M.: The global knowledge management framework: Towards a theory for knowledge management in globally distributed settings. Electron. J. Knowl. Manag. 10, 92–108 (2012)
19. Gregor, S., Hevner, A.R.: Positioning and presenting design science research for maximum impact 37, 337–355 (2013)
20. Cooper, H.: Organizing Knowledge Syntheses: A Taxonomy of Literature Reviews. SAGE (1998)
21. Vom Brocke, J., Simons, A., Niehaves, B., Reimer, K., Plattfaut, R., Cleven, A.: Reconstructuring the Giant: On the importance of rigour in the literature search process. In: ECIS 2009 Proc. Pap. 161 (2009)
22. Scholl, W., König, C., Meyer, B., Heisig, P.: The future of knowledge management: an international delphi study. J. Knowl. Manag. 8, 19–35 (2004)
23. Serenko, A., Bontis, N.: Meta-review of knowledge management and intellectual capital literature: citation impact and research productivity rankings. Knowl. Process Manag. 11, 185–198 (2004)
24. Bontis, N., Serenko, A.: A follow-up ranking of academic journals. J. Knowl. Manag. 13, 16–26 (2009)
25. Lee, M.R., Chen, T.T.: Revealing Research Themens and trends in knowledge management from 1995 to 2010. Knowledge-Based Syst. 28, 47–58 (2012)
26. Mcadam, R., Mccreedy, S.: A critical review of knowledge management models. Learn. Organ. 6, 91–101 (1999)

27. Kakabadse, N.K., Kakabadse, A., Kouzmin, A.: Reviewing the knowledge management literature: towards a taxonomy. J. Knowl. Manag. 7, 75–91 (2003)
28. Asl, N.S., Rahmanseresht, H.: Knowledge Management Approaches and Knowledge gaps in Organizations. Manag. Worldw. Oper. Commun. with Inf. Technol. 1427–1432 (2007)
29. Jafari, M., Akhavan, P., Mortezaei, A.: A review on knowledge management discipline. J. Knowl. Manag. Pract. 10, 1–23 (2009)
30. Moteleb, A.A., Woodman, M.: Notions of Knowledge Management Systems: A Gap Analysis. Electron. J. Knowl. Manag. 5, 55–62 (2007)
31. Probst, G., Raub, S., Romhardt, K.: Wissen managen - Wie Unternehmen ihre wertvollste Ressource nutzen. Gabler (2006)
32. Marwick, A.D.: Knowledge management technology. IBM Syst. J. 40, 814–830 (2001)
33. Nonaka, I., Toyama, R., Konno, N.: SECI, Ba and leadership: a unified model of dynamic knowledge creation. Long Range Plann. 33, 5–34 (2000)
34. Mayring, P.: Qualitative Inhaltsanalyse. In: Flick, A., et al. (eds.) Qualitative Forschung, 5th edn. Ein Handbuch. Rowohlt. Hamburg (2007)
35. Zack, M.H.: Managing codified knowledge. Sloan Manage. Rev. 40 (1999)
36. Park, H., Jeong, D.H.: Assessment of effective utilization of KM technologies as a function of organizational culture. In: Reimer, U., Karagiannis, D. (eds.) PAKM 2006. LNCS (LNAI), vol. 4333, pp. 224–233. Springer, Heidelberg (2006)

Combining Bottom-Up and Top-Down Generation of Interactive Knowledge Maps for Enterprise Search

Michael Kaufmann[1], Gwendolin Wilke[2], Edy Portmann[3], and Knut Hinkelmann[2]

[1] Lucerne University of Applied Sciences and Arts, Horw, Switzerland
`michael.kaufmann.01@hslu.ch`
[2] University of Applied Sciences North-Western Switzerland, Olten, Switzerland
`{gwendolin.wilke,knut.hinkelmann}@fhnw.ch`
[3] University of Bern, Switzerland
`edy.portmann@iwi.unibe.ch`

Abstract. Our research project develops an intranet search engine with concept-browsing functionality, where the user is able to navigate the conceptual level in an interactive, automatically generated knowledge map. This knowledge map visualizes tacit, implicit knowledge, extracted from the intranet, as a network of semantic concepts. Inductive and deductive methods are combined; a text analytics engine extracts knowledge structures from data inductively, and the enterprise ontology provides a backbone structure to the process deductively. In addition to performing conventional keyword search, the user can browse the semantic network of concepts and associations to find documents and data records. Also, the user can expand and edit the knowledge network directly. As a vision, we propose a knowledge-management system that provides concept-browsing, based on a knowledge warehouse layer on top of a heterogeneous knowledge base with various systems interfaces. Such a concept browser will empower knowledge workers to interact with knowledge structures.

Keywords: knowledge technology, knowledge engineering, concept extraction, enterprise ontology, enterprise search, concept browsing.

1   Introduction

Today's organizations are increasingly connected to knowledge work, dealing with new problems, complex decisions and unforeseen situations. Knowledge work requires creating, managing, exchanging and finding relevant information. Intranet search engines help make explicit information accessible. The information is stored e.g. in the file system of a company, in emails, in database systems or in corporate wikis. However, often, the relevant search terms are not known in advance, so the desired information cannot be found. Accordingly, much of the know-how and the stored information remain inaccessible.

When relevant keywords for a successful search are not known, it is often helpful to get a general idea of the higher-level concepts (i.e. the big picture), thereby creating semantic relationships between concepts and existing resources. A conventional search engine does not provide such an overview. This is because the relationships

R. Buchmann et al. (Eds.): KSEM 2014, LNAI 8793, pp. 186–197, 2014.
© Springer International Publishing Switzerland 2014

between concepts, as well as the relationships between concepts and resources, are not explicitly represented. This paper proposes an approach towards the development of a search engine that provides the following functionality: The user is able to navigate the conceptual level in a visual, interactive, auto-generated knowledge map.

1.1   Sublimating Data to the Conceptual Level

In the Stanford Encyclopedia of Philosophy (SEP), concepts are defined as "the constituents of thoughts" [1]. For knowledge extraction from text, for reasons of simplicity, we define concepts as equivalence classes of particular signs. Thus, two equal strings that occur in different locations are different particular pieces of information; nevertheless, they can be thought of as belonging to the same universal concept, defined as the class of not identical, but equal, and thus equivalent, signs. Essential for the representation of knowledge as abstract concepts is the structural embedding of concepts in an ontology [2], i.e. a *semantic network* that represents associations between concepts. In this sense, knowledge technology enables the user to interact with abstract concepts and their associations directly.

One possible solution is to automatically create the concept level, not by users, but by algorithms that provide automated knowledge extraction. Humans are unable to fully grasp the abstract implicit concept structure in the flood of available documents without manually working through them; this is not possible when dealing with big data. Therefore, our approach is to automatically extract a conceptual structure from the existing document base. This structure is intersubjective by nature and is based on machine-made belief-propositions and justifications from the given data (cf. section 3). As a result, the users can focus on creating linear, thematic documents. Meanwhile, a shared semantic concept structure is created by artificial intelligence as an emergent semantic network.

1.2   Vision

In the project Lokahi, we are developing a prototype of a semantic intranet search engine that allows users to browse and edit implicitly existing knowledge structures. The users will be able to get an overview of the relevant issues in the organization and corresponding documents and resources. Also, the users will be able to explicitly add and edit their own contextual knowledge in the concept network. As an integral part of the software, the search engine will provide the users with a knowledge map, which serves two purposes: (A) visualization of intranet knowledge and (B) interaction with this knowledge.

The concept of an intranet search engine is extended in order to allow the users to browse within concepts and their relationships. In this vision, semantic relationships between concepts, as well as between concepts and resources, are automatically extracted and visualized. The user can browse the knowledge landscape visually in a semantic network, and thereby find related resources from various sources of information.

1.3   The Lokahi Approach to Knowledge Engineering

The Hawaiian word "lokahi" stands for unity, unification and harmony. In the project Lokahi, knowledge technology brings together approaches for bottom-up and top-down knowledge engineering.

Top-down knowledge engineering applies existing ontologies to deduce structure from an ontology and apply it to existing data. These methods project a predefined structure from the concept level to the document level by deduction. For example, software tools, such as ontology editors, allow the user to model a semantic concept model and to allocate resources accordingly. However, these tools rely heavily on the manual entering of conceptual knowledge structures or deduction rules in order to assign data to concepts.

In contrast, bottom-up knowledge engineering generates ontologies from data. These methods induce a concept structure by sublimating concepts and their relationships from the document level to the concept level. These methods apply automated information extraction to draw conclusions about semantically relevant concepts and relationships through induction from the set of electronic information resources of an organization. However, pure bottom-up approaches do not incorporate readily available contexts and knowledge structures, such as employees, organizational units, or ongoing projects.

The Lokahi approach unifies and combines these two knowledge engineering methods to use synergies of the advantages of both methods. The aim is a partially automatic creation of a knowledge map. Automatically extracted knowledge is structured by manually created ontologies.

2   Related Work

Oesterle et al. [3] outlined the discipline of a *design-oriented information systems (IS) research*. Similar to behaviorist IS research, the object of investigation are sociotechnical information systems, per se, that involve humans, technology, and organizations. However, while the behaviorist approaches study IS empirically as given phenomena, the design-oriented approach aims at creating and evaluating innovative designs for information systems that provide some advantage or benefit. The research process involves analysis, artefact design, artefact evaluation, and diffusion. Accordingly, following a design-science oriented approach to informatics, we present a blueprint scheme that is being implemented in an ongoing research and development project. The field of application of our design is knowledge technology, by which we intend to blend emergent semantics and enterprise ontologies for knowledge engineering.

2.1   Knowledge Technologies

Preece et al. [4] proposed applying knowledge engineering methods to improve knowledge management. The authors describe *knowledge engineering* as a process that builds conceptual models and constructs a knowledge base. They suggest applying these processes to build a knowledge map that provides an organizational knowledge model. Knowledge engineering is embedded in the larger spectrum of *knowledge technologies* [5], defined by Milton et al. [6] as methods and tools that are specifically oriented towards knowledge, supporting creation, mapping, retrieval und use of knowledge, thereby acting as a bridge between people and information technology.

The *knowledge warehouse* has been proposed by Yacci [7] as a central repository for storing knowledge components, providing different views on knowledge and supporting queries for *knowledge retrieval*; this has been described as a specialized form of inference by Frisch [8]. Nemati et al. [9] described the knowledge warehouse as an extension of the data warehouse, which incorporates implicit knowledge that has not been captured explicitly in databases. The authors point out the use of artificial intelligence techniques to enhance knowledge creation, similar to the use of data mining in data warehouses. Our design envisions such a knowledge warehouse system; this system integrates heterogeneous knowledge sources, similar to a data warehouse, extracts implicit knowledge structures, and provides the knowledge content to the user.

Nilsson & Palmér [10] designed a *concept browser* as a paradigm for user interaction in knowledge-based systems. In a concept browser, the user can interact with knowledge structures in a graphical way. The browser presents a visual and interactive representation of ontology to the user, thereby facilitating knowledge retrieval. Accordingly, the user interface of our knowledge warehouse system will incorporate a concept browser that empowers the knowledge workers to query the knowledge components graphically.

2.2   Emergent Semantics for Knowledge Engineering

In knowledge engineering, as well as in robotics, there are deductive approaches, based on methods of formal reasoning. However, in robotics, there are also informal approaches involving *design for emergence* (see [11] pp. 124), in which the derivation of new knowledge from existing knowledge-structures emerges by an interaction of a computer with the environment. These approaches are based on the assumption that intelligence is not achieved by deducing from a rule base but by inducing a rule base from interactions with the real world. Inspired by emergent robotics, we propose a similar approach to knowledge engineering, by which ontologies are not entirely predefined by experts but are learned by interaction of the system with the dynamic knowledge base and its users.

Our research is based on a *connectionist* approach to cognitive science, namely, the view that knowledge is based on self-developing network structures (see [12], p. 299). Analogous to McClelland & Cleeremans [13], we hypothesize that complex knowledge representation and processing can be modeled with a very large number of very simple units and their connections. Learning and knowledge discovery can be implemented as changes in the association weights between the units. The working hypothesis is that meaningful knowledge automation, including semantics, can emerge from such adaptive network structures (*emergent semantics* [14]).

Portmann & Pedrycz [15] proposed to automatically extract a concept graph from data and presented to the user interactively. Approaches for extracting ontologies from data have been proposed e.g. by Maedche & Staab [16] and by Parameswaran et al [17]. Existing approaches for extracting concept graphs are e.g. latent semantic indexing [18], computing a distance between latent concepts in multidimensional space, and formal concept analysis [19], which can derive a hierarchy of concepts from a corpus. Portmann, Kaufmann & Graf [20] proposed a method of bottom-up knowledge extraction and visualization of knowledge for web-scale knowledge retrieval. This method inspired the design in this paper for the application to enterprise search.

2.3   Enterprise Ontologies

An enterprise ontology is a formal description of the organizational structure of an enterprise in machine readable format. Information in an organization is embedded in structures. Hinkelmann et al. [21] showed how semantic metadata, derived from the enterprise architecture description, can improve the exploitation of information. An attempt to directly link the world of enterprise ontologies and the world of enterprise architecture frameworks is ArchiMEO [22]; this is a formalization of a generic enterprise architecture in the form of an enterprise ontology, which is based on the Archi-Mate framework [23]. As an Upper Ontology, ArchiMEO can be adapted to specific organizational contexts and domains (c.f. e.g. [24] for its extension to software integration projects).

Since the manual extension and adaptation of ontologies to specific organizational contexts and domains is time-consuming and laborious, an automatic method is needed. In her dissertation, Thönssen [25] developed a method in order to automatically generate metadata for information objects from the information infrastructure of a company (such as registries or file system structures) in order to add it to the manually created ontology. In the Lokahi project, the resulting enterprise ontology is based on the ArchiMEO framework. The enterprise ontology is used to steer the bottom-up concept extraction. Thus, our research prototype incorporates an enterprise ontology as a backbone for bottom-up knowledge extraction.

3   Artefact Design

In the project Lokahi, a semantic intranet search engine with interactive visual concept browsing is being developed. The intention is to help the users quickly find meaningful relationships, thereby making it easier to find relevant resources and content in the intranet of an organization.

The goal of the knowledge extraction component is creating a weighted concept graph by automatic extraction of semantic relationships in existing documents and data. The extraction functionality analyzes the implicit and cross-document concept associations by (A) analyzing document contents (i.e. bottom-up approach) and (B) by incorporating the corporate context (i.e. top-down approach).

Methods of information extraction and semantic technologies are used to achieve the following:

- (A) In a bottom-up approach, substantive correlations between concepts (i.e. relevant n-grams as granules) are extracted from documents and data through an analysis of their contents. The result of this analysis is a *concept graph* that consists of concepts, concept associations, and concept-association weights.

- (B) In a top-down approach, enterprise ontologies are applied in order to deduce structure and project it onto the concept graph. Explicit top-down structuring of the knowledge landscape (e.g. organizational units, projects, employees, computer systems, etc.) is made available in order to complement the bottom-up extraction with a top-down backbone structure of the concept graph.

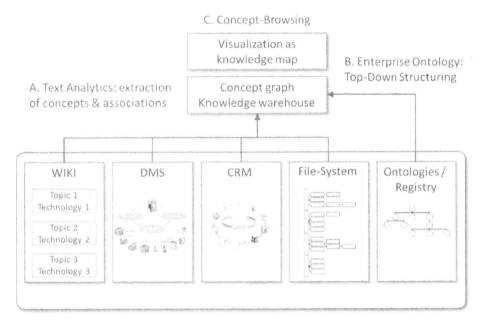

Fig. 1. Target vision of the proposed knowledge management system: automated knowledge extraction (A) structured by enterprise ontologies (B) enables concept browsing (C) on top of a heterogeneous enterprise knowledge base

The project focuses on the intranet file storage as a source of information. In case of a successful first prototype, the plan is to include further information, such as mail servers, customer relationship management systems (CRM), document management systems (DMS), or wikis. The final vision is a knowledge warehouse layer that integrates the whole of the heterogeneous enterprise-knowledge base of an organization (Figure 1). The knowledge warehouse will provide a standard interface for data indexing that is identical for all knowledge sources. For each one of the heterogenous knowledge sources, an individual adapter satisfying the knowledge warehouse interface will have to be implemented.

The user interface of the knowledge warehouse is an interactive visualization of the concept graph. This presents concepts and associations graphically to the users in the concept browser interface, allowing the users to interactively explore and edit semantic relationships between subject areas (concept browsing), as well as to display the resources (documents and data records) relevant to the respective concepts. This overview of the knowledge landscape enables the users to quickly identify resources that are associated with thematic concepts. The intent is to create a tool that empowers knowledge workers by giving them overview, guidance, structure and efficient access to available resources.

3.1   Bottom-Up Knowledge Extraction from Data

Although the definition of knowledge is controversial [26], Shadbolt [5] defined knowledge simply as *useful information*. According to Shi & Griffith [28], induction represents an adequate model for neuropsychological learning. In this sense, know-ledge (i.e. useful information) can be automatically extracted from data in the same way our brain does. That *biomimetic* stance is the main difference to existing methods such as latent semantic indexing (LSI) [18], which is rather algebraic and geometrical in nature. A drawback of LSI is that its concepts are represented as a vector matrix without labels. To address that, our approach determines concepts that have a speci-fied label in the form of a document keyword.

First, our procedure extracts the set of concepts. For every document, its keywords are extracted using Lucene [30] and its *MoreLikeThis* class. Every term (i.e. n-gram) that results as a document keyword is then considered important enough to be added to the set of concepts. Thus, the concept granularity is determined. The number of keywords that are extracted per document can be specified as a parameter. That mod-ulates the size of the set of concepts.

Secondly, for every concept pair, the degree of association is analyzed. After the extraction of the set of concepts from the corpus, sufficiently frequent concept pairs (A, B) are extracted as frequent 2-itemset in the sense of [31]. In his dissertation, Kaufmann [29] has developed a formula for inductive membership function induc-tion. We can measure the degree of membership of object x in y by the measure of *normalized likelihood ratio* (NLR In the project Lokahi, this measure is applied for calculating the weight of association between concepts. Fuzzy concept map mining is an instance of *inductive fuzzy classification* of concept pairs. Therefore, by counting the contingencies of keyword pairs in the corpus, the degree of membership of key-word A in keyword B, denoted as $\mu_B(A)$, can be induced. The degree of association of two concepts can then be computed by

$$\mu_B(A) \overset{\text{def}}{=} \frac{p(A|B)}{p(A|B)+p(A|\bar{B})} \ . \tag{1}$$

In Formula (1), $p(A|B)$ denotes the (sampled) probability that a document in the corpus contains term A under the condition that it contains term B, and $p(A|\bar{B})$ is the probability that a document contains A if it does not contain B. Thus, the strength of concept association is measured by a normalization of the likelihood ratio. This pro-cedure results in a bottom-up concept graph of document keywords, with concepts as vertices and association strengths as edges.

3.2   Top-Down Structuring of the Concept Graph Using Enterprise Ontologies

In a top-down structuring of the knowledge landscape of an organization, concepts, such as processes, organizational units, users, and contacts, are predefined (either by enumeration or by rule base) and are provided manually; existing data resources, such as documents and records are assigned to them a posteriori. The definition of concepts is usually done implicitly, e.g., by drawing organizational charts or by setting up the file system using descriptive filenames and semantically appropriate structures. Also, there are research approaches to model enterprise structures explicitly in the form of semantic enterprise ontologies [22].

Our approach utilizes enterprise ontologies to provide top-down structure to the bottom-up concept graph that has been extracted using the procedure described in the previous section. According to Thönssen [25], the upper ontology ArchiMEO is an enterprise ontology that is based on ArchiMate Standard [23], resulting in a formal, computer-readable description of the organizational context. The enterprise ontology is defined manually or semi-automatically by metadata extraction from databases. This representation of the enterprise ontology, adapted to the organization, serves as the backbone for the bottom-up extraction of the concept graph. In the user interface, this will provide a well-known structure of objects of the organizational context, such as employees, organizational units, and projects, to the user.

3.3   Integration of Bottom-Up and Top-Down Approaches to Knowledge Engineering

The aim of the project Lokahi is to bring together bottom-up and top-down approaches to knowledge engineering in order to generate a knowledge map of a company automatically. This can be used for concept browsing in a semantic intranet search engine. To utilize synergies, we propose a combined approach. On the one hand, the domain ontology is emergent and self-organizing, as it is extracted automatically from data (bottom-up), and thus, it is efficient. On the other hand, the extraction of the domain ontology is regulated and structured by a predetermined enterprise ontology (top-down), serving as the underlying structure of the knowledge graph; this incorporates existing organizational context in the emergent structure.

The enterprise ontology (EO) is intended to be used as the base structure or backbone. This backbone graph is expanded and refined by bottom-up extraction: In the EO, every concept is given by a plain text identifier, while concepts in the bottom-up concept graph are represented by extracted document keywords. A simple text matching provides the link between the unified resource identifier (URI) in the formal ontology and the keyword concept in the bottom-up concept graph. Thus, Formula (1) can provide a sorted list of bottom-up concepts associated with a top-down URI in the formal ontology.

To illustrate, an employee could be identified in the EO by the URI "http://example.com#Sandra+Meyer". If there are documents having the 2-gram "Sandra Meyer" as an extracted keyword, then there is a concept with the same name in the bottom-up concept graph. This graph then also contains concepts that are inductively associated with that keyword; e.g. the term "knowledge management", because Sandra Meyer is a knowledge manager, and she has more frequently written documents containing her name together with that keyword than with other keywords.

Thus, using a gradual strength of association, the bottom-up concept graph provides a fuzzy set [32] of associations to concepts in the predefined top-down ontology, thereby extending it gradually.

3.4   User Interface for Interactive Concept Browsing

Classic intranet search engines provide information-oriented user interfaces, which list the resources found for search terms, sorted by relevance. In contrast, the user interface in project Lokahi visualizes the knowledge structures of a company at the abstract concept level as a knowledge map. The user interface also displays the corresponding resources associated with the concepts. In addition, the user will be given the opportunity to browse this knowledge map; to search the knowledge map by keyword; to change the knowledge map; and to view the concept's associated resources.

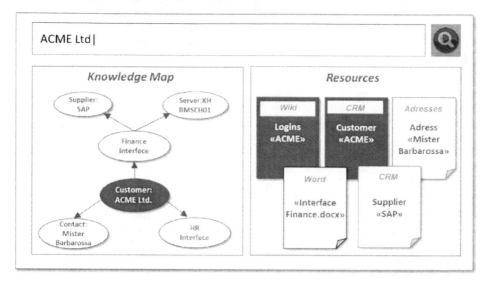

Fig. 2. The user interface includes a concept browser. The user can enter a search term, and the interface displays a knowledge map with associated concepts and their associations, together with corresponding resources from different data sources.

The graphical user interface of a concept browser provides both a visualization of the concept graph as a knowledge map and the ability for the user to interact with that knowledge map. The user interaction with the concept browser does not only concern the presentation of a knowledge map and accessing related resources (e.g. by browsing, zooming, panning), but also the ability for the user to edit the concept graph directly.

The resulting user interface for the search engine (Figure 2) will provide the user with a search field; a knowledge map that displays concepts and associations related to the search terms; and a resource display that provides documents and data records, related to the user's navigation in the knowledge map. Entering a search term provides a concept map with concepts associated to the search term based on our proposed methods; as well as resources that are relevant the search term (based on conventional Lucene indexing). Clicking on a concept in the concept browser results in a graph of associated concepts, as well as a list of associated resources, where link between the knowledge map and the resources is determined using classical full text search in the document base using the concepts as keywords.

In the Project Lokahi, the basic functionality of visualization of the concept graph, search functionality and visual resource allocation, as well as a simple browse functionality (e.g. by "panning") are implemented. The user interface is designed with focus on extensibility, so that it can be further developed after the successful completion of the project. The final vision of a concept browser will enable the following interactive browsing modes in the knowledge map:

- (A) The *panning* functionality allows the user to change the focus of the displayed section, thereby navigating in the map, e.g. by clicking on a concept node;
- (B) The *zoom* functionality makes it possible to change the scope and level of detail of the presentation of the knowledge map;
- (C) The *drill-down and roll-up* functionality allows the user to display more or less details of the concept map by changing the relevancy threshold set for the association weight;
- (D) The *edit* functionality allows the user to change the map by adding new or reviewing existing concepts and concept associations. Thus the user can extend the knowledge map with his or her tacit knowledge directly in the user interface.

4   Conclusions

We have presented a design for an enterprise search engine with an interactive concept browser. The technical solution implements a knowledge warehouse system that involves enterprise-ontology based knowledge extraction, as well as concept browsing, based on that system.

Automated knowledge extraction from data could become essential in coping with exponential information growth. According to a study by Hilbert & Lopez [33], the amount of stored information on earth has increased annually by 28% from 1986 to 2011, which corresponds to duplication every three years. A progression of this trend is likely and will demand creative solutions sooner or later. This trend means that knowledge in relation to the available information is becoming increasingly scarce, and knowledge is therefore becoming more and more valuable. To cope with the future information explosion, knowledge engineering can help the user with the automatic extraction of potential knowledge from the flood of information.

To evaluate our artefact design, we plan two complementary approaches: To evaluate the feasibility of the design, a demonstration in silico in the form of an implemented of a prototype, which is ongoing. To evaluate the usefulness of the design, a qualitative case study within a software enterprise will be conducted. The resulting prototype will be implemented and deployed at FIVE Informatik AG.

The strength of that approach is that the designed artifact can be tested in a real enterprise environment. However, it is difficult to formally evaluate the resulting artifact. So far, only qualitative evaluation of the resulting system is intended. The qualitative evaluation will be based on subjective feedback from the users of the system. If the results are promising, the research prototype could lead to product development.

As an outlook, we believe that knowledge extraction is essential for the future of knowledge technology research. The present approach is tested in a small data environment of a software company. If the approach is successful, we intend to apply a similar design to the Web as a whole (web knowledge engineering). In this case, a concept graph will be extracted from the Web, and presented to the Web user as a concept browser for Web content.

References

1. Margolis, E., Laurence, S.: Concepts. In: Zalta, E.N. (ed.) The Stanford Encyclopedia of Philosophy (2014)
2. Guarino, N.: Formal Ontology, Conceptual Analysis and Knowledge Representation. Int. J. Hum-Comput. Stud. 43, 625–640 (1995)
3. Österle, H., Becker, J., Frank, U., Hess, T., Karagiannis, D., Krcmar, H., Loos, P., Mertens, P., Oberweis, A., Sinz, E.J.: Memorandum on design-oriented information systems research. Eur. J. Inf. Syst. 20, 7–10 (2010)
4. Preece, A., Flett, A., Sleeman, D., Curry, D., Meany, N., Perry, P.: Better knowledge management through knowledge engineering. IEEE Intell. Syst. 16, 36–43 (2001)
5. Shadbolt, N.: Knowledge Technologies. Ingenia R. Acad. Eng. 58–61 (2001)
6. Milton, N., Shadbolt, N., Cottam, H., Hammersley, M.: Towards a knowledge technology for knowledge management. Int. J. Hum.-Comput. Stud. 51, 615–641 (1999)
7. Yacci, M.: The Knowledge Warehouse: Reusing Knowledge Components. Perform. Improv. Q. 12, 132–140 (1999)
8. Frisch, A.M.: Knowledge Retrieval as Specialized Inference. Ph.D Thesis, University of Rochester (1986)
9. Nemati, H.R., Steiger, D.M., Iyer, L.S., Herschel, R.T.: Knowledge warehouse: an architectural integration of knowledge management, decision support, artificial intelligence and data warehousing. Decis. Support Syst. 33, 143–161 (2002)
10. Nilsson, M., Palmér, M.: Conzilla - Towards a Concept Browser (No. CID-53, TRITA-NA-D9911). Stockholm: Centre for User Oriented IT Design, Dept. Computing Science, Royal Institute of Technology KTH (1999)
11. Pfeifer, R., Scheier, C.: Understanding Intelligence. MIT Press (2001)
12. Eysenck, M.W., Keane, M.T.: Cognitive Psychology: A Student's Handbook, 4th edn. Psychology Press (2000)
13. McClelland, J.L., Cleeremans, A.: Consciousness and Connectionist Models. In: McClelland, J.L., Bayne, T., and Wilken, P. (eds.) The Oxford Companion to Consciousness. Oxford University Press (2009)
14. Cudré-Mauroux, P., Liu, L., Özsu, M.T.: Emergent Semantics. Encyclopedia of Database Systems, pp. 982–985
15. Portmann, E., Pedrycz, W.: Fuzzy Web Knowledge Aggregation, Representation, and Reasoning for Online Privacy and Reputation Management. In: Papageorgiou, E.I. (ed.) Fuzzy Cognitive Maps for Applied Sciences and Engineering. ISRL, vol. 54, pp. 89–105. Springer, Heidelberg (2014)
16. Maedche, A., Staab, S.: Ontology learning for the Semantic Web. IEEE Intell. Syst. 16, 72–79 (2001)
17. Parameswaran, A., Garcia-Molina, H., Rajaraman, A.: Towards the Web of Concepts: Extracting Concepts from Large Datasets. Proc VLDB Endow. 3, 566–577 (2010)
18. Deerwester, S.: Improving Information Retrieval with Latent Semantic Indexing. Presented at the Proceedings of the 51st ASIS Annual Meeting (ASIS 1988) (October 23, 1988)
19. Ganter, B., Bock, H.H.: Software for formal concept analysis. Classification as a tool of research, pp. 161–167. North-Holland, Amsterdam (1986)

20. Portmann, E., Kaufmann, M.A., Graf, C.: A Distributed, Semiotic-inductive, and Human-oriented Approach to Web-scale Knowledge Retrieval. In: Proceedings of the 2012 International Workshop on Web-scale Knowledge Representation, Retrieval and Reasoning, pp. 1–8. ACM, New York (2012)
21. Hinkelmann, K., Merelli, E., Thönssen, B.: The Role of Content and Context in Enterprise Repositories. Presented at the 2nd International Workshop on Advanced Enterprise Architecture and Repositories (AER) (2010)
22. Thönssen, B.: An Enterprise Ontology Building the Bases for Automatic Metadata Generation. In: Sánchez-Alonso, S., Athanasiadis, I.N. (eds.) MTSR 2010. CCIS, vol. 108, pp. 195–210. Springer, Heidelberg (2010)
23. The Open Group: ArchiMate 2.1 Specification,
    `http://pubs.opengroup.org/architecture/archimate2-doc/`
24. Martin, A., Emmenegger, S., Wilke, G.: Integrating an enterprise architecture ontology in a case-based reasoning approach for project knowledge. In: Proceedings of the Enterprise Systems Conference, ES (2013)
25. Thönssen, B.: Automatic, Format-independent Generation of Metadata for Documents Based on Semantically Enriched Context Information. Ph.D Thesis, University of Camarino (2013)
26. Ichikawa, J.J., Steup, M.: The Analysis of Knowledge. In: Zalta, E.N. (ed.) The Stanford Encyclopedia of Philosophy (2014)
27. Hawthorne, J.: Inductive Logic. In: Zalta, E.N. (ed.) The Stanford Encyclopedia of Philosophy (2012)
28. Shi, L., Griffiths, T.L.: Neural implementation of hierarchical bayesian inference by importance sampling. Advances in Neural Information Processing Systems 22, 1669–1677 (2009)
29. Kaufmann, M.: Inductive Fuzzy Classification in Marketing Analytics. Springer (2014)
30. The Apache Software Foundation: Apache Lucene - Apache Lucene Core,
    `http://lucene.apache.org/core/`
31. Agrawal, R., Imieliński, T., Swami, A.: Mining Association Rules Between Sets of Items in Large Databases. In: Proceedings of the 1993 ACM SIGMOD International Conference on Management of Data, pp. 207–216. ACM, New York (1993)
32. Zadeh, L.A.: Fuzzy sets. Inf. Control. 8, 338–353 (1965)
33. Hilbert, M., López, P.: The World's Technological Capacity to Store, Communicate, and Compute Information. Science 332, 60–65 (2011)

A Modeling Procedure for Information and Material Flow Analysis Comprising Graphical Models, Rules and Animated Simulation

Christoph Prackwieser

University of Vienna, Research Group Knowledge Engineering
Währingerstrasse 29, 1090 Vienna, Austria
prackw@dke.univie.ac.at

Abstract. The domain-specific modeling method *SIMchronization* was originally developed to analyze and optimize industrial maintenance supply chains. The use of graphical models extended with behavior-describing rules and combined with discrete simulation, reveals the dynamic interactions of supply chain elements, IT-systems, and the resulting information and material flows.

This paper presents the basic meta-model and mechanisms of the modeling method and proposes a modeling procedure to achieve comprehensible supply chain models. Furthermore, this approach incorporates model verification, as the simulation algorithm requires syntactically correct and input-wise complete models, and the graphical flow animation enables practical verification by domain experts. A proof of concept illustrates the practicability of the approach which is discussed in a SWOT analysis.

Keywords: Domain Specific Modeling Language, Modeling Procedure, System Analysis, Simulation, Supply Chain.

1 Introduction

The work presented in this paper is based on the assumption that animated graphical models which show the flow of objects through the system are better understood than static diagrams alone by business oriented domain experts [1-3]. Animated models can help to foster system insights, especially by revealing "the minute spatial-temporal actions of components" [2]. Thus, the quality of system descriptions, the understanding of dependencies and chronological sequences within a system and the evaluation of change proposals could be improved.

With the development of the *SIMchronization* method [4, 5] we already followed this consideration and defined a basic notation and technique to analyze the information and material flows in supply chain networks. This method was initially developed to study the various effects of new technologies on maintenance supply chains [6, 7].

While evaluating this method in collaboration with maintenance practitioners, we recognized that the application of *SIMchronization* leads to a comprehensible presentation of supply chains which works well both for documentation and training purposes and deals furthermore as a robust basis for improvement initiatives.

R. Buchmann et al. (Eds.): KSEM 2014, LNAI 8793, pp. 198–209, 2014.
© Springer International Publishing Switzerland 2014

The structure of the paper is as follows: After this introduction, we provide background and the problem statement in Chapter 2. Chapter 3 outlines the conceptualization of the *SIMchronization* method, and Chapter 4 presents the concepts for the dynamization of the model. The modeling procedure is proposed in the next chapter, which is followed by a proof of concept. This paper closes with the results of a SWOT analysis in Chapter 7.

2   Problem Statement and Background

2.1   The Maintenance Supply Chain

Supply chain management focuses on the optimization of material and information flows through one or more companies to reduce costs and lead time. Additionally, in the maintenance domain a supply chain must ensure that a maintenance engineer with relevant knowledge, required tools and spare parts is on location in time. Furthermore, in the maintenance domain demands can arise from regular planned checkups or stochastically influenced failures. This results in a need for a supply chain that reacts immediately, in contrast to better predictable customer-induced demands for production or logistic processes. Therefore a highly efficient and synchronized coupling of information and material flows is required. A more detailed description of the domain-specific requirements and the reasons for the development of a new method can be found in [5].

2.2   Problem Positioning

The description of a system as a graphical model is helpful to explain and foster understanding of the basic relations between elements within the system. As many examples show this works very well for business processes as most modeling notations (BPMN, EPC, BPMS [8, 9]) use as the main flow-defining relation the logical sequence of tasks. However, in a supply chain network the interactions of supply chain elements, actors, IT-systems, and the resulting information and material flows can cause a highly dynamic interplay of actions and reactions which is due to the occurrence of stochastic events and changing demands not easily predictable. A static graphical model like a process model is not adequate to describe all the possible alternative system statuses. Obviously, beside the modelling notation's capability to represent such networks there is also a structured approach needed to acquire and formalize subject matter expert's knowledge in comprehensible models.

2.3   Proposed Solution

We are proposing a modelling procedure as an integral part of the *SIMchronization* method to utilize model simulation and animation as a mean to improve the quality of the documentation of maintenance supply chains. The use of behavior describing rules leads to a more precise documentation as text alone and can syntactically be checked by the simulation algorithm and semantically by the domain expert studying the animation.

2.4   Related Literature

This paper is part of our ongoing research to conceptualize a method for analyzing and optimizing maintenance supply chains. Related maintenance domain-specific literature can be found in [4, 5].

Our work shares some ideas with the Exemplary Business Process Modeling (eBPM) approach, which is described in Breitling et al. [10]. eBPM is a graphical modeling method that supports domain-specific information gathering of processes, objects, and involved actors, such as humans or IT-systems [11], for an object-oriented software application design. eBPM is usually used to depict business processes, whereas *SIMchronization* is intended for supply chains. Another important difference is that eBPM follows a scenario-based approach. Therefore, resulting models do not include case differentiation; they are exemplary, which means they describe just one specific scenario. A *SIMchronization* model describes the environment-related behavior of supply chain objects in a dynamic setting. Therefore, different scenarios can be incorporated into one model. The eBPM method is realized by using the meta-modeling platform ADOxx® [12]. The software provides a manually-triggered process stepper that allows a stepwise walk-through following the chronological control flow. An integral part of *SIMchronization* is discrete simulation and token flow animation, which visually display material and information flows.

The three major approaches in simulation engineering [13] are System Dynamics (SD), Discrete Event Simulation (DES), and Agent Based Simulation (ABS). There is no definite decision framework for selecting a suitable method, but Borshchev and Filippov [13] offer guidelines and Lorenz and Jost [14] developed an orientation framework. After applying these frameworks on our problem statement, we concluded that System Dynamics does not allow observing the flow of individual entities throughout the process and is therefore not suitable. Both DES and ABS provide this level of detail, but in ABS individual entities, called agents, are active, which means they control their behavior and way through the system independently. Maintenance supply chains are usually centrally planned and controlled and therefore Discrete Event Simulation suits best.

Numerous papers describe graphical modeling methods that support the design of Computation Independent Models in Model Driven Architecture (MDA) [15]; including: UML Activity Models, Interaction Diagrams, Use Case Models, Business Process Modeling Notation (BPMN), Event-Driven Process Chains (EPCs), and ADONIS Business Process Management Systems (BPMS) [8, 16], to name just a few. These modeling languages are widely used to model business process-oriented models, in order to foster insights or to define software requirements. However, they do not depict supply chain networks well because they are not able to model interactions between information and material flows expressed in the behavior of temporary objects, such as physical item movements or sent and received information objects. In a structured literature study, we analyzed 17 modeling methods [5], but just two of them - Petri Nets and Queuing Networks - were able to fulfill this requirement. However, as these methods are not domain-specific and therefore quite hard to understand for practitioners, we decided to develop a new modeling method for this approach.

3   Conceptualization of a Domain-Specific Modeling Method

Karagiannis and Kühn's approach [5] was chosen as the underlying framework for the development of the modeling method. Contrary to many other Method Engineering methods [17], their method comprises beside the *modeling language* and the *modeling procedure* also *mechanisms & algorithms* which are applied to the models. A core aspect of *SIMchronization* is the dynamization of static models using simulation and animation; thus, these algorithms are an integral component of our modeling method. Moreover, the approach supports for the operationalization of domain-specific methods using a meta-modeling platform, which is used to evaluate the research results by implementing a tool and testing practical examples. A *modeling procedure for SIMchronization* will be proposed later in this paper.

3.1   Modeling Language Requirements

In a supply chain items are transported between processing stations, stores, suppliers, and customers. To understand a system and its interactions, it is necessary to observe the status of production stations, inventory levels, fulfillment rates of customer orders, etc. On the other hand, since a specific order is necessary for each item movement and production run, and since coordination between this information and material flows is vital for an efficient system, it is important to track these flows over time. To get a full picture of the items in a system, the modeling method must combine static and dynamic views of the system.

While the static model contains all objects that exist during the entire observation or analysis period, the dynamic model depicts the motion and transformation of physical items and information objects. The static model contains resident entities, whereas the dynamic model adds transient entities, which are a temporary result of the simulation.

Resident entities do not change their position within the network and their relations to neighboring nodes are permanent throughout the entire analysis period [18], which is at least as long as the duration of the simulation. Even if temporary instances of resident entities can be created during the simulation the type of the entity stays resident as the inherited attributes and relations to other resident entities are the same. For example, resident entities could be modeling objects depicting production processes, storage processes, or IT-services. Each resident entity type has a type-specific-behavior, which can be influenced, to a certain extent, by the modeler.

Transient entities usually travel through the system and only stay for a certain period within the system. Transient entities get produced, transformed, and consumed by resident entities, and their properties and/or location are subject to change. Transient entities can be physical items, such as spare parts, or immaterial information objects, such as orders or inventory levels.

3.2   Metamodel

To describe the static model of a supply chain network, we adopt the 'Source,' 'Make,' 'Deliver,' and 'Plan' processes of the widely used Supply Chain Operations

Reference model (SCOR) [19] to name the respective modeling classes. Additionally, we studied a variety of supply chain specific ontologies [5] to define the metamodel for resources and components. To facilitate reusability we structured the metamodel using three model types:

— The *Supply Chain Network Model* depicts information and material flows
— The *Resource Model* describes resource objects as a repository
— The *Component Model* contains the definition of transient objects

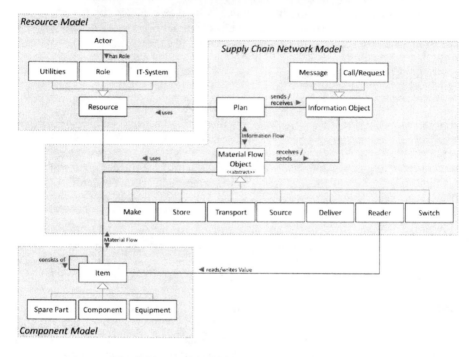

Fig. 1. Metamodel of *SIMchronization based on* [5]

3.3   Resided Modeling Classes and Relations

A graphical notation, syntax, and semantics define a modeling language [20]. Table 1 presents the visualization of all modeling classes for resident objects of Supply Chain Network Models. Each has a multitude of attributes containing quantitative and qualitative data that are useful for reports, more detailed requirement descriptions, references to supporting IT-systems, or rule expressions. The description of these attributes is not part of this paper.

The transient information object and its specializations - Messages and Calls - cannot be used directly by the modeler, but are a result of the dynamization of the model.

Table 1. Modeling classes and relation classes

Make *Modeling class*	MAKE	Plan *Modeling class*	Takt / Periode **0**
Produces, processes, or transforms physical items within a material flow		Controls and schedules the supply chain by receiving, processing and sending out information objects	
Store *Modeling class*	STORE	Transport *Modeling class*	TRANSPORT
Accepts, stores and provides transient objects (items) within the material flow		Transports transient objects (items) to a different geographical location	
Source *Modeling class*	SOURCE	Deliver *Modeling class*	DELIVER
Supplies the material flow with transient objects (items) Maps the supplier of the Supply Chain		Consumes transient objects (items) by removing them from the material flow. Maps/Stands for the customer	
Reader *Modeling class*	READER	Switch *Modeling class*	
Reads out or writes data onto transferring transient objects (items)		"Exclusive Or" in the material flow to divide the flow of transient objects (items) in different Material Flow Channels	
Information Flow Channel *Relation class*		Material Flow Channel *Relation class*	
Relation between two resident objects; shows the existence of a communication channel from one object to the other.		Relation between two resident objects; shows the existence of a transport connection from one object to the other.	

4   Dynamization of the Model

We use the term "Dynamization" of a model to describe the process of enriching a static graphical model with all information necessary to simulate and animate the model. This includes quantitative input data and behavior describing rules.

4.1   Rules Syntax

As stated earlier, we apply rules to determine the behavior of resident objects in the supply chain model and use the results of the rule evaluation as an input for the simulation algorithm.

However, the rules mechanism is not limited to representing a triggering logic for a production task; it can also send messages, describe demand functions, and formulate production scheduling programs.

To represent a rule corresponding with the *SIMchronization* method we use the OMG Production Rule Representation [21] as a basic syntax, in which a production rule is defined as a "statement of programming logic that specifies the execution of one or more actions in the case that its conditions are satisfied" [21]. A production rule consists of one or more conditions, in the form of logical formulas, and one or more produced actions, in the form of action expressions.

Example

As the simulation reaches the fifth period the rule shown in Fig. 2 evaluates to True and the expression GET(INFOIN1, 'Level') is evaluated. This term induces an information flow that uses the Information Flow Channel '1' and contains the current inventory level of the 'Raw Material Stock' object. As this rule is a value of the Order attribute the result is the number of production orders for the Make-object in the current period.

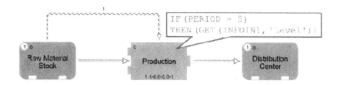

Fig. 2. Rule in attribute 'Order' of a Make-object

4.2   Simulation and Animation

In order to generate, consume, send, and receive parts and messages within a supply chain, a discrete simulation algorithm is applied on the model for a predefined number of periods. For all objects of the model, the simulation algorithm determines whose preconditions are met and evaluates their behavior-describing rules, in the context of the current input, model, and simulation data. Depending on the results, messages are sent to receiving objects by using specific Information Flow Channels, Web services are called, and production orders are given that subsequently lead to parts movements in Material Flow Channels.

As the simulation can also process stochastic input, the resulting system behavior can change over time; also, unexpected situations may occur. This is a major advantage of the approach because the identification of unforeseen events and the observation of the system's reaction is an important part of understanding how the system behaves.

5   Modeling Procedure

According to Karagiannis and Kühn [20], an integral component of a modeling method is the *modeling procedure*, which describes the application of the *modeling language* and the *mechanisms & algorithms* needed to achieve the desired results. The modeling procedure differs greatly, depending on the application field of the method. For example, to optimize spare part inventories with *SIMchronization*, the method has to be applied in combination with a meta-heuristic which gives guidance for changing influential input factors. As in this paper, the main focus lies on system design and analysis, we suggest the following modeling procedure. To put this approach into practice, there are at least two necessary roles: the role of a subject matter expert, who has deep knowledge and experience in the respective domain, and the role of a Business Analyst, who applies the method, processes the domain expert's input, and interprets and communicates the output.

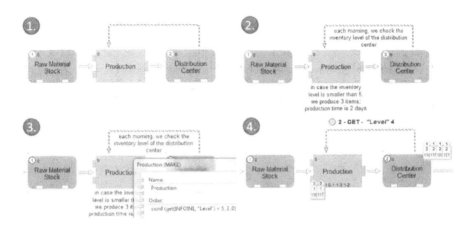

Fig. 3. Steps in modeling a supply chain network model

The following modeling procedure describes the basic steps in creating a supply chain network model for documentation and training purposes using *SIMchronization*. The most important steps of the modeling procedure are illustrated in Fig. 3.

1. **Identify System Boundaries:** Depending on the objective of the analysis boundaries of the supply chain, such as the start and end points of the material flow, the level of modeling detail and the simulation's time period have to be defined.
2. **Mapping the Static Models:** Supply Chain Network Models, Resource Models and Component Models. In a business process model, objects are arranged according to their chronological sequence, following a control flow. In contrast, a static supply chain model is built up by defining all scope-relevant resident objects, such as Store, Make, Plan, or Deliver, first and connecting them afterwards, according to their real ability to exchange items or information, by using respective channels (see Fig. 3, step 1). Whether these Material or Information Flow Channels are actually used during a simulation run depends heavily on the behaviors of each resident object, which is defined by its behavior-describing rules formulated in the next step of this approach.
3. **Enrich the Static Model by Adding Verbal Descriptions:** Subject Matter Experts describe verbally the interactions of supply chain elements, triggering events and dependencies. They use their domain-specific terminology and a business-oriented description. This information is stored in designated text attributes of objects and relations and can be visualized (see Fig. 3, step 2).

4. **Transform Verbal Descriptions in Behavior Describing Rules:** The Business Analyst transforms these non-formatted texts into production rules. For example, Fig. 3, step 2 describes inventory policy: "In case the inventory level is smaller than 5, we produce 3 items; ..." This corresponds to the following rule in the Order attribute of the Make-object (see Fig. 3, step 3):

```
cond (get(INFOIN1, "Level") < 5, 3, 0)
```

5. **Prepare Model for Simulation:** Ensure syntactically correct and input-wise complete models which are executable by the simulation algorithm.
6. **Run the Simulation and Animation:** The flow animation presents the material movements and exchanged information within the static supply chain network in chronological order to the domain expert (see Fig. 3, step 4). This provides a good overview of the dynamics within the network and potential effects of system changes. The animation supports the domain expert's understanding and allows to compare the animated model with the domain expert's practical experiences.
7. **Verify the Model:** In case a domain expert recognizes model deficits, necessary verbal descriptions and the associated rules can be adapted immediately or reveal the necessity of iteration. If the simulation of the model operates as conceived, the domain experts will approve the model. The Business Analyst uses the static and dynamic model and their rules for documentation and communication purposes or as a solid basis for supply chain improvement.

Especially the usage of domain specific terminology and modelling classes and the possibility to annotate verbal descriptions enable Subject Matter Experts to map, at least for other domain experts, comprehensible models using *SIMchronization*'s modelling notation (step 1-3). However, as soon as the models have to fulfill syntactical correctness as an input for a simulation algorithm and rules engine a modelling expert is necessary to ensure that the simulation is executable and delivers correct results.

6   Proof of Concept

To evaluate the approach, a prototype was implemented by using the meta-modeling platform ADOxx® (ADOxx® is a registered trademark of BOC Information Technologies Consulting AG). The modeling classes and their attributes were created within the software, and combined with a simulation and animation algorithm.

We tested the approach by modelling an extract of a supply chain which could be improved by the implementation of RFID (Radio-frequency identification) technology. After applying the proposed modeling procedure, simulation and animation we were able to stepwise follow the chronological movement of items and information objects within the system. The animation presented us with a full picture of resident and transient objects for each point of the simulation period. This snapshot could be easily transformed into UML Interaction Diagrams (Fig. 4) to provide IT specialists with a modeling notation they are used to work with.

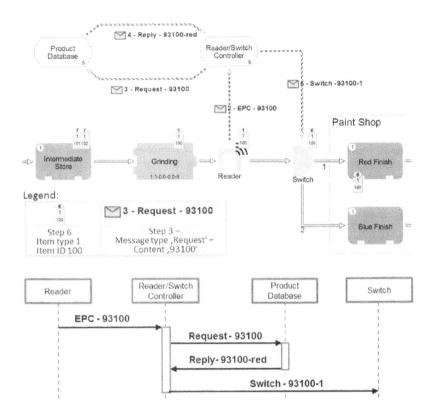

Fig. 4. Dynamic Supply Chain Network Model and Corresponding UML Interaction Diagram

The upper part of Fig. 4 shows a supply chain fragment of the item preparation for a paint shop. The depicted model shows interactions during a specific period of time. Each incoming item is equipped with an RFID tag; the central 'Product Database' stores the information on whether one specific item gets a red or blue finish.

After the grinding process, the item passes a `Reader` (Step 1) which reads the 'Electronic Product Code (EPC)' out of the RFID tag and sends this information to the 'Reader/Switch Controller' (Step 2). Next, the controller sends a 'Request' (Step 3) to the 'Production Database' and gets the ordered color for this specific item as an immediate 'Reply' (Step 4). The controller processes this information and sends a message to the `Switch` that the item has to be directed through `Material Flow Channel` '1' (Step 5). Subsequently the item will be put into the `Store` "Red Finish" (Step 6). The lower part of Fig. 4 shows the corresponding UML Interaction Diagram which depicts just the information flow and involved modelling objects.

7   Conclusive SWOT Analysis

In this paper, we presented the *SIMchronization* method and proposed a modeling procedure to create supply chain models. The application of the method leads to comprehensible supply chain network models which can be simulated and animated uncovering the interactions of information and material flows. A conducted SWOT analysis resulted in the conclusions below:

Strengths: The modeling procedure suggests the usage of the domain-specific modelling notation and to enhance the graphical model with unstructured, but for the business easy understandable, verbal descriptions. The outcome can therefore be re-used for documentation, communication, and training purposes. As the verbal descriptions get manually transformed in structured rules the approach incorporates model verification and applies integrity rules [22]. Both the simulation algorithm and the rules interpreter require syntactically correct and input-wise complete models and the graphical flow animation enables practical verification by domain experts [23].

Weakness: The usage of a dedicated rule language comes with the disadvantage that special knowledge is necessary to transform the verbal description in correct rules. For this task a specialist has to be involved which increases the effort of model creation. Further evaluation of the proposed approach is necessary. So far we have shown the practicability of this approach by implementing the method on a meta-modeling platform. Some practical examples were modeled, simulated, animated, and discussed with domain experts, Business Analysts and IT specialists. A major goal for the next research step is an evaluation based on bigger real-life examples.

Opportunities: Based on the method proposed in this paper many application scenarios of this approach are conceivable. For example, the defined rules could be directly re-used in productive rule engines which control the real supply chain. This would shift the focus of the method to an IT System Analysis Tool which could be used for prototyping, test case generation or e-learning.

Threats: To our knowledge *SIMchronization* is currently the only simulation algorithm in this domain, which shows the flow of the physical items **and** information objects such as messages or calls in an animation. Commercial simulation tools as Arena provide already powerful functionalities to influence the behavior of supply chain elements by implementing additional functions, but are currently focusing on the material and logical control flow.

References

1. Park, O.-C., Gittelman, S.: Selective use of animation and feedback in computer-based instruction. ETR&D 40, 27–38 (1992)
2. Tversky, B., Morrison, J.B., Betrancourt, M.: Animation: can it facilitate? International Journal of Human-Computer Studies 57, 247–262 (2002)
3. Thompson, S.V., Riding, R.J.: The Effect of Animated Diagrams on the Understanding of a Mathematical Demonstration in 11- to 14-Year-Old Pupils. British Journal of Educational Psychology 60, 93–98 (1990)
4. Prackwieser, C.: SIMchronization: a method supporting the synchronisation of information and material flows. In: Proceedings of the Winter Simulation Conference, Berlin, Germany, pp. 1–2 (2012)

5. Prackwieser, C.: SIMchronization: Eine Modellierungsmethode zur Synchronisation von Material- und Informationsflüssen in der Instandhaltung. In: 11th International Conference on Wirtschaftsinformatik, Leipzig (2013)
6. Fabry, C., Schmitz-Urban, A.: Maintenance Supply Chain Optimisation within an IT-Platform: Network Service Science and Management. In: 2010 International Conference on Management and Service Science (MASS), pp. 1–4 (2010)
7. Iung, B., Levrat, E., Marquez, A.C., Erbe, H.: Conceptual framework for e-Maintenance: Illustration by e-Maintenance technologies and platforms. Annual Reviews in Control 33, 220–229 (2009)
8. BOC Information Technologies Consulting AG: Method - BPMS (Business Process Management System), http://www.boc-group.com/products/adonis/method-bpms/
9. Prackwieser, C., Buchmann, R., Grossmann, W., Karagiannis, D.: Towards a generic hybrid simulation algorithm based on a semantic mapping and rule evaluation approach. In: Wang, M. (ed.) KSEM 2013. LNCS, vol. 8041, pp. 147–160. Springer, Heidelberg (2013)
10. Breitling, H., Kornstadt, A., Sauer, J.: Design Rationale in Exemplary Business Process Modeling. Rationale Management in Software Engineering 191 (2006)
11. Hofer, S.: Instances over Algorithms: A Different Approach to Business Process Modeling. In: Johannesson, P., Krogstie, J., Opdahl, A.L. (eds.) PoEM 2011. LNBIP, vol. 92, pp. 25–37. Springer, Heidelberg (2011)
12. Open Model Initiative - Open Model Project - eGPM, http://www.openmodels.at
13. Borshchev, A., Filippov, A.: From system dynamics and discrete event to practical agent based modeling: reasons, techniques, tools (2004)
14. Lorenz, T., Jost, A.: Towards an orientation framework in multi-paradigm modeling. In: Proceedings of the 24th International Conference of the System Dynamics Society, Nijmegen (2006)
15. OMG Model Driven Architecture, http://www.omg.org/mda/
16. Petzmann, A., Puncochar, M., Kuplich, C., Orensanz, D.: Applying MDA Concepts to Business Process Management. In: Fischer, L. (ed.) BPM AND Workflow Handbook, vol. 103-116, Future Strategies Inc. (2007)
17. Braun, C., Wortmann, F., Hafner, M., Winter, R.: Method construction - a core approach to organizational engineering. In: Proceedings of the 2005 ACM Symposium on Applied Computing, pp. 1295–1299. ACM, Santa Fe (2005)
18. Schruben, L.W., Roeder, T.M.: Fast Simulations of Large-Scale Highly Congested Systems. SIMULATION 79, 115–125 (2003)
19. Supply Chain Council Inc.: Supply Chain Operations Reference (SCOR®) model; Overview - Version 10.0 (2010)
20. Karagiannis, D., Kühn, H.: Metamodelling platforms. In: Bauknecht, K., Tjoa, A.M., Quirchmayr, G. (eds.) EC-Web 2002. LNCS, vol. 2455, p. 182. Springer, Heidelberg (2002)
21. Production Rule Representation (PRR). Version 1.0, vol. formal/2009-12-01 (2009)
22. Mylopoulos, J., Chung, L., Yu, E.: From object-oriented to goal-oriented requirements analysis. Commun. ACM 42, 31–37 (1999)
23. Sargent, R.G.: Verification and validation of simulation models. Journal of Simulation 7, 12–24 (2013)

A Practical Approach to Extracting Names of Geographical Entities and Their Relations from the Web

Cungen Cao, Shi Wang, and Lin Jiang

Key Laboratory of Intelligent Information Processing
Institute of Computing Technology
Chinese Academy of Sciences, Beijing 100080
{cgcao,wangshi}@ict.ac.cn

Abstract. Geographical information extraction is a special case of information extraction. In this paper, we present a practical method of extracting both names of geographical entities and their relations from the Web. The method is composed of three major phases. First, we manually designed a list of 493 Chinese lexico-syntactical patterns for matching Web page excerpts which contain names of geographical entities and their relations; second, we developed a knowledge extractor for extracting those names and relations to generate a geographical graph whose nodes are entities, and edges represent relations of the entities; third, we developed several methods for handling problems or errors in the generated graph. Experimental results show that the OMKast-Googling system has a satisfactory performance both in the entity name extraction and relation extraction.

Keywords: Geographical entity, geographical relation, knowledge acquisition from text, lexico-syntactical pattern.

1 Introduction

Geographical information extraction from text (GIET) is an important area in the current information extraction practice. In GIET, there are two basic tasks: extracting names of geographical entities (or geo-entities for short) and extracting their relations (or geo-relations for short).

In recent years, much work has been reported on the first task. For example, Nissim et al. presented an off-the-shelf maximum entropy tagger for recognizing location names for recognizing geo-entity name in Scottish historical documents [12]. Dutta et al. implemented a prototype for extracting geographical terms (i.e. geo-entity names) from Hindi text based on the Hindi syntax, linguistic rules, and a dictionary-based term normalization process [4]. Wang et al. proposed several methods for detecting provider, content and serving locations from Web resources to meet users' location-specific information needs [15].

As for the second task of GIET, to our knowledge, little attention has been paid on it. Fortunately, there is much work on general relation extraction [1,2,13]; and works in specific relation extraction such as is-a relations [9,10], part-whole relations [6,7,8], and coreferencing relations [14] offer some insights into extraction of geo-relations.

R. Buchmann et al. (Eds.): KSEM 2014, LNAI 8793, pp. 210–221, 2014.
© Springer International Publishing Switzerland 2014

In this paper, we introduce a pratical method of extracting both geo-entities and geo-relations from the Web. Geo-relations are a kind of common relations in real applications. For example, situated-in-the-south(v, v') represents a *location relation* that v is situated in the south part of v', and borders-in-the-south(v, v') represents a *neighborhood relation* that v borders v' in the south.

We developed a system called the OMKast-Googling system. It consists of two major subsystems: OMKast and Googler. The Googler is relatively simple, and it extracts Web page excerpts from search results by using Google and instantiated lexico-syntactical patterns as search queries. The OMKast provides a programming language, called EKEL (Executable Knowledge Extraction Language), for knowledge engineers to write knowledge-extraction agents to extract both geo-entities and their relations from Web page excerpts [16].

The paper is organized as follows. Section 2 outlines the OMKast-Googling system. Section 3 introduces our methods for collecting an initial set of geo-entities from Chinese publications and an initial set of geo-relations, and also describes methods for iteratively expanding this initial sets, based on an offline Chinese corpus. Section 4 focuses on methods for iteratively expanding neighorhood relations of geo-entities. Section 5 introduces the methods for checking possible problems in the generated geo-graph. Section 6 concludes the paper.

2   The OMKast-Googling System

The OMkast-Googling system consists of two subsystems: OMKast and Googler. The Googler is relatively simple, and it extracts Web page excerpts from the Web using concrere lexico-syntactical patterns. The OMKast provides a programming language (i.e. EKEL) for knowledge engineers to write knowledge extraction agents to extract geo-entities and their relations from the obtained excerpts, and it also provides checking functions which will be discussed in Section 5.

An EKEL knowledge-extraction agent has two major components: a task-specific syntax and semantic operations.

In this work, the task-specific syntax is for extracting geo-entities and their relations, and we designed 493 Chinese lexico-syntactical patterns for matching Web page excerpts. these patterns are classified into two categories: relation-specific patterns, and entity-specific patterns. Relation-specific patterns are further subclassified into neighbor-relation patterns which are specifically designed for extracting *neighborhood relations* of geo-entities, and location-relation patterns which are specifically designed for extracting *location relations* of geo-entities.

Two typical Chinese location-relation pattern are <geo-entity> 地处(is situated in) <loc>, and <geo-entity> 位于(is located in) <geo-entity>. With the input 南京地处华东地区 (Nanjing is situated in the Huadong Area), one parse tree can be generated, as depicted in Fig.1(a).

Two typical Chinese neighbor-relation pattern are <geo-entity> 东邻(borders in the east) <geo-entity>, and <geo-entity> 西邻(borders in the west) <geo-entity>. With the input 南京东邻苏州 (Nanjing borders Suzhou in the east), one parse tree can be generated, as depicted in Fig.1(b).

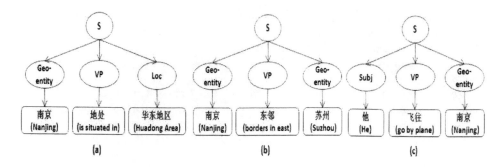

Fig. 1. Three Parse Trees

There are many geo-entity-specific patterns. For example, verbs of movement may be associated with a geo-entity. For example, In the input 他飞往南京 (He went to Nanjing by plane), Nanjing is a geo-entity, and the whole parse tree is depicted in Fig.1(c). We will further discuss lexico-syntactical patterns in Sections 3 and 4.

The OMKast subsystem is a parser with three additional functions. First, it generates parse trees, and three examples are depicted in Fig. 1.

Second, it uses a list of semantic operations to transform parse trees into a geographical graph (or geo-graph for short), and an example is depicted in Fig.2. This process makes use of the semantics of lexico-syntactical patterns; that is, the OMKast knows where the geo-entities and geo-relations are in the input text, and extract them properly (for more details see [16]).

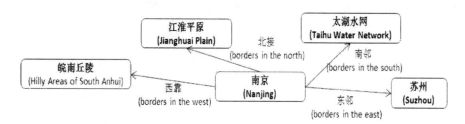

Fig. 2. An Example of Geo-graph

The third function is to check the geo-graph for various problems or errors, and we will discuss the checking function in detail in Section 5. In the next two sections, we will discuss how to use the OMKast-Googling system to iteratively extract geo-entities and their relations from the Web.

3   Building Initial Sets of Geo-Entities and Geo-Relations

In this section, we present methods for building an initial set of geo-entities (GV_0) and an initial set of geo-relations (GR_0). These are the basis for our iterative extraction. GV_0 is from two sources - Chinese publications and an offline Chinese corpus, and GR_0 is from the offline Chinese corpus.

3.1   Manually Collecting Geo-Entities from Chinese Publications

There are a lot of Chinese publications which contains geo-entities. We made use of four famous publications with different significance. The first publication is *Military Thesaurus of People's Liberation Army*, which contains 6,078 names of world-wide geo-entities with military significance.

The second publication is *The Dictionary of World-Wide Place Names*, which is downloadable from http://www.xzqh.org/suoyin/sjdm/, and it contains about 10,480 international geo-entity names.

The third publication is the column of *Foreign Administrative Zones* from the formal website of the Minitry of Foreign Affairs of China (http://www.fmprc.gov.cn), and it contains more 4000 items, including all names of countries and their first-level administrative zones.

The forth publication is *the Chinese Administrative Zones*, a Chinese standard published in 2002, which is downloadable from http://www.xzqh.org. It contains 3501 names of Chinese provinces, cities, counties, and urban districts.

From the above publications, we totally collected a set of 16,339 world-wide names of geo-entities, and we denote this set by GV_{init}.

3.2   Generating GV_0 and GR_0 from GV_{init} Using the OMKast-Googling System

In this subsection, we use GV_{init} as a starting point, and generate GV_0 and GR_0 using the OMKAST-Googling system on an offline Chinese corpus of 1.2TB Web pages.

Now, we introduce the four types of lexico-syntactical patterns we used in the experiments. The first type of patterns is related to verbs of movement, such as 飞往 (go to by plane), 前往 (head for) and 驻守 (garrison). The basic form of such patterns is <subject><verb-of-motion><geo-entity>, and we collected all the Chinese verbs of motion. The second type of patterns is related to phrases of Chinese adjectives which modify geo-entities. In the sentence 泰山风光秀丽, for instance, 风光秀丽(scenic) is an adjective, and 泰山 (Mt. Tai) is most likely to be a geo-entity.

The third type of patterns is related to terms of various social or political positions. For example, 市长 (mayor) is associated with a city or town, like in 北京市市长 (Mayor of Beijing). We have manualy summarized 31 common positions which are geographically meaningful. The form of these lexico-syntactical patterns is <geo-entity><position>, where <position> can be one of 总统 (president), 主席 (chairman), 总理 (PM), 市长 (mayor), 总裁 (CEO), and so on.

The final but more powerful type of patterns is related to location relations and neighborhood relations as mentioned in Section 2. The basic forms of location-relation patterns are <geo-entity><is situated in><geo-entity>, and <geo-entity> <is located in><geo-entity>. Of course, there are other forms of patterns which we omit here.

Neighbor-relation patterns describe neighborhood relations of geo-entities, and the basic form of such patterns is <geo-entity><direction><borders><geo-entity>, where <direction> can be 东 (east), 南 (south), 西 (west), 东南 (south-east), and so on, and <boders> can be 邻 (neighbors), 接 (borders), 靠 (borders), and so on.

We then used the OMKast-Googling system to use the patterns above to match the offline Chinese corpus, and obtained a set of 219,594 world-wide geo-entities. A manual examination of 1000 randomly selected geo-entities shows that the system achieves a precision rate of 94.3%.

Adding the 16,339 geo-entities from GV_{init} to the extracted geo-entities, and removing repetitions, we obtained an initial name set of 223,716 geo-entities[1]. This set is henceforth called GV_0. At the same time, we extracted a set of 78,650 location relations and neighbor relations (henceforth called GR_0). A manual examination of randomly selected 500 relations indicated that the OMKast-Googling system has a precision rate of 90.88%.

Table 1. GV0 and GR_0

	No. elements	Precision
GV_0	223,716	94.3%
GR_0	78,650	90.88%

3.3　Iteratively Expanding GV_0 and GR_0 by the OMKast-Googling System

To expanding GV_0 and GR_0, we designed a few types of search queries, let Googler use these queries to extract Web page excerpts from the Web, and then invoked the OMKast subsystem to extract the geo-entities and geo-relations from the excerpts. This process was iterated seven times, and terminated with a satisfactory result. In the following, we only discuss the ways of designing the search queries.

For each geo-entity p in GV_0, we defined two forms of queries: "p地处" (p is situated in) and "p位于" (p is located in). For example, suppose that p=南京(Nanjing). Using the queries "南京地处" or "南京位于", we can extract from the Web page excerpts like 南京地处华东地区 (Nanjing is situated in the Huadong Area).

Note that, in our experiments, the Googler retains, for each concrete query, the first 20 excerpts of Web pages returned by Google. Here, the number 20 was obtained through several preliminary tests. Generally, the first 10 excerpts suffice since the head word p and the relational words (i.e. 位于 and 地处) are fixed. However, in order to identify and handle possible hyponymic geo-entities, we decided to obtain more excerpts.

Another two forms of search queries are "地处p" (is situated in p) and "位于p" (is located in p). These forms of queries aim to extract Web page excerpts which tell what geo-entities *are* situated in p. For example, suppose p=华东地区 (Huadong Area). Using the concrete queries "地处华东地区" or "位于华东地区", the Googler can extract from the Web excerpts like 苏州地处华东地区 (Suzhou is situated in the Huadong Area).

For each concrete query, the Googler retains the first 300 excerpts of Web pages returned by Google. The reason is that in such search queries, p is usually a larger geo-entitiy (e.g. Huadong Area), and many smaller geo-entities may be described in relation to p.

---

[1] This means that 4,122 geo-entity names from the four publications were not extracted by the OMKast-Googling from the corpus.This justifies the significance of our manual labor.

We use $GV_{i+1}=OMKast(GV_i)$ to denote the result of the (i+1)-th iteration. After seven rounds of iteration, $GV_7=OMKast(GV_6)$ almost converses in the sense only 24 new geo-entity names were extracted based on GV_6. Table 2 summarizes the results of all the iterations. In Table 2, GR_0 contains only location relations,; that is, the neighborhood relations were not participated in the interactive process. Another point is that we did considered the 4122 geo-entities when calculating the sum of GV_0 to GV_7, because they were manually collected (see footnote 1).

Table 2. Interative Growth from GV_0 and GR_0

GV_i	No. new geo-entities extracted	GR_i	No. new relations extracted
0	223,716	0	39,977
1	231,494	1	209,095
2	95,271	2	81,692
3	16,376	3	9,498
4	2,085	4	1,050
5	376	5	309
6	80	6	96
7	24	7	8
Total	565,300		341,725

Now, we obtain a much bigger set of geo-entities and a big set of location relations. Based on Table 2, we define two new sets: $GV_7 = GV_0 \cup GV_1 \cup GV_2 \cup GV_3 \cup GV_4 \cup GV_5 \cup GV_6 \cup GV_7$, where $|GV_7| = 565,300$, and $GR_7 = GR_0 \cup GR_1 \cup GR_2 \cup GR_3 \cup GR_4 \cup GR_5 \cup GR_6 \cup GR_7$, where $|GR_7| = 341,725$.

Manually verifying 1000 geo-entities and geo-relations of GV_7 and GR_7, we obtained their precision rates, respectively (see Table 3).

Table 3. GV7 and GR_7

	No. elements	Precision
GV_7	565,300	93.11%
GR_7	341,725	89.68%

It is worthwhile to compare Table 1 and Table 3. When starting from GV_0 to extract geo-entities and their relations from the Web, more errors can be introduced to GV_7 and GR_7.

4   Extracting Neigborhood Relations for Geo-Entities in GV_7

In the previous section, we mainly focused on extracting and expanding location relations for geo-entities, and now we focus on extracting neighborhood relations for geo-entities in GV_7.

To describe the neighbors of a geo-entity in Chinese, one may use eight directions: 东(east), 西(west), 南(south), 北(north), 东南(southeast), 东北(northeast), 西南(southwest), and 西北(northwest). The use of directions in describing geo-entities in Chinese is quite different from English. For example, 东(east) can be used in "the east of" and "to the east" in English. But in Chinese, directional words are used as adverbs, as in 东连 (borders … on the east) and 南接 (adjacent to … in the south). We also must consider the types of geo-entities in designing search queries. For example, when q is a water body, we generally use "p东濒q", rather than "p东邻q".

A preliminary experiment showed that lexico-syntactical patterns, such as "p东邻" and "东邻q", work poorly in extracting neighborhood relations for geo-entities. One experimental result showed that when we used these patterns on 100,000 geo-entities, we generated 800,000 concrete queries, but only about 20,000 neighborhood relations were extracted from the Web. The reason is that when people describe a geo-entity in Chinese, they tend to first describe where it is situated using words like 位于 (be located), and then describe its neighbors using sentences in which the subject (i.e. the geo-entity name) is generally omitted.

After the considerations above, we designed, for each geo-entity name p in GV_7, a concrete query with three conjuncts:

(1) "p地处" OR "p位于"
(2) "东 OR 西 OR 南 OR 北 OR 东南 OR 东北 OR 西南 OR 西北"
(3) "濒 OR 邻 OR 临 OR 接 OR 靠"

Through a test, we found that the second conjunct above results in much redundancy. In fact, when a geo-entity p is described, some of its east, northeast, southeast, west, southwest, and northwest neighbors are generally described. We therefore used a simplified search pattern composed of "p地处" OR "p位于", "东 OR 西", and "濒 OR 邻 OR 临 OR接OR 靠".

南京地处长江下游的宁镇丘陵山区，总面积6597平方公里。南京东连富饶的长江三角洲，西靠皖南丘陵, 南接太湖水网, 北接辽阔的江淮平原。

(Nanjing is situated in the hilly areas of Nanjing and Zhenjiang in the lower reaches of Yangtze River, and its total area is 6597 square kilometers. It borders with the Yangtze River Delta to the east and the hilly areas of South Anhui to the west, adjacent to the water network of Taihu Lake in the south and the Jianghuai Plain in the north.)

Fig. 3. A Web Excerpt Using the Form of Query (8C) for Nanjing

Using the geo-entities in GV_7, we obtained 565,300 concrete queries. Running the Googler with such queries, we collected 291,748 excerpts from the Web. Fig. 3 gives an excerpt from http://www.cctv.com/science/special/C16334/20060906/103122. shtml.

From the 291,748 excerpts, 120,313 new relations and 73,022 new geo-entities are extracted using the OMKast-Googling system. The new relations are either location or neighbor relations of the geo-enities in GV_7. A manual examination of 500 relations showed that the OMKast-Googling system achieves a precision rate of 88.27% for relation extraction. Similarly, a manual examination of 1000 new geo-entity names shows that the OMKast-Googling system achieves a precision rate of 91.27%.

5   Geo-Graph Checking in the OMKast-Googling System

The GV and GR generated by the OMKast-Googling system are transformed into a geo-graph, where geo-entities are nodes and geo-relations are edges. In the following, we use G=(GV, GR) to denote the graph. In this section, we summarize a few types of common problems or errors in G, and present a few methods of handling them in the OMKast-Googling system.

5.1   Name Decontextualization

A common type of errors is that, in an extracted relation, we may find a generic name. These errors are hard to identify. In the relation borders-in-the-east(大学, 上海), for example, 大学 (the university) refers to a specific university from the context of the Web page it appears in, but the specificity may be lost during the extraction process.

To identify such errors, we used a heuristic rule and Google-based verification method. First, we checked whether a term is a suffix for at least α geo-entities. If so, we consider the term is generic. For example, 大学 is the suffix for many universities such as 上海大学 (Shanghai University) and 华东师范大学 (Huadong Normal University). In the OMKast-Googling system, we set α be to 50.

When the heuristic rule above fails for a geo-entity , we used a pattern "等name" (i.e. "name such as") as a search query, because the pattern indicates a potential hypernymy relation. If there are at least β Web pages returned by Google, we consider the name is generic. In our experiment, we chose β=1000 as the threshold.

Once generic names are identified, all relations contained these names are removed from the geo-graph. For example, the two relations borders-in-the-east(大学, 上海) and is-situated-in(大学, 南京) are removed from the geo-graph, since 大学 is a generic name.

5.2   Geo-Entity Name Verification

For extracting geo-entities from Web page excerpts, the OMKast-Googling system adopted a *general-purpose* term extraction system developed by Yu [17]. It segments and POS-tags input excerpts during syntactical analysis, and uses both statistical analysis and linguistic heuristics to extract and verify geo-entities. The system achieves precision rate of 86.1% on average. In this work, we enhanced Yu's system with a few *geography-specific* extensions.

One major extension is that we collected a list of 388 Chinese suffixes of geo-entities from a geographical standard of China (i.e. GB/T13923-92) and a geographical handbook [11], including 国(nation), 省(province), 州(state), 市(city), 县 (county), 大学(university), 中学(middle school), 小学(primary school), 公司 (company), 山(mountain), 沙漠(desert), and 岛(island). Although the list of suffixes might be not exhaustive, it covers almost all geo-entities in our experiments.

5.3   Sink Node Handling

In the geo-graph G=(GV, GR), a node v is a *sink node* if it points to no other nodes. In general, a sink node v in the geo-graph is problematic: The v is generally famous or with some significance, and therefore there should be Web pages which also offer geopraphical information about v. This implies v can not be a sink node.

We introduced a heuristic to handle sink nodes: If v is a sink node, and there are at least γ geo-entities which are described in relation to v, then we remove v and its associated relations from the graph. Based several tests, we set γ to 2 in our experiment.

5.4   Hyponym Handling

In the geo-graph **G**=(GV, GR), there may be hyponyms; that is, different geo-entities may have an identical name. For example, in relations is-situated-in(狮子山, 丽江) and is-situated-in(狮子山, 南京下关区), the two 狮子山 (Mt. Lion) actually refer to two different mountains. Hyponyms may be misleading in the geo-graph, and therefore need to be identified and split properly.

To facilitate hyponymic analysis, we need two important notions: *context of relations*, and *equivalence classes of relations*.

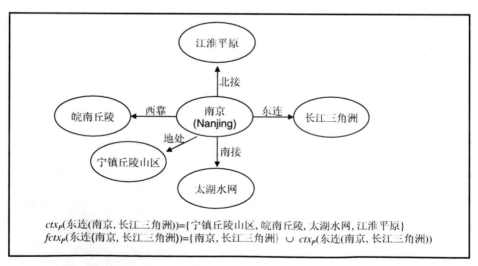

Fig. 4. Illustrating Context and Full Context of 东连(南京, 长江三角洲), i.e. borders-in-the-east(Nanjing, Yangtze River Delta)

First, we define the *context* of a relation $r(v_1, v_2)$ in $\mathbf{G} = (GV, GR)$. Suppose that $r(v_1, v_2)$ is extracted from a paragraph P (see Fig. 3 for an example). The context of $r(v_1, v_2)$ with respect to P, denoted as $ctx_P(r(v_1, v_2))$, is defined to be the set of geo-entity names in P with v_1 and v_2 being excluded. Let us take Fig.3 as an example. Suppose $r(v_1, v_2) =$ 东连(南京, 长江三角洲), i.e. borders-in-the-east(Nanjing, Yangtze River Delta). The context of $r(v_1, v_2)$ is $ctx_P(r(v_1, v_2)) = \{$宁镇丘陵山区, 皖南丘陵, 太湖水网, 江淮平原$\}$ ({Hilly Areas of Nanjing and Zhenjiang, Hilly Areas of South Anhui, Taihu Lake, Jianghuai Plain }).

For convenience, we also need the notion of *full context*. The full context of $r(v_1, v_2)$ with respect to P, denoted as $fctx_P(r(v_1, v_2))$, is the union of $\{v_1, v_2\} \cup ctx_P(r(v_1, v_2))$. For example, suppose $r(v_1, v_2) =$ borders-in-the-east(南京, 长江三角洲). Then, as shown in Fig. 6, $fctx_P(r(v_1, v_2)) = \{$南京, 长江三角洲, 宁镇丘陵山区, 皖南丘陵, 太湖水网, 江淮平原$\}$, i.e. $fctx_P(r(v_1, v_2)) = \{$Nanjing, Yangtze River Delta, Hilly Areas of Nanjing and Zhenjiang, Hilly Areas of South Anhui, Taihu Lake, Jianghuai Plain$\}$.

It is worth noticing that all relations extracted from the same paragraph P are assumed to have the same full context. For example, all relations extracted from Fig.3 have the same full context as shown above.

Now, we define a relation $\Diamond$ over the edge set GR: for $r(v_1, v_2) \in GR$ and $r'(v_3, v_4) \in GR$, $r(v_1, v_2) \Diamond r'(v_3, v_4)$ if the following two conditions hold:

(1) $v_1 = v_3$ or $v_2 = v_3$, and
(2) $|fctx_P(r(v_1, v_2)) \cap fctx_P(r'(v_3, v_4))| > 1$.

From the definition above, it can be verified $\Diamond$ is an equivalence relation over GR, and two relations $r(v_1, v_2)$ and $r'(v_3, v_4)$ in each equivalence class have at least two common elements in their full context. Therefore, the identical name in $r(v_1, v_2)$ and $r'(v_3, v_4)$, i.e. v_1 or v_2, very likely refers to the same geo-entitiy.

Based on the discussion above, we designed a method for identifying hyponyms in the geo-graph G. When hyponyms are identified, they are split into different copies, as described in step 4 in Fig. 5.

---

1. G=(GV, GR) be a geo-graph, and V=GV
2. If V is empty, stop;
3. Let v be any node in V, and $C_1, \ldots, C_k$ be the set of equivalence classes under $\Diamond$;
4. Make k copies of the node v in G, say $\underline{v}_1, \ldots, \underline{v}_k$, which have the label as v, and replace v in C_i with $\underline{v}_i$;
5. $V := V \setminus \{v\}$, and remove v from the node set GV;
6. Goto 2.

Fig. 5. Hyponym Identification and Splitting

5.5   Consistency Handling

Two edges or relations $r(v, v')$ and $r'(v, v')$ in the geo-graph may be inconsistent about v and v'. For practical reasons, we divided inconsistencies into two categories: strong inconsistencies and weak inconsistencies.

To check strong inconsistencies for the geo-graph, we defined for each relation a list of consistency checking rules. For example, for the relation borders-in-the-south(v, v'), we defined the following 5 rules, where the predicate s-incon(v,v') means strongly inconsistent:

(1) borders-in-the-south(v,v') & borders-in-the-east(v,v')→s-incon(v,v')
(2) borders-in-the-south(v,v') & borders-in-the-west(v,v')→s-incon(v,v')
(3) borders-in-the-south(v,v') & borders-in-the-north(v,v')→s-incon(v,v')
(4) borders-in-the- south(v,v') & borders-in-the-northwest(v,v')→s-incon(v,v')
(5) borders-in-the-south(v,v') & borders-in-the-northeast(v,v')→s-incon(v,v')

When the relations are found to be strongly inconsistent in the geo-graph, we remove both relations from the graph. There are two special cases of inconsistencies. As one may notice, each relation r in the geo-graph is asymmetric, and both r(v', v) and r(v, v') can not occur in the geo-graph at the same time. Therefore, we defined another rule, i.e. r(v, v') & r(v', v)→s-incon(v, v), for verifying such asymmetricity. When r(v, v') and r(v', v) appear in the geo-graph, we remove both edges from the graph.

Similarly, it is easy to verify that each relation r in the geo-graph is irreflexive. For example, we can not say borders-in-the-south(Nanjing, Nanjing). Therefore, we also defined a rule r(v, v)→s-incon(v, v) to handle these cases.

A weak inconsistency is a situation where two relations r(v, v') and r'(v, v') have certain "information" in common, while their other information is inconsistent. For example, borders-in-the-south(v, v') has the direction 南(south) in common with borders-in-the-southeast(v, v') and borders-in-the-southwest(v, v'), although the three relations are inconsistent about the neighborhood relations of v and v'.

For each relation r, we defined a few rules for checking weak inconsistencies. For example, for the relation 南邻 (borders-in-the-south), we defined the following 2 rules, where the predicate w-incon(v,v') means weakly inconsistent:

(6) borders-in-the-south(v,v') & borders-in-the-southeast(v,v')→w-incon(v,v')
(7) borders-in-the-south(v,v') & borders-in-the-southwest(v,v')→w-incon(v,v').

In the system, we retain weakly inconsistent relations.

6   Conclusion

The paper presented a practical method of extracting geo-entities and their geo-relations from the Web. The method contains more than 493 Chinese lexico-syntactical patterns.

Starting from an initial set of geo-entities, our knowledge extractor (i.e. OM-Kast-Googling system) uses the patterns to extract geo-entities and geo-relations from the matched Web excerpts, where the nodes represent geo-entities, and edges represent geo-relations of the entities. Large-scale comprehensive experiments show that 638,322 geo-entity names and 462,038 geo-relations are extracted from the Web. For extracting geo-entity names, the OMKast-Googling system achieved an average precision rate of 93.43%, where 93.43% = (94.3%+93.11%+92.87%)/3, and for geo-relations, the system achieved an average precision rate of 89.22%, where 89.22% = (90.88%+89.68%+88.27%)/3.

Acknowledgements. The work is supported by NSFC grants (No. 91224006, No.61035004, No.61173063, and No.61203284), and a MOST grant (No. 201303107).

References

1. Agichtein, E., Gravano, L.: Snowball: Extracting Relations from Large Plain-Text Collections. In: 5th ACM International Conference on Digital Libraries, pp. 85–94. ACM Press, New York (2000)
2. Blohm, S., Cimiano, P., Stemle, E.: Harvesting Relations from the Web–Quantifiying the Impact of Filtering Functions. In: 22rd Conference on Artificial Intelligence, pp. 1316–1321. AAAI Press, Menlo Park (2007)
3. Cao, C.G., Wang, H.T., Sui, Y.F.: Modeling and Acquisition of Traditional Chinese Drugs and Formulae. International Journal of Artificial Intelligence in Medicine 32, 3–13 (2004)
4. Dutta, K., Prakash, N., Kaushik, S.: Hybrid Framework for Information Extraction for Geographical Terms in Hindi Language Texts. In: 1st Interfnational Conference on Natural Language Processing and Knowledge Engineering, pp. 577–581. IEEE Press, New York (2005)
5. Gao, J.F., Li, M., Wu, A., Huang, C.N.: Chinese Word Segmentation and Named Entity Recognition: A Pragmatic Approach. Journal of Computational Linguistics 31, 531–574 (2005)
6. Girju, R., Badulescu, A., Moldovan, D.: Learning Semantic Constraints for the Automatic Discovery of Part-Whole Relations. In: 2003 Human Language Technology Conference Conference of the North American Chapter of the Association for Computational Linguistics on Human Language Technology, pp. 1–8. Association for Computational Linguistics, Stroudsburg (2003)
7. Girju, R., Badulescu, A., Moldovan, D.: Automatic Discovery of Part-Whole Relations. Journal of Computational Linguistics 32, 83–135 (2006)
8. van Hage, W.R., Kolb, H., Schreiber, G.: A Method for Learning Part-Whole Relations. In: Cruz, I., Decker, S., Allemang, D., Preist, C., Schwabe, D., Mika, P., Uschold, M., Aroyo, L.M. (eds.) ISWC 2006. LNCS, vol. 4273, pp. 723–735. Springer, Heidelberg (2006)
9. Hearst, M.: Automatic Acquisition of Hyponyms from Large Text Corpora. In: 14th International Conference on Computational Linguistics, pp. 539–545 (1992)
10. Liu, L.: Theories and Methods of Extracting Concept and Hyponymy Relations. PhD Thesis, Institute of Computing Technology, Chinese Academy of Sciences (2007)
11. Mao, H.Y., Liu, K.: Handbook of World Geography. Knowledge Publisher (1984)
12. Nissim, N., Matheson, C., Reid, J.: Recognising Geographical Entities in Scottish Historical Documents. In: Workshop on Geographic Information Retrieval at SIGIR (2004)
13. Pantel, P., Pennacchiotti, M.: Espresso: Leveraging Generic Patterns for Automatically Harvesting Semantic Relations. In: 21st International Conference on Computational Linguistics and the 44th Annual Meeting of the ACL, pp. 113–120. Association for Computational Linguistics, Stroudsburg (2006)
14. Tian, G.G.: Self-Supervised Knowledge Acquisition from Text from Constrained Corpora. PhD Thesis, Institute of Computing Technology, Chinese Academy of Sciences (2007)
15. Wang, C., Xie, X., Wang, L., Lu, Y.S., Ma, W.Y.: Detecting Geographic Locations from Web Resources. In: 3rd Workshop on Geographic Information Retrieval, pp. 17–24. ACM, New York (2005)
16. Wang, H.T., Cao, C.G., Gao, Y.: An Ontology-based System for Acquiring Knowledge from Semi-structured Text. Journal of Computers 28, 2010–2018 (2005)
17. Yu, L.: Acquisition and Verification of Terms from Large-Scale Chinese Corpora. MS Thesis, Institute of Computing Technology, Chinese Academy of Sciences (2006)

An Improved Backtracking Search Algorithm for Constrained Optimization Problems

Wenting Zhao[1], Lijin Wang[1,2], Yilong Yin[1,*], Bingqing Wang[1],
Yi Wei[1], and Yushan Yin[1]

[1] School of Computer Science and Technology,
Shandong University, Jinan, 250101, P.R. China
wenting.wentingzhaoid@gmail.com,
yilong.ylyin@sdu.edu.cn, bingqing.wangbqing@qq.com,
{yi.weiyi1991,yushan.yinyushande2012}@163.com
[2] College of Computer and Information Science,
Fujian Agriculture and Forestry University, Fuzhou, 350002, P.R. China
lijin.lijinwang@fafu.edu.cn

Abstract. Backtracking search algorithm is a novel population-based stochastic technique. This paper proposes an improved backtracking search algorithm for constrained optimization problems. The proposed algorithm is combined with differential evolution algorithm and the breeder genetic algorithm mutation operator. The differential evolution algorithm is used to accelerate convergence at later iteration process, and the breeder genetic algorithm mutation operator is employed for the algorithm to improve the population diversity. Using the superiority of feasible point scheme and the parameter free penalty scheme to handle constrains, the improved algorithm is tested on 13 well-known benchmark problems. The results show the improved backtracking search algorithm is effective and competitive for constrained optimization problems.

Keywords: constrained optimization, backtracking search algorithm, differential evolution algorithm, breeder GA mutation operator, mutation.

1   Introduction

Decision science and the analysis of physical system attach great importance to optimization techniques. Optimization problems can be mathematically formulated as the minimization or maximization of objective functions subject to constraints on their variables. Recently, nature-inspired meta-heuristic algorithms designed for solving various global optimization problems have been changing dramatically, e.g. genetic algorithm (GA) [1], differential evolution algorithm (DE) [2], ant colony optimization algorithm (ACO) [3], particle swarm optimization algorithm (PSO) [4,5], artificial bee colony algorithm (ABC) [6], social emotion optimization algorithm (SEOA) [7,8,9,10,11], bat algorithm (BA) [12], firefly algorithm (FA) [13], harmony search

---

* Corresponding author.

R. Buchmann et al. (Eds.): KSEM 2014, LNAI 8793, pp. 222–233, 2014.

algorithm (HS) [14], biogeography-based optimization algorithm (BBO) [15], group search optimizer (GSO) [16], and backtracking search optimization algorithm (BSA) [17].

BSA, a new nature-inspired algorithm proposed by Civicoglu, is effective, fast and capable of solving different numerical optimization problems with a simple structure. It has been proved that BSA can solve the benchmark problems more successfully than the comparison algorithms e.g. PSO, CMAES, ABC and JDE [17]. To our knowledge, no one has so far attempted making research on the BSA algorithm for constrained optimization problems. In light of this, we propose an improved BSA algorithm for constrained optimization problems, called IBSA. IBSA divides the evolutionary process into two phases. In the first phase, the proposed algorithm employs the mutation and crossover operators used in the standard BSA to take advantage of information gained from previous population. In the second phase, the mutation and crossover operators employed in the standard differential evolution algorithm is used to accelerate convergence and guide algorithm to find the optimal solution. In addition, the breeder genetic algorithm mutation is utilized to improve the population diversity with a small probability in the later phase.

The remainder of this paper is organized as follows. Section 2 describes general formulation of constrained optimization problem and constraint handling method. Section 3 introduces improved backtracking search algorithm. Results are presented in Section 4 and the concluding remarks are made in Section 5.

2 Constraint Problem and Constraint Handling Method

2.1 Constraint Problem

In the field of decision science and the analysis of physical system, there are a bundle of constrained optimization problems. Generally speaking, a constrained optimization problem can be described as follows (without loss of generality minimization is considered here).

$$f_{min} = \min\{f(x) \mid x \in \Omega\} \tag{1}$$

Feasible region:

$$\Omega = \{x \in \mathbb{R}^n \mid g_i(x) \le 0, h_j(x) = 0, l_m \le x_m \le u_m, \text{ for } i = 1,...p \;\; j = 1,...,q \;\; \forall m\} \tag{2}$$

In the above equations, $\vec{x} = (x_1, x_2, ..., x_D) \in \Omega \subseteq S$ is a D-dimensional vector. Each variable x_m subjects to Lower bound l_m and upper bound u_m. $f(x)$ is the objective function, $g_i(x)$ is the i-th inequality constraint, $h_j(x)$ is the j-th equality constraint. We divide constraints into four categories broadly, linear inequality constraints, nonlinear inequality constraints, linear equality constraints and nonlinear equality constraints. Most constraint handling techniques tend to deal with inequality constraints only. Consequently, we transform equality constraints into inequality constraints of the form $|h_j(x) - \delta| \le 0$, where δ is the constraint violation tolerance (a small positive value close to zero).

2.2   Constraint Handling Method

There are lots of constrained handling methods used in constrained optimization problems, but the penalty function has been used most widely. The basic penalty function can be formulated as follows:

$$\hat{f}(x) = f(x) + R \times G(x) \tag{3}$$

$$G(x) = R \sum_{i=1}^{s} \max[0, g_j(x)]^q \tag{4}$$

where R is the penalty parameter, and $\hat{f}$ is called an exact penalty function.

The superiority of feasible points (SFP) scheme is based on the static penalty method but includes an additional term in formulation (1). The purpose of this additional function is to ensure that infeasible points always have worst fitness values than feasible points. Eq.(1) can be rewritten as follows, where T_k is the population composed of trial individuals v_i at the k-th iteration.

$$\hat{f}(v_i^k) = f(v_i^k) + R \times G(v_i^k) + \Theta_k(v_i^k), \ v_i^k \in T^k \tag{5}$$

$$\Theta_k(v_i^k) = \begin{cases} 0 & \text{if } T^k \cap \Omega = \Phi \ \text{or} \ v_i^k \in \Omega \\ \alpha & \text{if } T^k \cap \Omega \neq \Phi \ \text{and} \ v_i^k \notin \Omega \end{cases} \tag{6}$$

The value α is calculated by:

$$\alpha = \max[0, \max_{v \in T^k \cap \Omega} f(v) - \min_{z \in T^k \setminus (T^k \cap \Omega)} [f(z) + R \times G(z)]] \tag{7}$$

The method of parameter free penalty (PFP) scheme is a modification of the SFP Scheme. The most significant feature is the lack of a penalty coefficient R. The fitness function in the PFP scheme is as follows:

$$\hat{f}(v_i^k) = f(v_i^k) + G(v_i^k) + \Theta_k(v_i^k), \ v_i^k \in T^k \tag{8}$$

$$\Theta_k(v_i^k) = \begin{cases} 0 & \text{if } v_i^k \in \Omega \\ -f(v_i^k) & \text{if } T^k \cap \Omega = \Phi \\ -f(v_i^k) + \max_{y \in T^k \cap \Omega} f(v_i^k) & \text{if } T^k \cap \Omega \neq \Phi \ \text{and} \ v_i^k \notin \Omega \end{cases} \tag{9}$$

3   Improved Backtracking Search Algorithm

3.1   BSA

BSA is a population-based iterative evolutionary algorithm designed to be a global minimizer. BSA maintains a population of N individual and D-dimensional members for solving bound constrained global optimization. Moreover, BSA possesses a memory in which it stores a population from a randomly chosen previous generation for use in generating the search-direction matrix [17]. To implement BSA, the following processes need to be performed.

BSA initials current population and history population according to Eq.(10) and (11) respectively where U is the uniform distribution.

$$P_{i,j} \sim U(l_j, u_j) \tag{10}$$

$$oldP_{i,j} \sim U(l_j, u_j) \tag{11}$$

At the start of each iteration, an $oldP$ redefining mechanism is introduced in BSA through the rule defined by Eq.(12) and (13), where $a, b \sim U(0,1)$ is satisfied.

$$oldP = \begin{cases} P, & a < b \\ oldP, & otherwise \end{cases} \tag{12}$$

$$oldP := permuting(oldP) \tag{13}$$

BSA has a random mutation strategy that uses only one direction individual for each target individual. BSA generates a trial population, taking advantage of its experiences from previous generations. F controls the amplitude of the search-direction matrix. The initial form of the trial individual u_i is created by Eq.(14).

$$u_i = P_i + F \times (oldP_i - P_i) \tag{14}$$

Trial individuals with better fitness values for the optimization problem are used to evolve the target population individuals. BSA generates a binary integer-valued matrix called map guiding crossover directions. Eq.(15) shows BSA's crossover strategy.

$$V_{i,j} = \begin{cases} P_{i,j}, & map_{i,j} = 1 \\ u_{i,j}, & otherwise \end{cases} \tag{15}$$

At this step, a set of v_i which has better fitness values than the corresponding x_i are utilized to renew the current population as next generation population according to a greedy selection mechanism as shown in Eq.(16).

$$x_i^{next} = \begin{cases} v_i & if \ f(v_i) \leq f(p_i), \\ x_i & otherwise. \end{cases} \tag{16}$$

3.2   Differential Evolution

Differential evolution (DE) is proposed by Storn and Price in 1995. So far, more than six mutation strategies have been proposed [18, 19] owing to its simple yet efficient properties. Compared with original DE mutation, "Rand-to-best" [18] mutation is able to improve population convergence, guiding evolution towards better directions. "Rand-to-best" mutation strategy is introduced as follows:

$$\vec{u_i} = \vec{x}_{best} + F \times (\vec{x}_{r_1} - \vec{x}_{r_2}) \tag{17}$$

Where r_1, r_2 are integers randomly selected from 1 to N, and satisfy $r_1 \neq r_2$. The scaling factor F is a real number randomly selected between 0 and 1. $\vec{x}_{best}$ is the best individual in the current population, and $\vec{u_i}$ is the mutant vector.

Subsequently, the crossover operation is implemented to generate a trial vector v_i shown by Eq. (18). Where I_i is an integer selected randomly from 1 to D, r_j is selected randomly from 1 to 0 and j denotes the j-th dimension. The index k is the number of iteration, and C_r is the crossover control parameter.

$$v_{i,j}^k = \begin{cases} u_{i,j}^k, & r_j \leq c_r \ or \ j = I_i \\ x_{i,j}^k, & otherwise \end{cases} \tag{18}$$

3.3   Breeder Genetic Algorithm Mutation Operator

$$v_{i,j} = \begin{cases} x_{i,j} \pm rang_i \times \sum_{s=0}^{15} \alpha_s 2^{-s} & rand < 1/D \\ x_{i,j} & otherwise \end{cases}, j = 1,...,D \tag{19}$$

$$rang_i = (u(i) - l(i)) \times (1 - current\_gen / total\_gen)^6 \tag{20}$$

Improved version of breeder genetic algorithm mutation operator proposed in [20, 21] intends to produce a highly explorative behavior in the early stage and ensures the global convergence in the later stage. Where $U(0,1)$ is the uniform random real number generator between 0 and 1. The plus or minus sign is selected with a probability of 0.5, and $\alpha_s \in \{0,1\}$ is randomly generated with expression $P(\alpha_s = 1) = 1/16$. Current generation number is denoted as $current\_gen$, and total generation number is denoted as $total\_gen$. Individuals in the interval $[x_i - rang_i, x_i + rang_i]$ are generated after IBGA mutation.

3.4   IBSA

BSA has a powerful exploration capability but a relatively slow convergence speed, since the algorithm uses historical experiences to guide the evolution. Focusing on excellent convergent performance of "Rand-to-best" mutation, it is combined with BSA. Meanwhile, IBGA is utilized to expand population diversity. Pseudo–code of IBSA can be present as follows:

Step 1: Initialize population size N, mutation probability pm, stage control parameter $rate$, crossover probability Cr, total number of iteration $IterMax$ and penalty coefficient R if SFP is used.

Step 2: Initialize population P, and historical population $oldP$ using Eq.(10) and Eq.(11), respectively.

Step 3: Evaluate the population P using Eq.(5) or Eq.(8).

Step 4: $k=0$;

Step 5: Update the historical population $oldP$ using Eq.(12) and Eq.(13).

Step 6: if $k<IterMax*rate$ then perform mutation and crossover operators according to Eq.(14) and Eq.(15), and goto Step 8.

Step 7: if $pm<=0.05$ then perform mutation operator using Eq.(19), else perform mutation and crossover operators according to Eq.(17) and Eq.(18), where the factor F is generated from the range of $[-1,-0.4]$ and $[0.4,1]$ uniformly.

Step 8: Evaluate the population P using Eq.(5) or Eq.(8), and select the best individual X_{best}.

Step 9: $k=k+1$, if $k<IterMax$ goto Step 5.

Step 10: output X_{best}.

4   Experiments

We use a set of 13 benchmark problems [22] in this paper to evaluate the performance of BSA, which were tested widely in evolution computation domain to show the performance of different algorithms for constrained optimization problems. The objective functions can be divided into 6 classes: quadratic, nonlinear, polynomial, cubic, linear, and exponential. Main characteristics of these functions are summarized in Tab.1.

Table 1. Main properties of benchmark functions(n: number of variables, |F|/|S|: the ratio of the feasible region to the given box constrained area, LI, NE, NI: number of linear inequality, nonlinear equality, and nonlinear inequality, a: number of active constraints at optimum)

	known optimal	n	Min/Max type	$f(x)$ type	|F|/|S|	LI	NE	NI	a
G01	-15	13	Minimum	quadratic	0.011%	9	0	0	6
G02	0.803619	20	Maximum	nonlinear	99.90%	1	0	1	1
G03	1	10	Maximum	polynomial	0.002%	0	1	0	1
G04	-30665.539	5	Minimum	quadratic	52.123%	0	0	6	2
G05	5126.4981	4	Minimum	cubic	0.000%	2	3	0	3
G06	-6961.8139	2	Minimum	cubic	0.006%	0	0	2	2
G07	24.306291	10	Minimum	quadratic	0.000%	3	0	5	6
G08	0.095825	2	Maximum	nonlinear	0.856%	0	0	2	0
G09	680.630057	7	Minimum	polynomial	0.512%	0	0	4	2
G10	7049.25	8	Minimum	linear	0.001%	3	0	3	3
G11	0.75	2	Minimum	quadratic	0.000%	0	1	0	1
G12	1	3	Maximum	quadratic	4.779%	0	0	9^3	0
G13	0.0539498	5	Minimum	exponential	0.000%	0	3	0	3

Additionally, for each problem, 30 independent runs were performed. Other parameters are given in Tab.2.

Table 2. Parameter values

	N	IterMax	C_r	p_m	rate	R	DIM_RATE
values	80	10000	0.9	0.05	0.6	10^{50}	1

Table 3. Results for IBSA using SFP

	optimal	best	mean	worst	std
G01	-15	**-15**	-15	-15	0
G02	0.803619	0.803614	0.788434	0.761742	0.009713
G03	1	**1.012555**	1.011447	0.992238	0.003785
G04	-30665.539	**-30665.539**	-30665.539	-30665.539	0
G05	5126.4981	**5126.484154**	5126.484154	5126.484154	0
G06	-6961.8139	**-6961.8139**	-6961.8139	-6961.8139	0
G07	24.306291	**24.306209**	24.306214	24.306279	0.000015
G08	0.095825	**0.095825**	0.095825	0.095825	0
G09	680.630057	**680.630057**	680.630057	680.630057	0
G10	7049.25	**7049.248021**	7049.248039	7049.248158	0.000037
G11	0.75	**0.7499**	0.7499	0.7499	0
G12	1	**1**	1	1	0
G13	0.0539498	**0.053942**	0.053942	0.053942	0

Table 4. Results for IBSA using PFP

	optimal	best	mean	worst	std
G01	-15	**-15**	-15	-15	0
G02	0.803619	0.803615	0.789926	0.75071	0.013549
G03	1	**1.01256**	1.010525	0.972599	0.007409
G04	-30665.539	**-30665.539**	-30665.539	-30665.539	0
G05	5126.4981	**5126.484154**	5126.484154	5126.484154	0
G06	-6961.8139	**-6961.8139**	-6961.8139	-6961.8139	0
G07	24.306291	**24.306209**	24.306213	24.306235	0.000007
G08	0.095825	**0.095825**	0.095825	0.095825	0
G09	680.630057	**680.630057**	680.630057	680.630057	0
G10	7049.25	**7049.248021**	7049.248051	7049.248432	0.000081
G11	0.75	**0.7499**	0.7499	0.7499	0
G12	1	**1**	1	1	0
G13	0.0539498	**0.053942**	0.06679	0.439383	0.070372

4.1   Results of the Proposed Method (IBSA)

Table 3 and Table 4 list the experimental results of the IBSA algorithm using SFP and PFP respectively. They include the known optimal solution for each test problem and the obtained best, mean, worst and standard deviation values. The best obtained by IBSA, which are same to or better than the known optimal, are marked in boldface.

As seen from Table 3, the IBSA using SFP algorithm can find the global optimal solutions for 12 problems except G02 in terms of the best results. However, result for problem G02 is very close to the optimal. The standard deviations for G01, G04, G05, G06, G08, G09, G11, G12 and G13 are zero, so we consider IBSA-PFP algorithm is under stable condition. G07 has found better solution than the known optimal solution. G03, G05, G11 and G13 are obtained better solutions as well, as a result of δ settings.

Table 4 shows the IBSA using PFP algorithm can also find the global optimization for 12 problems except G02 which is very close to the optimal. Within the allowed equality constraints violation ranges, better solutions have been found than the known optimal solutions such as G3, G5, G10, G11 and G07. As for problem G13, there are 29 out of 30 times finding better solution than the optimal.

Table 5. Results for BSA using SFP

	best of IBSA	best	mean	worst	std
G01	-15	-15	-15	-15	0
G02	**0.803614**	0.80358	0.79666	0.785462	0.00506
G03	**1.012555**	1.012298	1.002612	0.954234	0.016466
G04	-30665.539	-30665.539	-30665.539	-30665.539	0
G05	**5126.484154**	5126.4967	5198.319116	5415.4138	77.85568
G06	-6961.8139	-6961.8139	-6961.8139	-6961.8139	0
G07	**24.306209**	24.306242	24.321343	24.343854	0.018588
G08	0.095825	0.095825	0.095825	0.095825	0
G09	680.630057	680.630057	680.630057	680.630057	0
G10	**7049.248021**	7049.2555	7049.617975	7052.634	0.633024
G11	0.7499	0.7499	0.749901	0.749918	0.000003
G12	1	1	1	1	0
G13	**0.053942**	0.055766	0.633725	0.999993	0.305339

4.2   Comparison with BSA Using Penalty Function

We compare results of IBSA using penalty functions from Table 3 and Table 4 with BSA using penalty functions from Table 5 and Table 6 respectively. It can be found that IBSA performs far superior to BSA when solving problem G2, G3, G05, G07, and G10. When it comes to problem G13, IBSA successfully finds global optimization 29 times in 30 runs. However, there is only once that BSA finds the optimal. Improved strategies on BSA enhance the stability of the algorithm.

Table 6. Results for BSA using PFP

	best of IBSA	best	mean	worst	std
G01	-15	-15	-15	-15	0
G02	**0.803615**	0.803599	0.798024	0.792409	0.005266
G03	**1.01256**	1.012417	0.993189	0.899105	0.031189
G04	-30665.539	-30665.539	-30665.539	-30665.539	0
G05	**5126.484154**	5126.496714	5215.3329	5477.3847	114.543
G06	-6961.8139	-6961.8139	-6961.8139	-6961.8139	0
G07	**24.306209**	24.306244	24.323931	24.343783	0.018832
G08	0.095825	0.095825	0.095825	0.095825	0
G09	680.630057	680.630057	680.630057	680.630057	0
G10	**7049.248021**	7049.257983	7049.394	7049.7464	0.11844
G11	0.7499	0.7499	0.7499	0.749901	0
G12	1	1	1	1	0
G13	**0.053942**	0.069471	1.01215	13.782101	2.428104

Table 7. Comparison results with other algorithms using penalty function

	optimal	IBSA	MGSO	SAPF	OPA
G01	-15	-15	-15.000	-15.000	-15.000
G02	0.803619	0.803615	0.803457	0.803202	**0.803619**
G03	1	**1.012555**	1.00039	1	0.747
G04	-30665.539	**-30665.539**	**-30665.539**	-60,665.401	-30,665.54
G05	5126.4981	**5126.484154**	5,126.50	5,126.91	5,126.50
G06	-6961.8139	**-6961.8139**	**-6,961.8139**	-6,961.046	-6,961.814
G07	24.306291	**24.306209**	24.917939	24.838	24.306
G08	0.095825	**0.095825**	**0.095825**	**0.095825**	**0.095825**
G09	680.630057	**680.630057**	680.646	680.773	680.63
G10	7049.25	**7049.248021**	7,052.07	7,069.98	7,049.25
G11	0.75	**0.7499**	0.7499	0.749	0.75
G12	1	1	1	1	1
G13	0.0539498	0.053942	0.058704	**0.053941**	0.447118

4.3   Comparison with Other Algorithms Using Penalty Function

Constraint handling method highly affects the performance of intelligent algorithm for constrained optimization. To test the performance of the IBSA, we compare it with other algorithms.

The three algorithms are based on penalty function, such as modified group search optimizer algorithm with penalty function (MGSO) [23], genetic algorithm (GA) with a self-adaptive penalty function (SAPF) [24], over-penalty approach (OPA) [25]. Comparison results are summarized in Table 7. The best optimum values are listed.

As seen from Table 7, the performance of IBSA is much better than MGSO, SAPF, OPA on problems G03, G05, G07, G09 and G10 according to the best results obtained from algorithms with penalty function. Problem G01, G08, G11, G12 can be also solved successfully by IBSA. For problems g04, and g06, MGSO, OPA, and IBSA show the similar performance in terms of the best optimum, and better than SAPF. In all, using penalty function for handling constraint, IBSA shows better performance than other three algorithms for most problems.

4.4   Comparison with Other Algorithms Using Different Constraint Handling Methods

We employs three competitive evolutionary algorithms to further compare IBSA with other algorithms with different constraint handling methods, which were stochastic ranking (SR) method [22], improved stochastic ranking (ISR) [25] evolution strategy, and the simple multimember evolutionary strategy (SMES) [26]. Comparison results are summarized in Table 8.

Table 8. Comparison results with other algorithms using different constraint handling methods

	optimal	IBSA	SR	ISRES	SMES
G01	-15	**-15.000**	**-15.000**	**-15.000**	**-15.000**
G02	0.803619	0.803615	0.803516	**0.803519**	0.803601
G03	1	**1.012555**	1	1.001	1.000
G04	-30665.539	**-30665.539**	-30,665.539	-30,665.539	-30,665.54
G05	5126.4981	**5126.484154**	5,126.50	5,126.50	5,126.60
G06	-6961.8139	**-6961.8139**	-6,961.814	-6,961.814	-6,961.814
G07	24.306291	**24.306209**	24.306	24.306	24.327
G08	0.095825	**0.095825**	**0.095825**	**0.095825**	**0.095825**
G09	680.630057	**680.630057**	680.63	680.63	680.632
G10	7049.25	**7049.248021**	7,049.25	7,049.25	7,051.90
G11	0.75	**0.7499**	0.75	0.75	0.75
G12	1	**1**	1	1	1
G13	0.0539498	**0.053942**	0.053957	0.053957	0.053986

As seen from Table 8, Problem G01, G03, G04, G06, G08, G11 and G12 can be found optima by these four algorithms. IBSA performs better when dealing with G03, G05 and G13. For problems G07, G09 and G10, IBSA, SR, and ISRES show the similar performance in terms of the best optimum, and better than SMES.

5   Conclusion

This paper proposes an improved version BSA algorithm, called IBSA, which is very effective for solving constrained optimization problems. The proposed algorithm employs BSA operations, variant DE operations, and IBGA mutation at different stages. IBSA has better performance than BSA using PFP and SPF, and other three algorithms using penalty function to handle constraints according to experimental results. Comparing with other algorithms with different constraint handling methods, the results show that IBSA is comparable with its competitors. As a consequence, the effect of constraint handling methods on the performance of IBSA algorithm can be investigated in future works. Another direction is to improve algorithm to enhance accuracy of finding global optimum on some problems including equality constraint, such as G03 and G13.

Acknowledgement. This work was supported by the NSFC Joint Fund with Guangdong of China under Key Project U1201258, the Shandong Natural Science Funds for Distinguished Young Scholar under Grant No.JQ201316, and the Natural Science Foundation of Fujian Province of China under Grant No.2013J01216.

References

1. Holland, J.H.: Adaptation in natural and artificial systems. University of Michigan Press, Ann Arbor (1975)
2. Storn, R., Price, K.V.: Differential evolution: a simple and efficient adaptive scheme for global optimization over continuous spaces, Technical Report TR-95-012, Berkeley, CA (1995)
3. Dorigo, M., Maniezzo, V., Colorni, A.: The ant system: optimization by a colony of cooperating agents. IEEE Trans. on Systems. Man, and Cybernetics 26, 29–41 (1996)
4. Kennedy, J., Eberhart, R.C.: Particle swarm optimization. In: Proceedings of the IEEE International Conference on Neural Networks, pp. 1942–1948. IEEE Press (1995)
5. Eberhart, R.C., Kennedy, J.: A new optimizer using particle swarm theory. In: Proc. of the 6th International Symposium on Micro Machine and Human Science, pp. 39–43. IEEE Press (1995)
6. Karaboga, D.: An idea based on honey bee swarm for numerical optimization. Technical Report TR-06, Erciyes University, Engineering Faculty, Computer Engineering Department (2005)
7. Cui, Z.H., Cai, X.J.: Using social cognitive optimization algorithm to solve nonlinear equations. In: Proc. 9th IEEE Int. Conf. on Cog. Inf., pp. 199–203 (2010)
8. Chen, Y.J., Cui, Z.H., Zeng, J.H.: Structural optimization of Lennard-Jones clusters by hybrid social cognitive optimization algorithm. In: Proc. of 9th IEEE Int. Conf. on Cog. Inf., pp. 204–208 (2010)
9. Cui, Z., Shi, Z., Zeng, J.: Using social emotional optimization algorithm to direct orbits of chaotic systems. In: Panigrahi, B.K., Das, S., Suganthan, P.N., Dash, S.S. (eds.) SEMCCO 2010. LNCS, vol. 6466, pp. 389–395. Springer, Heidelberg (2010)
10. Wei, Z.H., Cui, Z.H., Zeng, J.C.: Social cognitive optimization algorithm with reactive power optimization of power system. In: Proc. of 2010 Int. Conf. Computational Aspects of Social Networks, pp. 11–14 (2010)

11. Xu, Y., Cui, Z., Zeng, J.: Social emotional optimization algorithm for nonlinear constrained optimization problems. In: Panigrahi, B.K., Das, S., Suganthan, P.N., Dash, S.S. (eds.) SEMCCO 2010. LNCS, vol. 6466, pp. 583–590. Springer, Heidelberg (2010)
12. Yang, X.S.: A new metaheuristic bat-inspired algorithm. Springer, Berlin (2010)
13. Yang, X.S.: Nature-Inspried Metaheuristic Algorithms, 2nd edn. Luniver Press, Frome (2010)
14. Geem, Z.W., Kim, J.H., Loganathan, G.V.: A new heuristic optimization algorithm: Harmony search. Simulation 76, 60–68 (2001)
15. Simon, D.: Biogeography-Based Optimization. IEEE Transactions on Evolutionary Computation 12, 702–713 (2008)
16. He, S., Wu, Q.H., Saunders, J.R.: A novel group search optimizer inspired by animal behavioral ecology. In: Proceedings of IEEE Congress on Evolutionary Computation, pp. 16–21 (2006)
17. Civicioglu, P.: Backtracking search optimization algorithm for numerical optimization problems. Applied Mathematics and Computation 219, 8121–8144 (2013)
18. Mezura-Montes, E., Miranda-Varela, M., Gómez-Ramón, R.: Differential evolution in constrained numerical optimization: An empirical study. Information Sciences 180, 4223–4262 (2010)
19. Neri, F., Tirronen, V.: Recent advances in differential evolution: a survey and experimental analysis. Artificial Intelligence Review 33, 61–106 (2010)
20. Wang, Y., Cai, Z., Guo, G., Zhou, Y.: Multiobjective optimization and hybrid evolutionary algorithm to solve constrained optimization problems. IEEE Transaction on Systems 37, 560–575 (2007)
21. Jia, G., Wang, Y., Cai, Z., Jin, Y.: An improved $(\mu+\lambda)$-constrained differential evolution for constrained optimization. Information Sciences 222, 302–322 (2013)
22. Runarsson, T.P., Yao, X.: Stochastic ranking for constrained evolutionary optimization. IEEE Transactions on Evolutionary Computation 4, 284–294 (2000)
23. Wang, L., Zhong, Y.: A modified group search optimiser for constrained optimisation problems. Int. J. Modelling, Identification and Control 18, 276–283 (2013)
24. Tessema, B., Yen, G.: A self-adaptive penalty function based algorithm for constrained optimization. In: Proceedings 2006 IEEE Congress on Evolutionary Computation, pp. 246–253 (2006)
25. Runarsson, T.P., Yao, X.: Search biases in constrained evolutionary optimization. IEEE Transactions on Systems, Man, and Cybernetics 35, 233–243 (2005)
26. Mezura-Montes, E., Coello Coello, C.A.: A simple multimembered evolution strategy to solve constrained optimization problems. IEEE Transactions on Evolutionary Computation 9, 1–17 (2005)

Sentiment Classification
by Combining Triplet Belief Functions

Yaxin Bi, Maurice Mulvenna, and Anna Jurek

School of Computing and Mathematics, University of Ulster
Newtownabbey, Co. Antrim, BT37 0QB, UK
{y.bi,md.mulvenna,a.jurek}@ulster.ac.uk

Abstract. Sentiment analysis is an emerging technique that caters for semantic orientation and opinion mining. It is increasingly used to analyse online product reviews for identifying customers' opinions and attitudes to products or services in order to improve business performance of companies. This paper presents an innovative approach to combining outputs of sentiment classifiers under the framework of belief functions. The approach is composed of the formulation of outputs of sentiment classifiers in the triplet structure and adoption of its formulas to combining simple support functions derived from triplet functions by evidential combination rules. The empirical studies have been conducted on the performance of sentiment classification individually and in combination, the experimental results show that the best combined classifiers made by these combination rules outperform the best individual classifiers over the MP3 and Movie-Review datasets.

Keywords: Sentiment analysis, opinion mining, triplet belief functions and combination rules.

1 Introduction

In a sense sentiment analysis can be regarded as a special text categorization technique that focuses on formulating a label assignment process as a binary decision making problem - identifying the polarity of opinions or attitudes to products. Views and opinions on the quality, services and various offerings about some products are currently proliferated on e-commerce web sites, they collate timely feedback about products and they are a key source of helping companies monitor customers' opinions and business behaviors in addition of aiding customers to make purchasing decision. A popular product could normally receive reviews in the hundreds and in various forms of expressions which are often affected by domain authorities. As a result opinions contained in the reviews might be difficult to precisely explain and identify, there is uncertainty in determining the decision boundary of sentiment polarity. This study develops an evidential reasoning approach to modelling such uncertainty and incorporates it into the process of sentiment polarity classification.

Text categorization methods are usually built on supervised machine learning approaches, which are focused on aspect/topic-based multi-class classification

R. Buchmann et al. (Eds.): KSEM 2014, LNAI 8793, pp. 234–245, 2014.

tasks. A typical approach is to use labeled text corpora to train classification models and then apply them to assign new text documents with the pre-defined topic/class labels. Sentiment analysis is aimed at determining a sentiment polarity against the opinions to aspects/topics in the positive or negative form. It can take a supervised learning approach, but the requirements on labeling opinionated texts and classification decision are different from the conventional supervised learning. Aspects or topics in text categorization are often identifiable by keywords alone, in contrast sentiments are expressed in a more subtle manner that requires more semantics and context understanding in addition to that the polarity labels classified may be used simply for summarizing the content of opinionated texts on an aspect/topic - whether it be positive or negative.

In recent years online reviews are rapidly growing in social media, much work on sentiment classification has been published in the literature [1]. The three supervised machine learning methods of Naive Bayes, Maximum Entropy and Support Vector Machines (SVM) were employed to sentiment classification based on traditional topic-based categorization in [2]. The empirically comparative studies demonstrated that accuracies on the sentiment classification problem were not comparable to those reported in standard topic-based categorization. In [3], a classifier for product feature extraction and binary classification was developed by applying various machine learning methods. The authors compared a uni-gram model with more complex models exploiting linguistic substitutions, greater n-grams, thresholding, smoothing of feature probabilities, scoring, and reweighting. The results showed greater n-grams outperform uni-grams. As labelled review data is not always available, an unsupervised approach was used to identify sentiment in [4] [5]. The authors determined the morphological characteristics of words and semantic orientations of sentences through the part-of-speech tagging in conjunction with sets of seed words. The advantage of this approach is that it makes use of the context of sentences, thereby gaining a better understanding of particular aspects reflected in the reviews.

Although over 7,000 articles have been published on the sentiment analysis related topics in the past decade [1], research on sentiment classification by information fusion based approaches is still very limited. In [6] the authors discussed domain adaptation problems that involve two domains: source domain and target domain. They used opinionated texts from the source domain to train a classification model and applied it for classifying texts from the target domain. A similar work was documented in [8], in which an online algorithm was proposed for multi-domain learning that could be directly applied to multi-domain sentiment classification tasks. A more recent work was concentrated on investigating how to make multiple domains aid each other in sentiment classification when each domain only has partially labeled training data in [7]. In that study each domain data were used to train one classifier and the results on test data drawn from one domain were combined by using fixed rules in terms of *sum* and *product* rules in conjunction with two dynamic *meta-learning* rules. The experimental results demonstrated that the two simple fixed combination rules give comparative performance, and the weighted sum rule significantly outperforms the simple fixed rules and the meta-learning rule.

Building on our previous studies [9], this work develops an evidential ensemble approach by formulating sentiment classification results in the triplet evidence structure [10]. A sentiment polarity classification normally involves predicting opinionated texts to a category in the form of positive, negative or neutral. However the neutral category may not be a case in which customers are interested for online reviews. Meanwhile when a classifier is trained by using the labelled opinionated text corpora, the neutral category cannot be directly outputted by standard supervised machine learning algorithms, but alternatively it can be handled through designing a membership function.

In this study, we define positive and negative categories as binary mutually exclusive propositions making up a frame of discernment and use the frame to represent the neutral situation in terms of *ignornace*. In such a way we can employ the triplet function to represent sentiment polarity classifier outputs. Furthermore we tailor the computational formulas derived for combining multiple triplet functions by the three combination rules of Demspter's rule of combination [11], and adapt them for the Transferable Belief Model(TBM) conjunctive rule [12] and the cautious conjunctive combination rule [13]. To evaluate the performance of the proposed approach, we carry out two experiments by using eight machine learning algorithms in conjunction with the three combination rules over the MP3 and Movie-Review review datasets, the experimental results show that on both of the datasets, the best combined classifiers by the three rules outperform the best individual classifier. Due to the limited space, more mathematic formulas and comparative studies with the normalized cautious conjunctive and linear sum rules have been included in the sister paper [14].

2   Basics of Belief Functions

This section briefly introduces some essential concepts and formulism used in this study. Given a frame of discernment, denoted by Θ that consists of mutually exclusive propositions, for any subset $A \subseteq \Theta$, the Dempster-Shafer (DS) theory of evidence uses a numeric value within the range of $[0, 1]$ to represent the degree of support for A, denoted by $m(A)$, which is called a *Basic Belief Assignment* (BBA) [11]. It has two conditions:

$$(1)\ m(\emptyset) = 0; \quad (2) \sum_{A \subseteq \Theta} m(A) = 1$$

where A is called a focal element or focus if $m(A) > 0$, and it is called a singleton if A contains only one element with $m(A) > 0$.

Notice that $m(A)$ represents the measurement of a support degree that one commits exactly to the subset A, rather than the total support committed to A. To obtain the measurements of the total support committed to A and of other situations, the DS provides three other evidential functions in terms of *belief function (bel)*, *plausibility function (pls)*, and *commonality function (q)*.

Particularly the computation of a commonality function for subset $A \subseteq \Theta$ by a mass function is given as follows:

$$q(A) = \sum_{B \supseteq A} m(B) \tag{1}$$

In practice, evidence sources may not be entirely reliable. Support degrees derived from such sources need to be discounted to truly reflect the reliability of the sources. The DS theory provides a discount mechanism by the following formula.

$$m^r(A) = \begin{cases} (1-r)m(A), & \text{if } A \subset \Theta \\ r + (1-r)m(\Theta), & \text{if } A = \Theta \end{cases} \tag{2}$$

where r represents the discounting rate within the range of $[0,1]$. When an evidence source is completely reliable, r takes on the value of 0. As oppose to this, if the source is completely unreliable, r takes the value of 1.

Definition 1. Let m_1 and m_2 be two mass functions on the frame of discernment Θ, and for any subset $A \subseteq \Theta$, the *orthogonal sum* of two mass functions on A, denoted by $\oplus$, is defined as follows:

$$m_1 \oplus m_2(A) = \frac{\sum_{X \cap Y = A} m_1(X) m_2(Y)}{1 - \sum_{X \cap Y = \emptyset} m_1(X) m_2(Y)} \tag{3}$$

where $X, Y \subseteq \Theta$ and $1 - \sum_{X \cap Y = \emptyset} m_1(X) m_2(Y)$ is a normalization factor, denoted by K. The orthogonal sum is also called Dempster's rule of combination, it allows two mass functions to be combined into a third mass function. The above definitions are based on the notations given in [11]. In the framework of the Transferable Belief Model (TBM) [12], Smets lifts the restriction on $m(\emptyset) = 0$ and removes the normalization operation in Equation (3), resulting in a TBM conjunctive combination rule (Smets's rule for short), denoted by $\bigcirc$. Both combination rules have been employed in the cautious conjunctive combination rule recently developed in [13].

Definition 2. Let m_1 and m_2 be two non dogmatic mass functions on the frame of discernment Θ. The cautious conjunctive combination of m_1 and m_2, denoted by $m_{1 \bigotimes 2} = m_1 \bigotimes m_2$, is defined on the basis of a weight function below:

$$w_{1 \bigotimes 2}(A) = w_1(A) \wedge w_2(A), \forall A \subset \Theta \tag{4}$$

and then

$$m_1 \bigotimes m_2 = \bigcirc_{A \subset \Theta} A^{w_1(A) \wedge w_2(A)}. \tag{5}$$

where $A^{w(A)}$ denotes a simple mass function m such that $m(A) = 1 - w(A)$ and $m(\Theta) = w(A)$, and the weight $w(A)$ can be obtained from the commonality values by the following formula:

$$w(A) = \prod_{B \supseteq A} q(B)^{(-1)^{|B|-|A|+1}} \tag{6}$$

if the conjunctive operation $\cap$ in Equation (5) is replaced by the Dempster orthogonal sum $\oplus$, then the revised Equation (5) has to be divided by the normalization factor of $K = 1 - m_1 {\textstyle\bigwedge} m_2(\emptyset)$, resulting in a normalized version of the cautious rule as follows:

$$m_1 {\textstyle\bigwedge} m_2 = \frac{\bigoplus_{\emptyset \neq A \subseteq \Theta} A^{w_1(A) \wedge w_2(A)}}{K}. \tag{7}$$

As indicated in [13] that $m_1 {\textstyle\bigwedge} m_2 = 1$ never holds as 'the cautious combination of two non dogmatic BBAs can never be dogmatic (i.e. Θ is not a focal set)'.

3   Triplet and Binary Structures

This study treats sentiment polarity classification simply as a special case of topic-based categorization with the polar topics of positive and negative sentiments. In such a case, a categorization algorithm is provided with a training data set made up of $D \times C = \{\langle d_1, c_1 \rangle, \cdots, \langle d_{|D|}, c_q \rangle\}$ $(1 \leq q \leq |C|)$ for deriving a classifier denoted by φ. Instance $d_i \in D$ is characterized by the bag of words of $(\omega_{i_1}, \cdots, \omega_{i_n})$ where ω_{i_j} is a score of keyword, C is composed of positive sentiment (c) and negative sentiment $(\tilde{c})$, i.e. $C = \{c, \tilde{c}\}$, and a classifier output on a new text document d is denoted by $\varphi(d) = C \times [0, 1]$. Given classifier output $\varphi(d)$, we can formulate it as a piece of evidence in the form of triplet below.

Definition 3. Let $\Theta = \{x_1, x_2, ..., x_n\}$ be a frame of discernment and $\varphi(d) = \{m(\{x_1\}), m(\{x_2\}), ..., m(\{x_n\})\}$ be mass probabilities derived from classifier outputs, where $n \geq 2$. An expression in the form of $A = \langle A_1, A_2, A_3 \rangle$ is defined as a *triplet*, where $A_1, A_2 \subset \Theta$ are singletons, A_3 is the whole set Θ and they satisfy

$$m(A_1) + m(A_2) + m(A_3) = 1$$

These elements are obtained by a focusing operator σ on m, and denoted by m^σ as follows:

$$A_1 = arg \max m(\{x_1\}), m(\{x_2\}), ..., m(\{x_n\}) \tag{8}$$

$$A_2 = arg \max m(\{x\} \mid x \in \{x_1, ..., x_n\} - A_1) \tag{9}$$

$$A_3 = \Theta, m^\sigma(\Theta) = 1 - m^\sigma(A_1) - m^\sigma(A_2) \tag{10}$$

We refer to m^σ as a *triplet mass function* or as a *two-point mass function*. When $n = 2$, the frame of discernment Θ is composed of only two focal elements, denoted by $\{x\}$ and $\{\tilde{x}\}$, i.e. $\Theta = \{x, \tilde{x}\}$, such that a triplet mass function defined on Θ also satisfies the following condition:

$$m(\{x\}) + m(\{\tilde{x}\}) + m(\Theta) = 1$$

the new support degrees for $\{x\}$, $\{\tilde{x}\}$ and Θ can be obtained by using the discounting Equation (2).

4   Combinations of Triplet Mass Functions by Using the Evidential Combination Rules

In this section, we formulate the combinations of two mass functions committed to binary focal elements by using Dempster's rule, Smets's rule and the cautious conjunctive rule, which can be tailored to compute the combinations of any number of triplet functions.

Definition 4. Let $\Theta = \{x, \tilde{x}\}$ be a frame of discernment and $m_1, \ldots, m_n$ be triplet mass functions with the following condition:

$$m_i(\{x\}) + m_i(\{\tilde{x}\}) + m_i(\Theta) = 1$$

for any two triplet mass functions m_i and m_j, by using Smets's rule, we have a new mass function below:

$$m(\{x\}) = m_i(\{x\})m_j(\{x\}) + m_i(\{x\})m_j(\Theta) + m_i(\Theta)m_j(\{x\}) \qquad (11)$$

then Equation (11) can be rewritten as Equation (12) with a normalization factor as combined by Dempster's rule below

$$m(\{x\}) = 1 - (1 - m_i(\{x\}))(1 - m_j(\{x\}))/K \qquad (12)$$

where $K = 1 - (1 - m_i(\{x\}))(1 - m_j(\{x\})) + (1 - m_i(\{\tilde{x}\}))(1 - m_j(\{\tilde{x}\})) - m_i(\Theta)m_j(\Theta)$.

Likewise we have new mass functions for $\{\tilde{x}\}$ and Θ:

$$m(\{\tilde{x}\}) = 1 - (1 - m_i(\{\tilde{x}\}))(1 - m_j(\{\tilde{x}\}))/K \qquad (13)$$

$$m(\Theta) = m_i(\Theta)m_j(\Theta)/K \qquad (14)$$

In respect of combining any two triplet mass functions m_i and m_j by using the cautious conjunctive rule, it involves three steps [13]. The first step is to compute the commonality functions q_i and q_j from m_i and m_j by Equation (1); the second is to compute the weight functions w_i and w_j using Equation (6) and then generate (inverse) simple mass functions in the form of $A^{w_1(A) \wedge w_2(A)}$, for all $A \subset \Omega$ such that $w_1 \wedge w_2(A) \neq 1$; and finally compute $m_{1 \textcircled{\wedge} 2} = m_1 \textcircled{\wedge} m_2$ on these simple mass functions by using Smets's rule. Following the first two steps, we thus have,

$$\{x\}^{w_i(\{x\}) \wedge w_j(\{x\})} = \{x\}^{min(\{\frac{m_i(\Theta)}{m_i(\{x\}) + m_i(\Theta)}, \frac{m_j(\Theta)}{m_j(\{x\}) + m_j(\Theta)}\}) = w}$$

$$\{\tilde{x}\}^{w_i(\{\tilde{x}\}) \wedge w_j(\{\tilde{x}\})} = \{\tilde{x}\}^{min(\{\frac{m_i(\Theta)}{m_i(\{\tilde{x}\}) + m_i(\Theta)}, \frac{m_j(\Theta)}{m_j(\{\tilde{x}\}) + m_j(\Theta)}\}) = \tilde{w}}$$

with w and $\tilde{w}$, we can generate two (inverse) simple mass functions as follows,

$$m_{ij}^1(\{x\}) = 1 - w; m_{ij}^1(\Theta) = w.$$

$$m_{ij}^1(\{\tilde{x}\}) = 1 - \tilde{w}; m_{ij}^1(\Theta) = \tilde{w}.$$

From the last step above, we have

$$m(\{x\}) = m_{ij}^1(\{x\})m_{ij}^2(\Theta). \tag{15}$$

$$m(\{\tilde{x}\}) = m_{ij}^2(\{\tilde{x}\})m_{ij}^1(\Theta). \tag{16}$$

$$m(\Theta) = m_{ij}^1(\Theta)m_{ij}^2(\Theta). \tag{17}$$

Repeatedly using Equations (12)-(17), we can combine all multiple triplet mass functions defined on the binary frame of discernment.

5   Empirical Study

To evaluate the effectiveness of sentiment ensemble classifiers made by using Dempster's rule of combination, the TBM conjunctive combination rule and the cautious conjunctive rule, we conducted the experiments with eight machine learning algorithms for generating base classifier over the MP3 and Movie-Review datasets. These learning algorithms are directly taken from the Waikato Environment for Knowledge Analysis (Weka) version 3.6 and briefly described in Table 1 and detailed description of these algorithms can be found in [15]. For our experiments parameters used for each algorithm were set at the default settings.

The MP3 dataset contains MP3 digital camera and player reviews collected in Amazon.com [16]. Each of review consists of short sentences and labeled with a five star scale. The reviews with 1 and 2 stars are considered very negative and negative, respectively, whereas reviews with 4 and 5 stars are considered positive and very positive, and reviews with 3 stars are regarded as neutral. On the other hand, the Movie Review dataset were drawn from the IMDB's archive of rec.arts.movies.reviews newsgroup [17]. The reviews were rated in either a five star scale or numerical scores and were converted into one of three categories: positive, negative, or neutral. For the work described in this paper, we concentrated only on discriminating between positive and negative polarity sentiment and the reviews with neutral are discarded for our experiments. As a result the MP3 dataset used in our experiments contains 21519 positive and 6390 negative text reviews and the Movie-Review consists of 1000 positive and 1000 negative movie reviews.

To reflect the performance of the ensemble sentiment classifiers faithfully and to avoid overfitting to some extent, the experiments were conducted using a ten-fold cross validation. The performance of classifiers in individuals and combinations was measured by the F-measure that is commonly used in text categorization area. For construction of ensemble classifiers by the three combination

Table 1. A brief description of eight machine learning algorithms

No	Classifier	Description
1	NaiveBayes	The Naive Bayes classifier using kernel density estimation over multiple values for continuous attributes, instead of assuming a simple normal distribution
2	IBk	A instance-based learning algorithm. It uses a simple distance measure to find the training instance closest to the given test instance, and predicts the same class as this training instance
3	KStar	The K instance-based learner using all nearest neighbors and an entropy-based distance
4	DecisionStump	Building and using a decision stump, but it is not used in conjunction with a boosting algorithm
5	J48	Decision tree induction, a Java implementation of C4.5
6	RandomForest	Constructing random forests for classification
7	DecisionTable	A decision table learner
8	JRip	A propositional rule learner - a a Java implementation of Ripper. It repeats incremental pruning to produce error reduction

Table 2. The accuracies of best individual classifier, best ensemble classifiers constructed by Dempster's rule, Smets's rule and the cautious conjunctive rules on MP3 and Movie-Review (MR) datasets

Datasets	Best individual	Demspter's rule	Smets's rule	Cautious rule
MP2	79.52%	81.99%	82.32%	81.74%
MR	75.17%	77.61%	76.49%	76.57%
Av	77.34%	79.8%	79.41%	79.16%

rules, all combinations of eight classifiers were exhaustively permuted. In other words, we first combine any two classifiers, denoted by $C2$, and combine the resulting combination of two classifiers with a third classifier, denoted by $C3$, and the result with a fourth classifier, denoted by $C4$, until combine all eight classifiers, denoted by $C8$.

Table 1 presents the accuracies of the best individual and best ensemble classifiers constructed by the three combination rules on the two datasets. On average, the best accuracies increase by 2.5% compared with the best individual on the MP3 dataset, and increase by 1.72% on Movie-Review. These give a totally averaged 2.1% increase on these datasets. The details of each of the experimental results are depicted below.

Figure 1 shows the averaged accuracies of different ensemble classifier groups on MP3, where $C1$ on the x-axis represents the averaged accuracy of eight classifiers, $C2$ represents an averaged accuracy of all accuracies of the ensemble classifiers that are comprised of two classifiers by using three combination rules respectively, $C3$ represents an averaged accuracy of all accuracies of the ensemble classifiers made up of three classifiers, and so forth. This figure illustrates

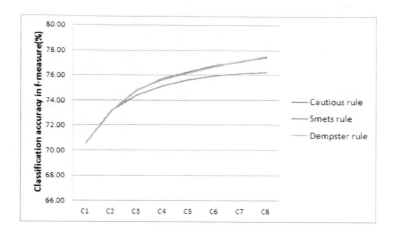

Fig. 1. Averaged accuracies of different groups combination of classifiers using the three combination rules over the MP3 dataset

that the averaged accuracies of different groups of ensemble classifiers gradually increase with adding more classifiers into classifier ensembles. It can be seen that the averaged performance of the combined classifiers by Smets's rule is very similar to that by Dempster's rule, and the ensemble classifiers made by both rules outperform those constructed by the cautious rule. Specifically compared with the averaged accuracy of eight base classifiers, the averaged accuracy of all ensemble classifiers made by the cautious rule is 4.69% better, by Dempster's rule 5.32% better and by Smets's rule it is 5.36% better.

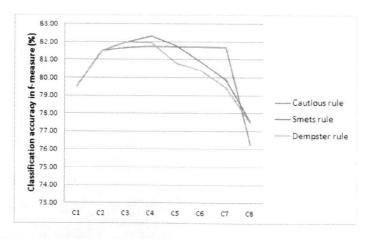

Fig. 2. The best accuracies of different groups combination of classifiers using the three combination rules over the MP3 dataset

Figure 2 shows the accuracies of the best ensemble classifiers among the respective groups of ensemble classifiers on MP3, where $C1$, $C2$, ..., $C8$ on the x-axis represent the best individual accuracies among each of the ensemble classifier groups, respectively, which are contrast with the averaged accuracies shown in Figure 1. Unlike the trends embodying in Figure 1, the best accuracies increase from the combination of two classifiers to four classifiers and then drop down with including more classifiers into ensemble classifiers. It can be found that the ensemble classifiers made by Smets's rule perform best, Dempster's rule perform in the second position and both rules outperform the cautious rule. The best accuracy drawn from all the ensemble classifiers made by Smets's rule is 2.8% better than the accuracy of best individual classifier, and the accuracy of combined classifier by Dempster's rule is 2.47% better than that of the best base classifiers, the cautious rule is 2.22% better than the best individual classifier.

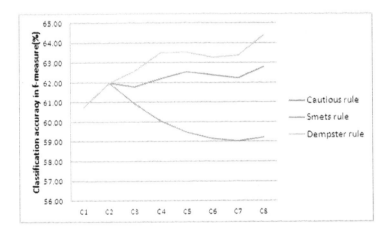

Fig. 3. Averaged accuracies of different groups combination of classifiers using different combination rules over the Movie Review dataset

Figure 3 presents a similar analysis as depicted in Figure 1 on Movie-Review instead. Compared with the analysis results with those in Figure 1, the general performance trend of the ensemble classifiers built by Dempster's and Smets's rules increases with some fluctuations, which is broadly similar to that reflected on MP3, that is the averaged accuracies increase with more classifiers being combined. However the trend of the ensemble classifiers made by the cautious rule is opposite to what the cautious rule conducts on MP3. Specifically the averaged accuracy of all ensemble classifiers constructed by the cautious rule is 0.79% less than that of the averaged eight base classifiers, but by Smets's rule it is 1.51% better and by Dempster's rule it is 2.46% better than that of the averaged eight base classifiers. Additionally, the difference margins of the averaged accuracies on Movie-Review is smaller than those on MP3.

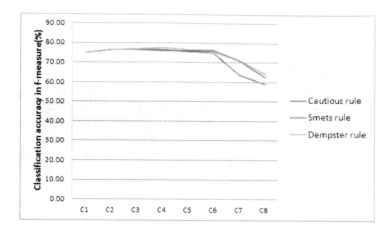

Fig. 4. The best accuracies of different groups combination of classifiers using different combination rules over the Movie Review dataset

In correspondence with the results shown in Figure 2, Figure 4 shows the best accuracies among each of the ensemble classifier groups denoted on the x-axis for the Movie-Review dataset. The performance trends of the ensemble classifiers made by the three combination rules are roughly embodied in a decreasing manner. The best accuracies achieved are among the combinations of two or three classifiers cross three groups of ensemble classifiers made by the three combination rules. From the experimental results, it is found that Dempster's rule performs best, the accuracy of best ensemble classifier among seven groups of ensemble classifiers is 2.45% better than that of the best individual classifier, for Smets's rule and the cautious rule they are 1.33% and 1.4% better than the best individual classifier, respectively.

6   Conclusions

In this paper we propose to use the triplet function to represent polarity sentiment classification outputs and tailor the formulas developed for triplet functions to combining multiple triplets in the form of simple support functions by using Dempster's rule of combination, Smets's rule and the Cautious rule. The evaluation demonstrates that Smets's rule performs better than Dempster's and the Cautious rules on the MP3 dataset, whereas Dempster's rule performs better than Smets's and the Cautious rules on Movie-Review, but on average Dempster's rule performs the best over these two datasets. From a further examination on the characteristics of MP3 and Movie-Review, it is noticed that the former consists of more instances than the latter, but the latter is a balanced class dataset being made up of the same number of positive and negative instances. Therefore these results conjecture that the ensemble classifiers made by Dempster's rule prefer more over balanced datasets than unbalanced datasets. More analysis results and theoretical aspects can be refereed to the sister paper in [14].

References

1. Feldnan, R.: Techniques and Applications for Sentiment Analysis. Communications of the ACM 56(4), 82–89 (2013)
2. Pang, B., Lee, L., Vaithyanathan, S.: Thumbs up? Sentiment Classification using Machine Learning Techniques. In: ACL 2002 Conference on Empirical methods in Natural Language Processing, vol. 10, pp. 79–86. Association for Computational Linguistics, Morristown (2002)
3. Jijkoun, V., de Rijke, M., Weerkamp, W.: Generating Focused Topic-specific Sentiment Lexicons. In: Annual Meeting of the Association for Computational Linguistics (ACL 2010), pp. 585–594 (2010)
4. Hu, M., Liu, B.: Mining and summarizing customer reviews. In: 2004 ACM SIGKDD International Conference on Knowledge Discovery and Data Mining, pp. 168–177. ACM, New York (2004)
5. Kim, S., Hovy, E.: Determining the Sentiment of Opinions. In: The 20th International Conference on Computational Linguistics Association for Computational Linguistics, Morristown, NJ, USA (2004)
6. Blitzer, J., Dredze, M., Pereira, F.: Biographies, Bollywood, Boom-boxes and Blenders: Domain adaptation for sentiment classification. In: ACL 2007 (2007)
7. Li, S., Huang, C., Zong, C.: Multi-domain Sentiment Classification with Classifier Combination. Journal of Computer Science and Technology 26(1), 25–33 (2011)
8. Dredze, M., Crammer, K.: Online Methods for Multi-Domain Learning and Adaptation. In: EMNLP 2008 (2008)
9. Burns, N., Bi, Y., Wang, H., Anderson, T.: Sentiment Analysis of Customer Reviews: Balanced versus Unbalanced Datasets. In: König, A., Dengel, A., Hinkelmann, K., Kise, K., Howlett, R.J., Jain, L.C. (eds.) KES 2011, Part I. LNCS, vol. 6881, pp. 161–170. Springer, Heidelberg (2011)
10. Bi, Y., Guan, J.W., Bell, D.: The combination of multiple classifiers using an evidential approach. Artificial Intelligence 17, 1731–1751 (2008)
11. Shafer, G.: A Mathematical Theory of Evidence, 1st edn. Princeton University Press, Princeton (1976)
12. Smets, P., Kennes, R.: The transferable belief model. Artificial Intelligence 66, 191–243 (1994)
13. Denoeux, T.: Conjunctive and disjunctive combination of belief functions induced by nondistinct bodies of evidence. Artificial Intelligence 172(2-3), 234–264 (2008)
14. Bi, Y.: Evidential Fusion for Sentiment Polarity Classification. In: The 3rd International Conference on Belief Functions, Oxford, UK (2014)
15. Witten, I.H., Frank, E.: Data Mining: Practical machine learning tools and techniques, 3rd edn. Morgan Kaufmann, San Francisco (2011)
16. Kim, S., Pantel, P., Chklovski, T., Pennacchiotti, M.: Automatically assessing review helpfulness. In: The Conference on Empirical Methods in Natural Language Processing (EMNLP 2006), Sydney, Australia, pp. 423–430 (2006)
17. Internet Movie Database (IMDb) archive, http://reviews.imdb.com/Reviews/

Relating the Opinion Holder and the Review Accuracy in Sentiment Analysis of Tourist Reviews

Mihaela Colhon[1], Costin Bădică[2], and Alexandra Şendre[2]

[1] Computer Science Department, University of Craiova
Str. A. I. Cuza 13, 200585 Craiova, Romania
[2] Computer and Information Technology Department, University of Craiova
Bvd. Decebal 107, 200440 Craiova, Romania

Abstract. In this paper we propose a sentiment classification method for the categorization of tourist reviews according to the sentiment expressed. We also give the results of the application of our sentiment analysis method on a real data set extracted from the AmFostAcolo tourist review Web site. In our analysis we were focused on investigating the relation between the opinion holder and the accuracy of the review sentiment with the review score. Based on our initial experimental results we concluded that specific characteristics of the opinion holder, like for example his or her reputation, might relate to the accuracy of the opinions expressed in his or her reviews.

Keywords: sentiment analysis, natural language processing, intelligent tourism.

1 Introduction

A great interest was shown during the last 15 years to the application of intelligent information technologies for the development of intelligent or smart tourism business [1, 2]. Recent developments in semantic technologies, dynamic pricing, information extraction, recommender systems and sentic computing are at the forefront of the field. A lot of efforts are currently spent on the application of text mining and natural language processing for improving the quality of tourism information services. Tourists will benefit of advanced IT systems for knowledge and information management to assist them in making decisions with less effort and in shorter time.

There are so many information sources containing tourist reviews and opinions about tourist destinations like for example: post-visit experiences, tourist advertisements, descriptions of tourist attractions, tourist highlights and advices, recommendations, photos, etc, addressing various aspects like accommodation, trips, historical places, landscape, sightseeing, food, shopping, entertainment, local attractions, a.o. Tourist information is most often presented as reviews or comments expressed in natural language that describe customer opinions or experiences about various tourist destinations. This information is usually poorly structured, can be more or less focused on a tourist entity or aspect, and can be multi-lingual, Collecting, aggregating and presenting it in a meaningful way can be very difficult by posing cognitive challenges to the users, as well as technical challenges to the computational methods employed.

R. Buchmann et al. (Eds.): KSEM 2014, LNAI 8793, pp. 246–257, 2014.

Following [3], "sentiment analysis or opinion mining is the computational study of people's opinions, appraisals, attitudes, and emotions toward entities, individuals, issues, events, topics and their attributes.". Typical usage scenarios of opining mining are: (1) detection of good or bad aspects about a certain target object that needs to be evaluated, (2) detection of sudden changes in sentiment, or (3) automatic interpretation of large amounts of opinionated data.

Our initial task was to extract and classify tourist reviews according to the sentiment expressed in three categories: positive, negative or neutral. We employed an unsupervised sentiment classification method and we present the results of its application results to tourist reviews. We focused on a real data set that we extracted from AmFostAcolo[1] – a Romanian Web site having a similar purpose as IveBeenThere[2].

We have developed an enhanced term-counting method built primarily on exploiting parsing techniques from natural language processing for detecting dependency links between the words of a text, as well as considering the contextual valence shifters [4]. Our proposed term-counting method has the advantage that it does not require training, so it can be applied to reviews where training data is not available.

In this paper we present the results of the application of our sentiment analysis method on a real data set extracted from the AmFostAcolo tourist review Web site. In particular, our analysis was focused on investigating the relation between the opinion holder and the accuracy of the review sentiment with the review score. Based on our initial experimental results we concluded that specific characteristics of the opinion holder, like for example his or her reputation (expressed as user class on AmFostAcolo Web site), relate to the accuracy of the opinions expressed in his or her reviews.

The paper is structured as follows. We start with an overview of related works in Section 2, then in Section 3 we present the data set preparation and preprocessing. We follow in Section 4 with the introduction of our sentiment analysis method. Then we present and discuss the experimental results of the application of our method on the real data set extracted from AmFostAcolo Web site. In the last section we conclude and point to future research and developments.

2   Related Works

Sentiment analysis and opinion mining, in particular focused on the analysis of customer reviews in the tourism industry, attracted a lot of research interest during the last decade. The efforts on mining opinions can be roughly divided into two directions: sentiment classification and sentiment related information extraction. The first task is to identify positive and negative sentiments from a text, while the second focuses on extracting the parts composing a sentiment text. Most of the researches reported in the literature were devoted to results obtained on the sentic analysis of documents expressed in widely spread natural languages, for example English, Chinese and Spanish, while less efforts were dedicated to other languages, like for example Romanian [5, 6].

A number of works take the effort of firstly creating an opinion lexicon [7–9]. A simple method is to manually decide the degree of positivity and negativity of words,

---

[1] http://amfostacolo.ro/

[2] http://www.ive-been-there.com/

and then to define a procedure to calculate the sentiment for each sentence or whole text based on the values of the words [4]. This can of course be tricky because it is difficult to set a sentiment value on a simple word, since many words can be both positive and negative depending either on the context or situation or on their Part-Of-Speech (POS). This is a problem of text mining rather than pure opinion mining. Considering word phrases and some forms of word disambiguation, as well as capturing syntactic relations and dependencies between words, can improve the results of sentiment analysis.

Reference [10] reports results obtained by applying machine learning classifiers for opinionated sentence detection and opinion target extraction based on manually annotated blog articles in the tourism domain. So, this work is more related to opinion-related information extraction, rather than opinion polarity detection. Moreover, the approach is different, as it involves manual text annotation which is a tedious activity.

Reference [11] describes the Nebular system that addresses the problem of classifying tourist reviews extracted from Web 2.0 sites related to tourism, as well as the detection of their sentiment polarity – positive or negative.

Reference [12] presents the BESAHOT system that helps hoteliers by providing them with summaries of the textual comments posted by users about their hotel(s) on the web. This system performs a mixed processing of the analyzed comments for text segmentation, statistical polarity detection of text segments and linguistic information extraction of review topics and their aspects.

Reference [5] introduces Sentimatrix – a system for extracting opinions with regard to named entities. The system uses language resources for name entity recognition and extraction, as well as for sentiment identification.

Reference [13] describes the Post-via system that combines semantic technologies and recommender systems for improving Customer Relationship Management to enhance tourist loyalty by capturing and management of after-visit tourist experiences.

Reference [14] is focused on enhancing opinion mining of tourist reviews with intuitive visualizations using Google Maps. The paper is more focused on human-computer interaction aspects, rather than the computational methods employed.

References [15, 16] introduce the OpinionZoom system for the analysis of opinions expressed by tourists on various Web sites. The system was designed to experiment with aspect-oriented opinion mining algorithms inspired by Liu's approach [3]. The authors also proposed an extension of Liu's approach to be applied to the tourism domain.

Last but not least, paper [6] addresses opinion identification from news and social media. The proposed method uses machine learning and a manually annotated corpus.

3   Data Set: Preparation and Preprocessing

3.1   Data Set Preparation

The initial source that we chose as basis for building our data set was the AmFostAcolo Web site. It provides a large semi-structured database with information describing post-visit tourist reviews about a large variety of tourist destinations covering specific aspects of accommodation, as well as general impressions about geographical places.

There are basically two different formats in which reviews are given on tourist sites. The format "pro and cons" describes separately the positive ("pro") and negative

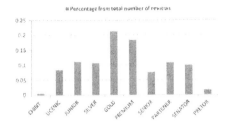

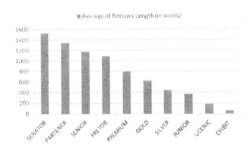

Fig. 1. AmFostAcolo – Percentage of the reviews number posted by users of a given type from the total number of reviews

Fig. 2. AmFostAcolo – Average number of words per review for each user type

("cons") opinions (Booking.com[3] uses this format). The other "free format" does not separate pros and cons, and the reviewer can write freely his or her comments. This format is used by several sites, including AmFostAcolo.

The information of the data source can be conceptualized as a tree-structured index according to the destination, region, section and location. Most often a destination represents a country, for example Romania, while sometimes it can be a continental region including several countries, for example Caucaz or Central America. Each destination contains several regions. For example, a region of Romania is Oltenia. A section most often represents a locality (for example Craiova or Runcu) or a more general category of locations, like for example "MĂNĂSTIRI din Oltenia" (En. "Monasteries of Oltenia"). A location can represent an accommodation unit or general impressions about the locality and surroundings, for example "Descoperă zona Runcu (Gorj)" (En. "Discover Runcu area (Gorj)"). Finally, each location is a container of tourist impressions or reviews written by users registered at AmFostAcolo. Each user has associated a trust score that is calculated based on his or her activity and feedback received on the site.

Our data set contains: 423 male users and 662 female users; 2521 reviews; 45 countries destinations; 161 regions among which there are 16 country subregions and 145 other regions (not country subregions); 529 sections among which 489 localities (cities, towns or villages) and 40 other sections (not localities); 1420 tourist locations among which 534 accommodation units (cottages, pensions, hotels, houses or villas) and 886 sections (not accommodation units).

When crawling AmFostAcolo, 2521 reviews were extracted and the accuracy of the original reviewers scores for each topic of interest was calculated based on a positive / negative / neutral term-counting method for sentiment analysis. From the total number of reviews we found 2280 positive reviews, 234 negative reviews and 7 neutral reviews.

The data set contains also information about the users that posted reviews on the AmFostAcolo Web site. This information contains: the user identifier, the age interval (one of 20-30 years, 30-40 years, 40-50 years, and 50-60 years), sex, geo-location, and user type (also called "statut" (in Romanian).

---

[3] http://www.booking.com/

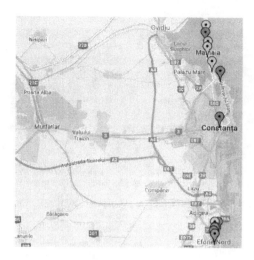

Fig. 3. Hotels on Romanian seaside with positive (in green), negative (in red) or neutral (in yellow) reviews

Very interesting is the *user type* that actually represents a qualitative score that in our opinion characterizes the user reputation [17]. It is determined by a method that is succinctly described on the AmFostAcolo Web site. The method is based on defining a user activity portfolio (including posting tourist impressions, uploading holiday photos, posting replies to comments, and answering questions posted by other users) for which the user receives a number of points. Clearly, the larger and better is the user portfolio, the higher is the number of points received by the user. Based on the number of points, the user is assigned into one of the following ordered set of user types: $UCENIC < JUNIOR < SILVER < GOLD < PREMIUM < SENIOR < PARTENER < SENATOR < PRETOR$.

Moreover, based on the data extracted from the Web site, we also identified another user type called $CHIBIT$ and we suspect that this is a transition type. The user will receive the lowest type $UCENIC$ only after he or she acquired a number of points larger than a given minimum threshold. Figure 1 displays the amount of reviews in percents that were posted by each set of users of a given type in our data set while Figure 2 presents the partitioning of reviews depending on their words size. It can be seen that users of highest types ($SENIOR$, $PARTENER$, $SENATOR$, $PRETOR$) posted the largest reviews (more than 1000 words).

We have also developed an intuitive visualization of the data set items. For example, Figure 3 presents the result of visualizing hotels on the Black Sea Coast region according to the explicitly provided user review scores.

3.2 Data Set Preprocessing

The text content of the reviews extracted from the Web site was preprocessed and organized as a corpus by adopting a simplified form of the XCES standard [18]. The words of the resulted corpus were annotated with syntactic data as it will be further detailed.

```
<S id="1">
  <W DEPREL="det." HEAD="2" ID="1" LEMMA="un" POS="ARTICLE">un</W>
  <W DEPREL="ROOT" HEAD="0" ID="2" LEMMA="hotel" POS="NOUN">hotel</W>
  <W DEPREL="a.subst." HEAD="2" ID="3" LEMMA="cu" POS="ADPOSITION">cu</W>
  <W DEPREL="det." HEAD="5" ID="4" LEMMA="un" POS="ARTICLE">o</W>
  <W DEPREL="prep." HEAD="3" ID="5" LEMMA="grădină" POS="NOUN">grădină</W>
  <W DEPREL="a.adj." HEAD="5" ID="6" LEMMA="superbă" POS="NOUN">superbă</W>
</S>
```

Fig. 4. Corpus snapshot

By running an automatic processing chain that includes sentence segmentation, to-kenisation, POS-tagging and lemmatization [19], the boundaries of the reviews sen-tences were marked and each word got attached its part of speech data and lemma. We need lemmas as the method of counting positive and negative terms requires the mapping of comments words into their base forms. The word tokens of the reviews sentences were automatically annotated for their head-words and the corresponding de-pendency relations by running a Dependency Parser[4].

The corpus resulted from preprocessing of the text extracted from the tourist reviews contains 91966 sentences and 2242841 word tokens. The corpus notations are expressed in XML, as shown in Figure 4. Basic layers of annotation include: borders of each sen-tence (marked as <S></S> elements, and identified by unique IDs) and words (marked as <W></W>) and including unique IDs, part of speech (POS), lemma (LEMMA) and de-pendency information (dependency relation DEPREL and head - HEAD).

4   Sentiment Analysis Use Case

Here we present our proposed sentiment analysis method. We conclude with discussing experimental results obtained on a realistic data extracted from AmFostAcolo.

4.1   The Sentiment Analysis Method

In our approach we rely on lexical information combined with the syntactical infor-mation obtained by running a Romanian automatic pre-processing chain available as a Web Service[5] and a Dependency Parser on the textual reviews contents. However, for this study, lexical information implies a list of keywords that must be lookup in the re-views and based on which, the analysis can be implemented, as well as a list of negative and positive words upon which the sentiment value of a review can be "calculated".

The features that contribute to increase the accuracy of sentiment analysis are the words in the lists of positive and negative terms [8]. The idea of counting positive and negative terms or expressions was firstly proposed by Turney [21]. We note that the term-counting method has the advantage that it does not require training, so it can be applied to reviews where training data is not available.

---

[4] At the Faculty of Computer Science of the "Alexandru Ioan Cuza" University of Iaşi, a De-pendency Treebank for the Romanian language was built by the NNLP Group [20]. We have used this resource in the training and evaluation stages of the Dependency Parser.

[5] The WSDL specifications are available at
http://nlptools.infoiasi.ro/WebPosRo/

We used an English lexicon of positive and negative terms[6] that was firstly reported in [9]. In order to be used on Romanian texts, we translated it using an English-Romanian dictionary[7]. By removing duplicates and by taking care that the intersection of the two lists of terms must be empty (otherwise the senses cancel each-other), we obtained 1943 terms with positive senses and 4678 terms with negative senses. Most of them are adverbs and adjectives in base form with no plurals and other inflected forms.

We augment the considered term-counting method by taking into account contextual valence shifters. Valence shifters are terms that can change the semantic orientation of another term, by increasing or decreasing its qualification or even by changing the sentiment of positive or negative terms in the sentence into its contrary.

As most of the terms used to define the sentiments in human communication are adjectives, we must pay attention to the modifiers that can affect the qualification expressed by this kind of words. The morphological means used for intensification are adverbs but not necessarily. Following [4], we consider as modifiers the negatives and intensifiers terms. In [8] also diminishers were taken into account (the language under their study is English), but we limited the valence shifters to negations and intensifiers as we did not get any improvements by taking also the diminishers into account.

Negations are terms that complement the sentiment of the word they modify. Examples are: "nu" (En. "not"), "nici" (En. "nor", "neither"), "niciodată" (En. "never"), "nicidecum" (En. "noway"), "deloc" (En. "none"), "nimeni" (En. "nobody"), "niciunul" (En. "none"), "niciun" (En. "no"), "nimic" (En. "nobody"), "nicăieri" (En. "nowhere").

Intensifiers increase the intensity of a positive/negative term. Examples of intensifiers: "super", as in "super frumos" (En. "super nice"), "extrem de" (En. "highly"), "extraordinar de" (En. "extraordinary"), etc.

Our proposed approach is based on counting the positive and negative terms in a review that are related to *aspects* of the object under discussion (these *aspects* or *facets* are presented in what follows). In this approach, a review is considered positive if it contains more positive than negative terms, and negative if there are more negative than positive terms. A review is neutral if it contains (approximately) equal numbers of positive and negative terms.

4.2 Knowledge Representation

In many applications, sentiment opinions are related to entities, but more frequently to particular aspects of the entities under discussion. For example, a negative review does not necessarily mean that everything mentioned in the comment is negative. There can be some positiveness regarding a particular aspect or facet. Likewise for a positive review. To obtain such detailed results we have to deepen our analysis at the sentence level and extract the relevant features.

We analyzed and evaluated the tourist reviews based on the aspects that are usually debated in such comments: *accommodation, kitchen, services*, etc. We performed an accurate sentiment analysis of the tourist opinions based on their specific aspects. The goal of this level of analysis was to discover general sentiments on entities based on

---

[6] The lexicon is currently maintained at address
http://www.cs.uic.edu/~liub/FBS/sentiment-analysis.html
[7] The dictionary can be found at www.mcolhon.ro

the sentiments of their specific aspects. The more positive (or negative) aspects has an entity, the stronger will be the positive (or negative) general opinion for that entity.

For example, the sentence "The hotel is good, but the restaurant is poor" evaluates two aspects, the accommodation quality and the kitchen quality, of the entity "hotel". The sentiment on hotel quality is positive, but the sentiment on its restaurant is negative. The accommodation and kitchen qualities of the entity under discursion (that is, the hotel) are called aspects or facets. Because we have a positive opinion and a negative opinion, the overall sentiment opinion on the hotel entity is neutral.

In the evaluation process we exploit the fact that all the reviews on the AmFostAcolo Web site can be uploaded only by registered users. Thus, we could focus also on analyzing the possible correlation between the source of an opinion, i.e. the user itself and the accuracy of the reviews posted by this user (by comparing the scores obtained in the evaluation of the proposed SA method with the user provided scores).

Following [3] in our proposed sentiment analysis method, the entities are formalized as tuples of the form (*entity, facet, sentiment, opinion holder*) where:

- an *entity* can be an accommodation unit (hotels, apartments, villas, etc.), as well as a tourist place or region. According to the *entity* type, some facets can not be used to properly describe an entity aspect[8].
- the *facets* under discussion on AmFostAcolo are *services, accommodation, kitchen, landscape, entertainment* and *GENERAL*[9]. The reviewers can grade them separately.
- *sentiment* can be *positive, negative* or *neutral*.
- *opinion holder* denotes the *user_id* that posted the review.

With this representation, a structured opinion analysis about entities and their aspects can be produced, which turns unstructured text to structured data which can be further used for all kinds of qualitative and quantitative analysis.

The evaluation of the opinion sentiment towards a specific facet rather than the whole review is not an easy task. To determine a sentiment towards a specific facet of an entity, we have to find correlations between the corresponding facet and the textual content of the review. According to this, we built five sets of so-called *seeds*, each set being determined to belong to a certain facet enumerated before.

We relate the positive and negative terms with the seeds under investigation by means of the grammatical relations generated by the Dependency Parser on the reviews sentences and by investigating the terms present around seeds using the following context window-based approach. We lookup to match the reviews words with the selected seeds. For each occurrence of seed s in the text:

- We select all positive and negative terms that are in a dependency relation with s.
- By applying the *bag-of-words* principle, we consider a fixed-size context-window around the seed[10]. We select all the positive and negative terms within the window that are not in a dependency-based relation with other seeds.

---

[8] For example, hotels as customer-based units are greatly described by the *services* scores. This is not the case for tourist regions for which, usually, the *services* scores have 0 as value.

[9] The *GENERAL* aspect denotes the entity itself as a whole and not a particular facet.

[10] The window is considered to include maximum 8 words immediately before and after the seed, being properly resized not to exceed the seed's sentence.

- The positive/negative score for s is set to the number of selected positive/negative terms. Then we map the obtained sentiment scores of s to the corresponding facet.

One argument to support the correctness of the scores resulted by the application of this proposed Sentiment Analysis algorithm (SA in what follows) is that the positiveness and negativeness of the terms found in reviews are considered only if they affect the facets' seeds of the evaluated entities.

4.3   Results and Discussions

On the AmFostAcolo Web site, the tourist entities under discussion (hotels, regions, etc.) are described by five facets which receive scores ranging from 0 (maximum negative) up to 100 (maximum positive). Even though it is not explicitly specified on the site, we suspect that sometimes the 0 score assigned to a facet in a review means that its text does not describe anything related to that facet rather than assigning the maximum negative opinion to the facet. But there is another issue related to this score: we suspect that the default value for a score is 0 and if the user forgets to explicitly set it, the score remains assigned with the erroneous 0 value. The 0 scores of reviews that do not match with 0 scores obtained by the application of our SA algorithm (i.e. our algorithm obtains nonzero positive and/or negative values) will be named in what follows 0-$scores$.

We evaluated the resulted sentiment analysis scores by dividing the data set into positive, negative and neutral rankings. For a scale from 0 to 100, we consider that:
- A negative comment on a certain topic means that the score given by the user is between $[0, 40)$. In order to be considered correct, our algorithm must obtain a greater negative score than the positive one: negative score > positive score
- A positive comment on a certain topic means that the score given by the user is between $(60, 100]$. In order to be considered correct, our algorithm must obtain a greater positive score than the negative one: positive score > negative score;
- A comment is considered neutral if the user's score is between $[40, 60]$. In order to be considered correct, our algorithm must obtain approximately equal positive and negative scores, which means | positive score - negative score | $\leq threshold$[11].

Evaluation is performed by comparing the initial user reviews scores with the scores calculated by our SA algorithm. We evaluate all the reviews and separately the reviews about accommodation units (called in what follows as *hotel reviews*) in terms of the standard measures of *Precision* (P), *Recall* (R) and $F - score$.

In what follows $\mathcal{R}$ is the set of reviews, and $\mathcal{R}_{0-scores-facet}$ is the set of reviews with 0-$scores$ for the $facet$ category. The evaluations are considered in several cases:

– upon each specific $facet$s:

$$P(facet) = \frac{matched(\mathcal{R}_{facet-related})}{|\mathcal{R}_{facet-related} \setminus R_{0-scores-facet}|} ; R(facet) = \frac{matched(\mathcal{R}_{facet-related})}{|\mathcal{R}|}$$

where $\mathcal{R}_{facet-related}$ is the set of facet-related reviews and $facet \in \{$ services, accommodation, kitchen, landscape, entertainment $\}$.

---

[11] Following [8], for the neutral comments we allow a maximum *threshold* value for the difference between positive and negative scores instead of asking for a strict equality between these values (in this approach we take *threshold* = 1).

Table 1. Evaluation of reviews per facets and per sentiments

Facets	Precision			Recall			F-score
accommodation	0.90			0.65			0.75
entertainment	0.70			0.58			0.63
kitchen	0.81			0.65			0.72
landscape	0.75			0.68			0.70
services	0.72			0.61			0.66
Reviews	**Precision**			**Recall**			**F-score**
	Corrects	Total	Score	Corrects	Total	Score	
Positive	1981	2280	0.87	1981	2280	0.87	0.87
Negative	11	18	0.61	43	234	0.18	0.27
Neutral	2	7	0.29	2	7	0.29	0.29
ALL	1994	2305	0.87	2026	2521	0.80	0.83

Table 2. Evaluation of reviews facets upon user types

User type	Precision			Recall			F-score
	Corrects	Total	Score	Corrects	Total	Score	
CHIBIT	27	74	0.36	27	75	0.36	0.36
UCENIC	667	993	0.67	667	1025	0.65	0.66
JUNIOR	946	1256	0.75	946	1355	0.70	0.72
SILVER	884	1166	0.76	884	1300	0.68	0.72
GOLD	1725	2209	0.78	1725	2615	0.66	0.72
PREMIUM	1392	1773	0.79	1392	2250	0.62	0.70
SENIOR	527	682	**0.77**	527	910	**0.58**	**0.67**
PARTENER	789	963	**0.82**	789	1315	**0.60**	**0.69**
SENATOR	704	838	**0.84**	704	1215	**0.58**	**0.68**
PRETOR	105	132	**0.80**	105	195	**0.54**	**0.64**

– upon each sentiment category

$$P(sent) = \frac{matched(\mathcal{R}^{sent})}{|\,\mathcal{R}^{sent} \setminus \mathcal{R}_{0-scores-GENERAL}\,|}; \ R(sent) = \frac{matched(\mathcal{R}^{sent})}{|\,\mathcal{R}^{sent}\,|}$$

where $sent \in \{pozitive, negative, neutral\}$. We have:

$$P(ALL) = \frac{matched(\mathcal{R})}{|\,\mathcal{R} \setminus \mathcal{R}_{0-scores-GENERAL}\,|}; \ R(ALL) = \frac{matched(\mathcal{R})}{|\,\mathcal{R}\,|}$$

because $\mathcal{R} = \mathcal{R}^{pozitive} \cup \mathcal{R}^{negative} \cup \mathcal{R}^{neutral}$. In this case, the facet under investigation is *GENERAL*.

– *F-score* in all cases is calculated with the formula: $F\text{-}score = (2 \times Precision \times Recall)/(Precision + Recall)$.

where $matched(\mathcal{R})$ is a function that returns the number of scores obtained by the *S A* method by processing the reviews of $\mathcal{R}$ for which the scores obtained by our algorithm match the reviewers scores provided by users.

Generally, we achieved a *Precision* of 87% for positive reviews, i.e. a very good score. Our algorithm performs slightly worse for negative reviews ($\approx$ 60%) and neutral reviews ($\approx$ 20%). Nevertheless, the overall *Precision* score remains 87% as the number of positive comments is larger compared to the negative and neutral comments.

It is well known that generally NLP is a difficult task. Moreover, analyzing texts for the facets level of detail is harder. Still, if we look in Table 1 at the *Precision* scores of the facets classification one can note that it varies from 70% to 90% (i.e. high values), while values of *Recall* are much smaller (from 58% to 68%). The big difference between *Precision* and *Recall* means that there are many reviews with 0-scores for facets that do not match with the scores obtained by the algorithm for those facets.

We grouped the evaluation scores according to the *user type* of the opinion holder. We noticed that the highest *Precision* scores are obtained for the reviews of the users classified higher in the hierarchy: *SENIOR, PARTENER, SENATOR* and *PRETOR* while the smallest scores matched with the "beginners" – users of *CHIBIT* type.

Another interesting observation results by analyzing the scores of Table 2: the reviews posted by more reputable users got also the smallest *Recall* scores. We believe that the difference between *Precision* and *Recall* scores can be explained by the fact that the largest reviews posted on the Web site correspond to the categories of more reputable users: *SENIOR, PARTENER, SENATOR* and *PRETOR* (as it is clearly illustrated in Figure 2 from Section 3.1). This is the reason why the *Recall* scores are so small: more reputable users tend to provide larger descriptions in their comments with many positive or negative terms that are not well related to the facets seeds. This type of behavior can be interpreted as the result of these users getting more "freedom" in writing their comments, according to their higher reputation value.

5   Conclusion and Future Work

In this paper we investigated the application of opinion mining on real data extracted from a travel review site. We presented the results of a sentiment analysis algorithm for the textual reviews contents at several levels of details (facets). In our analysis we focused on investigating the relation between the opinion holder and the accuracy of the review sentiment with the review score. Our initial experimental results showed that specific characteristics of the opinion holder, like his or her reputation, can be related to the accuracy of the opinions expressed in user reviews. Note that our proposed approach is domain independent. The utilized sets of features can be configured for use with data sets from other application domains. On the short term we plan to strengthen our results by investigating the correlation between other user characteristics (like sex, age interval, and geo-location) and the accuracy of the review sentiment. Moreover, we intend to investigate the application of our proposed opinion mining techniques to the analysis of changes and evolution of opinions.

References

1. Stabb, S., Werther, H., Ricci, F., Zipf, A., Gretzel, U., Fesenmaier, D.R., Paris, C., Knoblock, C.: Intelligent systems for tourism. IEEE Intelligent Systems 17(6), 53–66 (2002)
2. Becheru, A.: Agile development methods through the eyes of organisational network analysis. In: Akerkar, R., Bassiliades, N., Davies, J., Ermolayev, V. (eds.) 4th International Conference on Web Intelligence, Mining and Semantics (WIMS 14), p. 53. ACM (2014)
3. Liu, B., Zhang, L.: A survey of opinion mining and sentiment analysis. In: Aggarwal, C.C., Zhai, C. (eds.) Mining Text Data, pp. 415–463. Springer US (2012)
4. Polanyi, L., Zaenen, A.: Contextual valence shifters. In: Shanahan, J.G., Qu, Y., Wiebe, J. (eds.) Computing Attitude and Affect in Text: Theory and Applications. The Information Retrieval, vol. 20, pp. 1–10. Springer (2006)

5. Gînscă, A.-L., Boroş, E., Iftene, A., Trandabăţ, D., Toader, M., Corîci, M., Perez, C.-A., Cristea, D.: Sentimatrix: Multilingual sentiment analysis service. In: Proceedings of the 2nd Workshop on Computational Approaches to Subjectivity and Sentiment Analysis (WASSA 2011), pp. 189–195. Association for Computational Linguistics (2011)
6. Cardei, C., Manisor, F., Rebedea, T.: Opinion mining for social media and news items in romanian. In: 2nd International Conference on Systems and Computer Science (ICSCS 2013), pp. 240–245. IEEE (2013)
7. McCallum, A., Nigam, K.: Text classification by bootstrapping with keywords, em and shrinkage. In: Proceedings of ACL 1999 - Workshop for Unsupervised Learning in Natural Language Processing (1999)
8. Kennedy, A., Inkpen, D.: Sentiment classification of movie reviews using contextual valence shifters. Computational Intelligence 22(2), 110–125 (2006)
9. Hu, M., Liu, B.: Mining opinion features in customer reviews. In: Proceedings of Nineteeth National Conference on Artificial Intellgience (AAAI 2004) (2004)
10. Lin, C.J., Chao, P.H.: Tourism-related opinion detection and tourist-attraction target identification. IJCLCLP 15(1) (2010)
11. Palakvangsa-Na-Ayudhya, S., Sriarunrungreung, V., Thongprasan, P., Porcharoen, S.: Nebular: A sentiment classification system for the tourism business. In: Proc.8th Int. Joint Conference on Computer Science and Software Engineering (JCSSE 2011), pp. 293–298 (2011)
12. Kasper, W., Vela, M.: Sentiment analysis for hotel reviews. In: Proc.of the Computational Linguistics-Applications Conference, Polskie Towarzystwo Informatyczne, pp. 45–52 (2011)
13. Colomo-Palacios, R., Rodríguez-González, A., Cabanas-Abascal, A., Fernández-González, J.: Post-via: After visit tourist services enabled by semantics. In: Herrero, P., Panetto, H., Meersman, R., Dillon, T. (eds.) OTM-WS 2012. LNCS, vol. 7567, pp. 183–193. Springer, Heidelberg (2012)
14. Bjørkelund, E., Burnett, T.H., Nørvåg, K.: A study of opinion mining and visualization of hotel reviews. In: Proceedings of the 14th International Conference on Information Integration and Web-based Applications and Services (iiWAS2012), pp. 229–238. ACM (2012)
15. Marrese-Taylor, E., Velásquez, J.D., Bravo-Marquez, F., Matsuo, Y.: Identifying customer preferences about tourism products using an aspect-based opinion mining approach. In: 17th International Conference in Knowledge Based and Intelligent Information and Engineering Systems – KES 2013. Procedia Computer Science, vol. 22, pp. 182–191. Elsevier (2013)
16. Marrese-Taylor, E., Velásquez, J.D., Bravo-Marquez, F.: Opinion zoom: A modular tool to explore tourism opinions on the web. In: Proc. IEEE/WIC/ACM Int. Conf. on Web Intel.and Intel.Agent Technology 2013 Workshops, pp. 261–264. IEEE Computer Society (2013)
17. Jøsang, A., Ismail, R., Boyd, C.: A survey of trust and reputation systems for online service provision. Decis. Support Syst. 43(2), 618–644 (2007)
18. Nancy, I., Bonhomme, P., Romary, L.: Xces: An xml-based encoding standard for linguistic corpora. In: Proceedings of the Second International Conference on Language Resources and Evaluation, LREC 2000. European Language Resources Association (2000)
19. Simionescu, R.: Hybrid pos tagger. In: Proceedings of Language Resources and Tools with Industrial Applications Workshop (Eurolan 2011 Summerschool) (2011)
20. Perez, C.A.: Casuistry of romanian functional dependency grammar. In: Moruz, M.A., Cristea, D., Tufiş, D., Iftene, A., Teodorescu, H.N. (eds.) Proceedings of the 8th International Conference "Linguistic Resources and Tools for Processing of the Romanian Language" (ConsILR 2012), pp. 19–28. Alexandru Ioan Cuza University of Iaşi Publishing House (2012)
21. Turney, P.: Thumbs up or thumbs down? semantic orientation applied to unsupervised classification of reviews. In: Proceedings of the 40th Annual Meeting of the Association for Computational Linguistics (ACL 2002), pp. 417–424 (2002)

Formal Modeling of Airborne Software High-Level Requirements Based on Knowledge Graph

Wenjuan Wu, Dianfu Ma, Yongwang Zhao, and Xianqi Zhao

School of Computer Science and Engineering,
Beijing University of Aeronautics and Astronautics, Beijing, 100191
{wuwj,dfma,zhaoyw,zhaoxq}@act.buaa.edu.cn

Abstract. Airborne airworthiness certification DO-178C software release proposes a higher safety and reliability demands of airborne software. This raises great challenges to airborne software modeling and verification. In order to achieve airborne software high-level requirements objectives, we propose a formal method of modeling high-level requirements based on knowledge graph. The method gives a formal language to describe knowledge graph and constructs knowledge graph collaboratively. Then we represents high-level requirements by causal model and formal modeling of high-level functional requirements and non-functional requirements by knowledge graph. These improve the requirement traceability, namely these are helpful to trace the high-level requirements to system requirements so as to achieve high-level requirements' traceability objective that DO-178C demands. Additionally, we provide the modeling tool for domain experts to construct knowledge graph collaboratively and realize their high-level requirements modeling. We also give some high-level requirements verification. These are significant to generate safe, reliable, accurate and high-quality airborne software.

Keywords: Formal modeling, High-level requirements, Airborne software, Knowledge graph, DO-178C.

1   Introduction

With sharply increasing software scale and complexity, development of avionics software faces huge challenges including rising safety requirement, increasing verification cost and shortening time to market demands [1]. In order to ensure airborne software safety, it is necessary to provide adequate safety certification for software before put into use. The verification criteria currently used are aviation airworthiness certification standards DO-178B [2] and DO-178C [3] which are made by The Radio Technical Commission for Aeronautics, RTCA. DO-178C [3] proposes that system requirements, hardware interfaces, system architecture from system life cycle process and software development plan, software requirement standard from software plan process are inputs of the model. When

R. Buchmann et al. (Eds.): KSEM 2014, LNAI 8793, pp. 258–269, 2014.

conversion rules are determined, these inputs will be used to develop high-level requirements. The output of the process is software requirement data.

Current main modeling languages of airborne software are AADL [8], UML, UML MARTER [9] and so on. The main requirement modeling approach is use case diagram and use case description of UML. However, UML has some drawbacks [4]. First, UML lacks of process guidance. It is not a method but a modeling language and does not define process guidance. Then we cannot develop a really good system and guarantee the quality of the software only by UML. Second, UML is too complicated. Third, there is not an effective and rigorous method to verify and test software system modeled by UML. At the same time, lack of accuracy will reduce software quality. Also, requirement documents are tedious in traditional and high-level requirements traceability is hard to achieve. Thus leads workload and can't guarantee accurate traceability to meet the need of high quality software.

We apply knowledge graph approach to high-level requirements modeling of airborne software. The inputs of system requirements are perception concepts and the outputs of system requirements are actuation concepts. How to get actuation concepts from perception concepts is a black box problem. Knowledge graph turn the black box problem into white box problem and get high-level requirements. Traceability can be improved and formal modeling of high-level requirements is helpful for us to verify our high-level requirements.

This paper makes the following contributions:

a. It proposes a new formal method of high-level requirements modeling of airborne software called RMKG(Requirement Modeling based on Knowledge Graph).

b. It displays requirements traceability which is beneficial to trace the system requirements from high-level requirements .

c. Its visual knowledge graph modeling is intuitive and helpful for domain experts to communicate when constructing knowledge graph.

The rest of this paper is organized as follows. Some related work is given in Section 2. We give detailed description of knowledge graph in section 3. Section 4 describes the high-level requirements modeling and verification. Section 5 contains experimental results to evaluate our modeling tool. We draw conclusions and propose our future work in section 6.

2   Related Work

Knowledge graph is used to represent an idea, event, situation or circumstance described by a trend graph, which consists of nodes to represent concepts and links to represent the conceptual relationships which is an instrument that represents some knowledge as a way to represent the logical structure of knowledge described in natural language [5]. Nguyen-Vu Hoang proposes a representation of the knowledge on relationships existing between symbolic objects in a collection of images. They present a graph based representation of this knowledge and its associated operations and properties [6]. While we apply knowledge graph and knowledge inference to our requirements modeling. Harry S. Delugach apply conceptual graphs to acquiring software requirements. They use conceptual graphs to represent requirements knowledge, repertory grids to acquire requirements knowledge and formal concept analysis to form requirements concept [7]. But they left some problems such as requirement traceability which we'll improve in this paper.

3   Knowledge Graph

3.1   Formal Description of Knowledge Graph

Definition 1: *Causal model* summarizes the interaction between environment and computer system, identifies the relationship between the initial concept and other concepts based on the first-order logic and collections to get the architecture of a system.

Definition 2: *Causality* is defined as $< x_i, x_{i+1}, ..., x_j, y_i >$. Output y_i is calculated by internal behavior of the system based on the inputs $x_i, x_{i+1}, ..., x_j$. Such even forms a causality.

Definition 3: In airborne system, output data must be calculated by input data and intermediate data got by input data. Therefore, there exists a *causal chain* like $< x_i, x_{i+1}, ..., x_m, r >< x_{m+1}, ..., x_j, r, s >< s, t >< t, y >$ which can describe how to get output data from input data and intermediate data. The *causal chain* and *causality* $< x_i, x_{i+1}, ..., x_j, y_i >$ satisfy dependency. For each causality in the causal chain, there exists a new causal chain satisfying its dependency, or the causality can be mapped to a precise mathematical formula or algorithm.

Domain knowledge of airborne software can be expressed as $< x_1, ..., x_n, y_m >$ by mathematical formula or algorithm. x_k and y_m are concepts. $< x_1, ..., x_n, y_m >$ represents that $< x_1, ..., x_n >$ has relationship with y_m. $< x_1, ..., x_n, y_m >$ is a knowledge unit satisfying causality, that is we can get y_m from $x_1, ..., x_n$ through mathematical formula or algorithm. In airborne software domain knowledge, the attribute of things can be expressed as a concept and the relation between concepts can be expressed as a relationship. In this way, airborne software domain knowledge can be formed into knowledge graph.

Definition 4: We introduce the triple N=(C,F,R) known as a directed graph. Where C represents knowledge concept set and its attributes such as performance, safety and security. F represents rule such as mathematical formula or algorithm. R represents the knowledge relationship. The structure is shown in Figure 1.

For example, velocity $v = at$, mileage $s = vt$. Then the concept v is generated by the acceleration concept a and time concept t, the concept s can be generated by concept v and time concept t. These based on causal model.

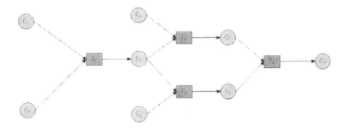

Fig. 1. Knowledge graph structure, Sequence $< C1, C2, F1, C3 >$ represents C3 is generated by C1 and C2 according to F1 and it is a relationship

Such model has the following properties:

1. Reachability. If you can reach a concept from initial concept, then the concept is reachable. N=(C,F,R), if there exists $f \in F$ makes $C[f] \rightarrow C_1$, then C_1 can be reached from C directly. If there exists a sequence $f_1, f_2, ..., f_{k-1}$ and $C_1, C_2, ..., C_k$ makes $C_1[r_1] \rightarrow C_2[r_2]...C(k-1)[r_{k-1}] \rightarrow C_k$, then C_k can be reached from C_1.

2. No Deadlock. This model is established on the causality . Perform a sequence can ultimately lead to any other sequence, then it will not have deadlock.

3. Reversibility. Any concept reached from the initial concept can go back to the initial concept.

There are some advantages when adopt this model. First, model is described in a graphic way and it is easy to understand. At the same time, it supports mathematical analysis to improve the accuracy of semantic description. Second, the formal method can realize concurrency, synchronization, resource sharing modeling of the system although not discussed in this paper. Third, it is helpful to analysis and verify requirements. This can be analyzed in section 4.

Although the model represented by figure is intuitive, it is poor normative and not identified in other systems. Then we propose XML to represent domain knowledge model in order to make up the defects.

3.2   XML Representation Framework of Knowledge Graph

Our knowledge graph includes three model elements: concept, relationship and rule. The concept describes attributes of knowledge. It has an identified ID and Name, performance attributes, safety-related attributes and interface description. The specific model structure of concept is shown in the left of Figure 2. The relationship describes the relation between concepts. Each relationship has an identified ID and Name, inputs section Variables, outputs section DependentVariables and corresponding rule section Description. The specific model structure of relationship is shown in the right of Figure 2. The rule is mainly formulas. We adopt MathML tool to express it. Each rule corresponds to an XML document. An example is shown in Figure 3.

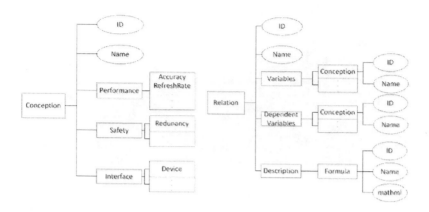

Fig. 2. Left is concept model structure, right is relationship model structure

```
<math>
  <mrow>
    <mfrac>
      <mrow>
        <mtext mathvariant='normal'>d</mtext>
        <mi>H</mi>
      </mrow>
      <mrow>
        <mtext mathvariant='normal'>d</mtext>
        <mi>t</mi>
      </mrow>
    </mfrac>
    <mrow>
      <mo>=</mo>
      <mi>v</mi>
    </mrow>
    <mtext mathvariant='normal'>d</mtext>
    <mo> v </mo>
  </mrow>
</math>
```

Fig. 3. An example of the rule represented by MathML

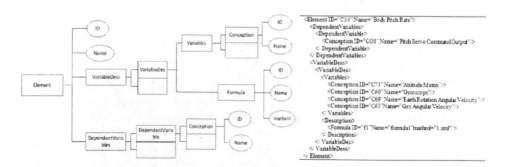

Fig. 4. Left is the complete domain knowledge graph model structure of airborne software, right is a detailed XML description of element 'Body Pitch Rate'

Finally, the complete knowledge graph model structure is shown in the left of figure 4. Element represents a knowledge. Each Element has an identified ID and Name. VariableDess represents a collection of knowledge and correspond rule which the knowledge depends. VariableDes represents single knowledge and correspond rule it depends. Variables represents knowledge which the knowledge depends. Conception represents a single knowledge. Formula represents correspond rule like mathematical formula or algorithm. DependentVariables represents a collection of knowledge which depends on the Element. DependentVariable represents single knowledge which depends on the Element. An example of specific knowledge graph XML description is shown in the right of Figure 4.

4 Formal Modeling of High-Level Requirements Based on Knowledge Graph(RMKG)

Due to the high safety and reliability features of airborne software, coupled with the importance of requirement analysis, high-level requirements modeling and verification is critical. The paper describes high-level requirements formally in order to verify conveniently and it generates high-level requirements by knowledge reasoning. Through an effective mechanism it converts non-functional requirements such as safety, security and real-time into related non-functional requirements of high-level requirements to ensure airborne software's safety and reliability at high-level.

4.1 Formal Description of High-Level Requirements

Definition 1: *High-level requirements representation* is a functional assertion structure formed by matching knowledge graph based on initial concept and terminate concept sets. It can be expressed as $< x_1, ..., x_n, r, y_m >$ formally and it means y_m is calculated by formula r based on the inputs $x_1, ..., x_n$. Single high-level requirements XML representation is shown in Figure 5.

```
<?xml version="1.0" encoding="UTF-8"?>
<Function>
  <TaskOutput>
   <Conception ID="C22" Name="Dynamic Pressure"/>
  </TaskOutput>
  <TaskInput>
    <Conception ID="C1" Name="Total Pressure Input"/>
    <Conception ID="C2" Name="Impact Pressure Input"/>
  </TaskInput>
  <Rule>
    <Formula ID="f1" Name="formula1" mathml="1.xml"/>
  </Rule>
</Function>
```

Fig. 5. An example of XML description of a single high-level requirements

The TaskOutput indicates the generated knowledge, TaskInput represents knowledge inputs and Rule represents rules such as mathematical formulas and so on which the generated knowledge depends on. Each high-level requirements only has one clear and unified interpretation and it is unambiguous and not in conflict with others. Using our formal representation, High-level requirements possess accuracy, consistency and concurrency.

4.2   Generation of High-Level Requirements

After the knowledge graph data structure has been determined, domain experts build knowledge graph by input knowledge through modeling tool. When the knowledge graph is generated, find and match knowledge graph. We use knowledge reasoning techniques to get all paths sets from the initial inputs to terminate outputs. Then we get high-level requirements. Shown in Figure 6. We can see given the initial inputs C_1, C_2, C_3, terminate outputs C_8, C_9, C_10 and knowledge graph, we'll get high-level requirements $< C_1, F_1, C_4 > < C_4, F_6, C_8 >, < C_2, F_2, C_4 > < C_4, F_6, C_8 >, < C_1, F_1, C_4 > < C_4, F_7, C_9 >, < C_2, F_2, C_4 > < C_4, F_7, C_9 >, < C_2, F_3, C_5 > < C_5, F_7, C_9 >$ and $< C_3, F_4, C_5 > < C_5, F_7, C_9 >$. $< C_3, F_5, C_6 >$ means initial input C3 can't get terminate outputs. $< C_7, F_8, C_{10} >$ means terminate output C_{10} can't be derived from initial inputs. So they are not included into high-level requirements.

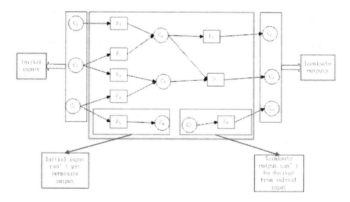

Fig. 6. High-level requirements formation process according to initial inputs and terminate outputs in system requirements

Requirement conversion is querying and inferring knowledge graph to obtain high-level requirements when given initial inputs and terminal outputs in the system requirements. We first use backward reasoning to get the causal chain of high-level requirements based on given outputs. For the redundant inputs we apply forward reasoning to achieve the causal chain of high-level requirements. Then we get all high-level requirements sets when given system requirements. On the one hand we reuse domain knowledge by querying and inferring mechanism.

On the other hand, we extend and improve knowledge graph while querying and inferring. Therefore, as the domain knowledge become diverse increasingly, the success rate of subsequent requirements matching can improve significantly.

We adopt improved breadth-first traversal in our backward and forward reasoning. According to our given complete knowledge graph model structure in section 3.2, our algorithm's variables correspond to the element name of the structure. *mapVariable* is the knowledge and its variabledess mapping. *mapdependVariable* is the knowledge and its dependvariables mapping. In backward reasoning, for each given output, we first judge that the dependent inputs variabledess of given output are all initial inputs or initial inputs and intermediate nodes or all intermediate nodes. If they are initial inputs, then we determine whether they are in given inputs, if they are intermediate nodes, we will recursively call backward reasoning algorithm. The backward reasoning algorithm is as follows.

Algorithm 1. BackwardTraversal Algorithm

1: *Input* ⇐ *outputknowledge, variabledess*
2: *Output* ⇐ *stack*
3: **for** each *variables* ∈ *variabledess* **do**
4:    **if** judge(*variables*)==1 **then**
5:       **if** IsInGivenInput(*variables*) **then**
6:          addToStack(*variables*)
7:          continue
8:       **end if**
9:    **else if** judge(*variables*)==2 **then**
10:       **for** each *variable* in *variables* **do**
11:          **if** IsInitialInput(*variable*) **then**
12:             **if** IsInGivenInput(*variable*) **then**
13:                addToStack(*variable*)
14:             **end if**
15:          **else**
16:             addToStack(*variable*)
17:             BackwardTraversal(*variable,mapVariable[variable],stack*)
18:          **end if**
19:       **end for**
20:    **else**
21:       **for** each *variable* ∈ *variables* **do**
22:          addToStack(*variable*)
23:          BackwardTraversal(*variable,mapVariable[variable],stack*)
24:       **end for**
25:    **end if**
26: **end for**

In forward reasoning, for each remaining input and its dependent output sets dependvariables which is a collection of arrays, we get the knowledge variabledess for each element conception in array of dependvariable, an element in dependvariables. For each element variable in array variables which is an the element in variabledess, we judge whether it is an intermediate node or an initial input. If it is an intermediate node and not in the intermediate nodes which backward reasoning generates, we call BackwardTraversal algorithm. If it is an initial input and not in the given inputs, then the path is interrupted. If all the element in array variables are in given inputs or in the generated nodes, we add it to stack. Then judge whether the conception is final output or not. If conception is not final output, we will recursively call forward reasoning algorithm. The forward reasoning algorithm is as follows.

Algorithm 2. ForwardTraversal Algorithm

 1: *Input* ⇐ *inputknowledge, dependvariables*
 2: *Output* ⇐ *stack*
 3: **for** each *dependvariable* ∈ *dependvariables* **do**
 4:    **for** each *conception* ∈ *dependvariable* **do**
 5:       *variabledess* = mapVariable[*conception*];
 6:       **for** each *variables* ∈ *variabledess* **do**
 7:          **for** each *variable* ∈ *variables* **do**
 8:             **if** IsIntermediateNode(*variable*)&&!InGeneratedNodes(*variable*) **then**
 9:                BackwardTraversal(*variable,mapVariable[variable],stack*)
10:             **else if** IsInitialInput(*variable*)&&!InGivenInput(*variable*) **then**
11:                break
12:             **end if**
13:          **end for**
14:          **if** IsAll(*variables*) **then**
15:             addToStack(*variables*)
16:          **end if**
17:       **end for**
18:       *arr* = mapdependVariable[*conception*]
19:       **if** *arr*!=null or *arr*.length!=0 **then**
20:          ForwardTraversal(*conception,arr,stack*);
21:       **end if**
22:    **end for**
23: **end for**

The improved breadth-first traversal can get all high-level requirements from system requirements. Because of reachability of our knowledge graph model, we can judge whether the initial inputs can reach the terminate outputs. As the model is alive, there's no dead loop in our algorithms. The model is reversible, so we can use backward reasoning to judge whether the terminate outputs can be got from initial inputs.

High safety and reliability is significant to airborne software. As the performance, safety and other attributes of concepts in knowledge graph may not meet the demands proposed in system requirements, We propose a reliable conversion mechanism to transfer non-functional requirements such as safety, security and real-time of system requirements into corresponding non-functional high-level requirements.

Non-functional Requirements Conversion Mechanism: For airborne software safety, assign the highest safety level to each node of high-level requirements according to output safety level given by system requirements. Namely, if the highest safety level of the given output is 5, then the safety level of each node in the high-level requirements is 5. Only when the safety level of each node in the high-level requirements is 5, the safety level of output we obtain can be 5. For airborne software security, assign the highest security level to each node of high-level requirements according to input security level given by system requirements. Namely, if the highest security level of the given input is 7, then the security level of each node in the high-level requirements is 7. Because the security level of input is the highest, then the security level of each node in the high-level requirements generated based on the input must be the highest. For airborne software real-time, assign the highest real-time to each node of high-level requirements according to system requirement. If the highest real-time in system requirements is sample once every 10ms, then each node of the high-level requirements also sample once every 10ms. Only in this way can we get the highest real-time system requirements.

Our conversion mechanism allocate safety, security and real-time according to the highest level, so the high-level requirements we get own the highest safety, security and real-time. Ensure the related requirements such as safety, security and real-time of airborne software are developed into high-level requirements, we start to monitor non-functional requirements from high-level requirements, which plays an important role in forming high safe and reliable airborne software finally.

5 Experimental Results

In this section, we present some experiments. The knowledge graph is built based on atmospheric data calculation, flight control and flight management which airborne software use. There are 73 concepts, 37 relationships and 47 mathematical formulas. We provide modeling tool for domain experts to construct knowledge graph collaboratively. When domain experts input xml files of concepts or relations, knowledge graph built according to the input xml files. If they search knowledge, the knowledge graph of the relevant knowledge will be shown. When expert modifies the knowledge graph, other experts will receive the modified message in the Latest News column. Experts can also decide whether to modify the knowledge by online conversation, you can see in the Figure 7.

When experts input system requirements files, the system will presents a graphical display of high-level requirements to experts. Experts can see how system requirements are turned into high-level requirements clearly. They will eventually get written documents of high-level requirements. We give some verification information from the aspect of integrity, consistency, correctness and traceability. When conflicts are detected among the graphs, the source can be identified and the problem can be clearly specified. Some verification information is given in the Latest News column finally. We give two examples in the Figure 8, when input different system requirements, the results are different.

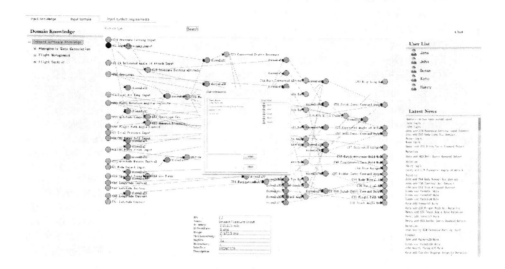

Fig. 7. An example of collaboration based on online conversation

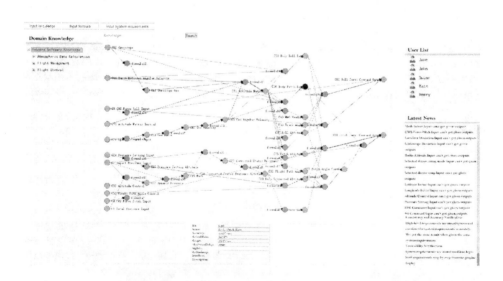

Fig. 8. Examples of high-level requirements formation and verification

6 Conclusions and Future Work

In this paper we have presented a new approach of high-level requirements modeling of airborne software. Formal description of knowledge graph is helpful to analysis and verify requirements. Formal description of high-level requirements makes it have accuracy, consistency and concurrency. Ensure the safety, security and real-time of airborne software at high-level requirements makes it important to form high safety and reliability software. We realize a modeling tool for experts to construct knowledge graph collaboratively, and complete the conversion from system requirements to high-level requirements based on knowledge graph. Our tool realize traceability automatically which can only be achieved manually in traditional methods and we obtain pretty good results in our experiments. Further, we firstly do more theoretical research. Moreover, we will improve our experiment to support the following airborne software architecture design.

References

1. Camus, J.L.: The Airborne Software Development Challenge. White Paper, Esterel Technologies (2010)
2. RTCA Inc.:RTCA/DO-178B: Software Considerations in Airborne Systems and Equipment Certification, Washington D.C(1992)
3. RTCA Inc.:RTCA/DO-178C: Software Considerations in Airborne Systems and Equipment Certification. Washington D.C(2011)
4. Zhong, S.W., Hong, M.E.I.: Review of the unified modeling language (UML). Journal of Computer Research and Development (1999)
5. Sowa, J.F.: Conceptual graphs as a universal knowledge representation. In: Computers & Mathematics with Applications, pp. 75–93 (1992)
6. Hoang, N.V., Valerie, G.B., Marta, R.: Object detection and localization using a knowledge graph on spatial relationships. In: 2013 IEEE International Conference on Multimedia and Expo (2013)
7. Delugach, H.S., Brian, E.L.: Acquiring software requirements as conceptual graphs. In: Fifth IEEE International Symposium on Requirements Engineering (2001)
8. David, S.: AADL Display System Model Description. Rockwell Collins, Inc. (2004)
9. OMG:UML Profile for MARTE: Modeling and Analysis of Real-Time Embedded Systems. Version 1.0 (2009)
10. Bouquet, P.: Theories and uses of context in knowledge representation and reasoning. Journal of Pragmatics, 455–484 (2003)
11. Yao, Y.L.: A Petri net model for temporal knowledge representation and reasoning. IEEE Transactions on Systems, Man and Cybernetics, 1374–1382 (1994)
12. Chen, S.M.: Fuzzy backward reasoning using fuzzy Petri nets. IEEE Transactions on Systems, Man, and Cybernetics, Part B: Cybernetics, 846–856 (2000)

Scalable Horn-Like Rule Inference
of Semantic Data Using MapReduce

Haijiang Wu[1,3], Jie Liu[2,3], Dan Ye[3], Jun Wei[2,3], and Hua Zhong[3]

[1] University of Chinese Academy of Sciences
[2] State Key Laboratory of Computer Science, Institute of Software
[3] Institute of Software, Chinese Academy of Sciences
{wuhaijiang12,ljie,yedan,wj,zhongh}@otcaix.iscas.ac.cn

Abstract. Semantic data analysis tasks benefit much from rule inference, which derives implicit knowledge from explicit information. Recently, available semantic data from the Web, sensor readings, semantic databases and ontologies exploded drastically. However, most of the existing approaches for semantic rule inference are either centralized, which cannot scale out to infer big semantic data; or rule-specific, which hinder their wildly use. In this paper, we propose a scalable approach for Horn-like rule inference of semantic data based on MapReduce, which can evaluate domain- and application-specific rules, and can be easily extended to evaluate RDFS and OWL ter Horst semantic rules. We first introduce a general rule-evaluation mechanism, which translates a Horn-like rule to one or more MapReduce jobs. To improve rule-evaluation performance, two optimization policies job-parallelization and job-reusing are then introduced. Using a large semantic data set generated by the LUBM benchmark, we give a detailed experimental analysis of the scalability and efficiency of our approaches.

Keywords: Sematic Web, Horn-Like Rule, MapReduce, Semantic Inference.

1 Introduction

Semantic inference, deriving implicit knowledge from explicit information, is one of the key steps of semantic data analysis tasks, such as automatic question-answering [1], consistency checking of semantic data [2], quality assessing of knowledge base [3], and so on. Over the recent years, large volumes of data have been published in Semantic Web format , including data in many different fields. Moreover, for the high computational complexity of semantic inference, even the state-of-the-art single-node inference engine will spend more than tens of seconds in reasoning less than one million facts [4]. Such big semantic data raise the challenges of scalable inference. Scalable rule-based inference is one of the best approach for reasoning Web-scale semantic web data [5], by expressing latent semantic with rules, matching the facts in a cluster with rules, and conducting new facts with matched facts.

A MapReduce-based RDFS rule inference method is first introduced in [9]. The authors exploit the fact that all the RDFS rules require a join between one schema triple and one instance triple, it is reasonable to load all the schema triples into the memory of each computing node in advance, to make their algorithms scalable. While

R. Buchmann et al. (Eds.): KSEM 2014, LNAI 8793, pp. 270–277, 2014.

some rules in OWL ter Horst rule set do not respect to this pattern; e.g., rules that contain "owl:sameAs", "owl:someValuesFrom" and "owl:allValuesFrom", the authors then design special algorithms to cover these rules in [5]. Unfortunately, the approaches introduced in [5],[9] are not applicable for domain- and application-specific rule inference, because these rules may contain no schema triple and it is unreasonable to design MapReduce algorithms for each rule. Domain- and application-specific rules can be naturally expressed as Horn-Like rules [10]. While there are two significant challenges in scalable Horn-Like rule inference based on MapReduce: (1) One is generalizing the rule evaluation mechanism of Horn-Like rules and separating rule representation and rule evaluation. (2) The other is optimizing the job execution.

In this paper, we focus on automatic scalable Horn-Like rule inference using MapReduce, and propose a novel approach based on three observations. First, According to the Horn-Like rule evaluation process, inference result is generated by joining all the facts that matches the antecedents of the rule. For special rules that contains "owl:allValuesFrom", "owl:sameAs" and "owl:allValuesFrom" in OWL ter Horst rule set, we can use the optimized methods introduced in [5] in preparing stage. Second, considering the evaluation of a rule may not achieve in one MapReduce job. A naive approach to translate it to MapReduce jobs is joining the antecedents one by one. Then all the facts will be loaded from the disc for each join. Obviously, duplicated data-loading can be avoided. Finally, it is common that different rules may share same antecedents, and then different rules can share the result of the same MapReduce job.

The contributions of this paper include: we first introduce a scalable Horn-Like rule evaluation mechanism based on MapReduce, which can be easily extended to inference RDFS, OWL ter Horst semantic rules. Add then we design several policies to optimize the rule evaluation. Based on the policies we proposed, less MapReduce jobs are generated, and less time is used. Finally, we implement our approach on a 10-nodes Hadoop cluster, and give detailed experimental analysis.

The rest of this paper is organized as follows. In section 2, we give a formal definition of Horn-Like rule and describe the basic process of rule-evaluation using MapReduce. Section 3 gives the overall process flow of our approach, and depicts Horn-Like rule inference using MapReduce in detail. In section 4, we propose the job-reduction and job-reusing algorithms to optimize the rule evaluation. We analyze the scalability and efficiency experimentally in section 5 and review the related work in section 6; Section 7 concludes the paper with a summary and outlines the future work.

2   Preliminaries

2.1   Horn-Like Rule

Definition 1. [Horn-Like rule, HLR] A Horn-Like rule is a definite Horn clause [6], in which all the propositions are in the form of RDF triples and only allow variables in subject position and object position.

$$HLR = \{\{antecedent_1, antecedent_2, \dots, antecedent_n\}, \{consequence\}\} \quad (1)$$

$$antecedent_i \in (US \cup V) \times US \times (US \cup LS \cup V) \quad (2)$$

$$onsequence \in (US \cup V) \times US \times (US \cup LS \cup V) \qquad (3)$$

Here, V refers to a set of variables. US refer to the URL set and LS to the literal set.

2.2 Rule Evaluation on MapReduce Framework

When programming on MapReduce, users specify a map function that processes a key/value pair to generate a set of intermediate key/value pairs, and a reduce function that merges all intermediate values associated with the same intermediate key. To evaluate a rule on MapReduce framework, all the triples in the input data set should be taken to match each antecedent of the rule. Taking rule '$?s\ rdf : type\ ?X \& ?X\ rdfs : subClassOf\ ?Y?s\ rdf : type\ ?Y$' as an example, and giving three triples: $tr1\langle c1rdf : typeC\rangle, tr2\langle c2rdf : typeC\rangle, tr3\langle Crdfs : subClassOfS\rangle$, tr1 and tr2 will match the first antecedent, tr3 will match the second antecedent. When tr1 joins tr3, a rule execution result will be produced. When tr2 joins tr3, another result will be produced.

3   Horn-Like Rule Inference Using MapReduce

3.1   Process Flow Overview

The overall process flow of our approach is summarized in Fig. 1. It comprises four components: rule parser, rule-job translator, rule-evaluation job scheduler and rule evaluator. Rule parser parses a raw rule, stored in a file as a string, to antecedents and consequences. For the results, we identify all the consequence-variables and antecedent-variables, as well as the dependency between them. In the rule-job generator, variables that appear in more than one antecedent will be taken as MapReduce keys, which correspond to MapReduce jobs. When a MapReduce job is executing, the input facts match each antecedent of the rule in the map stage. If all of the triples succeed in matching, the rule is triggered, and then intermediate result is generated. Once the last job of a rule is finished, new facts are produced. The output facts are merged into the input facts and taken as the input of the succeeding rule evaluation. More details will be given in the next several sections respectively.

3.2   Rule-Job Translation

Definition and Notation. Before evaluating rules using MapReduce as described in section 2.2, map keys should be identified in advance to emit key-value pairs in map stage, and then consequence-antecedent dependency information is used to generate results in reduce stage. For better illustration, several notations are defined:

Definition 2. [Rule-Key, RK] Given a rule R with more than one antecedent, the antecedent set of R is denoted as AS_R, and then a antecedent-group(AG) of R refers to a subset of AS_R. Rule-Key RK includes a variable that exists in more than one antecedent, and a set of antecedent-groups which have the variable in their antecedents. $RK = (v, \{ag_1, ag_2, , ag_n\})$, v will taken as map output key when rule is evaluated on a Mapreduce framework.

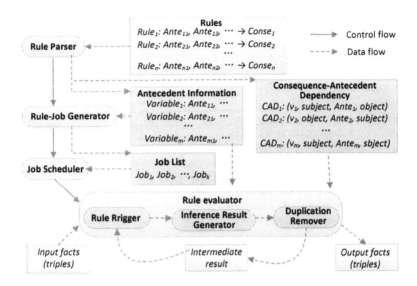

Fig. 1. Overall process flow of Horn-Like rule inference

Definition 3. [Consequence-Antecedent Dependency, CAD] A CAD is a 4-tuple that comprises information a consequence-variable depends: $CAD_i = (V_i, VPosition_i, DT_i, DPosition_i)$. Here, V_i is a variable in the consequence of the rule; $VPosition_i$ indicates the position where the variable sites in the consequence; DT_i indicates which antecedent does the variable depends on, and $dPosition_i$ indicates the position where the dependent variable sites in the antecedent.

Definition 4. [Rule-Evaluation Job, REJob] A Rule-Key RK_i corresponds to a MapReduce job, the output of the MapReduce job is produced using the CAD information of $RK_i : REJob_i = (RK_i, CAD_{RK_i})$

Rule-Job Translation Algorithm. Rule-Job translation aims at transforming a Horn-Like rule to a set of REJobs. To identify the rule-keys of each REJob, all the variables in the rule are extracted, and then the variables just exist in one antecedent are filtered out, the rest variables are used to generate RKs. The complete picture is shown in Algorithm. 1.

3.3   Rule Evaluation

One-REJob Rule Evaluation. With the rule parsing result and the REJob information, facts will be split to computing nodes to trigger rules and conduct new facts. For rules translated into one REJob, rule evaluation will be finished in two stages. In map stage, input facts are taken to compare with the antecedent of the AGs. If all the corresponding elements, except for variables, between an input fact and an antecedent are equal, the map function will emit a key-value pair $(element_{RK}; oldFact, AGid)$. $element_{RK}$ refers to the element of the input fact that corresponds to the variable of the RK. In

---

Algorithm 1. Rule-Job Translation

Require:
    The set of rule antecedents, RAS; The set of Rule-Evaluation jobs, RCS;
Ensure:
    set of Rule-Evaluation jobs, RJS;
 1: $RKS = \Phi, CADS = \Phi$;
 2: **for** each $RA \in RAS$ **do**
 3:    **if** RA.element is variable **then**
 4:       **if** RA.element is variable && RKS contains RA.element **then**
 5:          RKS.get(RA.element).addAG(RA)
 6:       **else if** RA.element is variable **then**
 7:          create a new RK and add it to RKS; RKS.get(RA.element).addAG(RA)
 8:       **end if**
 9:    **end if**
10: **end for**
11: **for** each $RC \in RCS$ **do**
12:    **if** RC.element is variable **then**
13:       create a new CAD, and add it to CADS
14:    **end if**
15: **end for**
16: **for** each $RK \in RKS$ **do**
17:    create a new REJob for RK if $RK.getAG().size > 1$, and add the REJob to RJS
18: **end for**
19: **return** RJS;

---

reduce stage, values with same $element_{RK}$ will be sent to the same reduce node. The reduce function of the REJob groups these values into Value-Groups according to their AGid. If the size of AG is less than the size of the Value-Groups, no inferred fact will generated for $element_{RK}$. Otherwise, an inferred fact can be produced by taking one value from each Value-Group, according to the CADs.

Multi-REJob Rule Evaluation. Rules that translated to more than one REJob cannot be finished in one shot. In any single REJob of such rule, only part of the antecedents are covered, then the reduce function will not conduct any new fact. The output of such REJob is temporary data, which is in the form: $Tuple = (newAGid, \{ant_1, ant_2, \ldots, ant_n\})$. Here, newAGid is the id of a new AG that produced by the rule-result generator. This AG contains all the antecedents in the rule-key of the REJob. After the REJob completes, the rule-key of the REJob will be removed from the RK set, the AGs of the rest RKs will be replaced by the new AG, on condition that their antecedents are included in the new generated AG.

4 Job Parallelization and Reusing

4.1 Job Parallelization

In naive multi-REJob rule evaluation, a REJob is generated to produce temporary data, for each RK in RK set. In the last REJob, new facts are produced according to the

CADs. In this way, Rule1 can be abstracted as a REJob chain. In this REJob chain, not all the REJobs have data dependency on their pre-REJobs. It is possible to explore the potential of this REJob chain to improve the rule inference performance, by transforming it to a DAG and parallelizing the REJob evaluation. Data dependency is caused by sharing RK between antecedents. In other words, it is caused by sharing antecedents between RKs. Then to parallelize the REJobs is to find out an RK set, the intersection of which is a empty set.

$$Rule_1 : ?x \ rdf : type \ Professor, \ ?x \ worksFor \ ?d, \ ?y \ rdf : type \ Professor,$$
$$?y \ worksFor \ ?u, ?d \ subOrgnizationOf \ ?u \rightarrow ?x \ workmateOf \ ?y$$
$$Rule_2 : ?x \ rdf : type \ Professor, \ ?x \ worksFor \ ?d, \ ?d \ subOrgnizationOf \ ?u$$
$$\rightarrow ?x \ worksFor \ ?u \quad (4)$$

4.2 Job Reusing

In some use-cases, two Horn-Like rules may share antecedents. After the Rule-Job Translation stage, different REJobs in these rules can be matched by the same facts and output the same intermediate result or inference result. Obviously, only one of these REJobs is necessary to be executed. We identify these duplicated REJobs by comparing RKs in different rule. If all the AGs of a RK_{K_i} are contained by another RK_{K_j}, the REJob corresponding to Ki will be removed from the job schedule list, and the output result of K_j will be used as that of K_i. Taking $Rule2$ as an example, antecedents in $Rule2$ are completely included in $Rule1$. Definitely, all the RKs are contained in the RK set of $Rule1$. When executing the REJobs of Rule1, inferred fact of $Rule2$ can be conducted by pass.

5 Evaluation

5.1 Experiment Setting

We have implemented our approach on a 10-nodes Hadoop cluster (1 master, 9 slaves). Each node has 4 cores and 16G memory. LUBM is a widely used benchmark for semantic reasoning, which can generate semi-realistic data sets of arbitrary size. For experiments, we generate more than 200 million RDF triples. The semantic rule set of specific-application is in constant evolution. Thus, to evaluate our approach of scalable Horn-Like rule inference, we create a rule set(Table. 1) according to LUBM ontology.

5.2 Experiment Result

Scalability. We evaluate the scalability of the approach introduced in this paper in two indices: (1) elapsed time increase linearly as the linear increasing of the input data size; (2) speedup with increasing number of computing nodes. We generated 10 data sets using LUBM benchmark, the size of which ranges from 20 million to 200 million. After evaluating the rules in Table. 1, we show our experimental results in (a) and (b) of Fig. 2. We can see that our approach can scale linearly as the size of the input data increase, and speed up when adding computing nodes.

Table 1. Rule set

ID	Antecedents	Consequence
R_1	?x rdf:type FullProfessor; ?x worksFor ?d; ?d subOrganizationOf ?u	?x worksFor ?u
R_2	?x rdf:type FullProfessor; ?x worksFor ?d; ?d subOrganizationOf ?u; ?y rdf:type FullProfessor; ?y worksFor ?u	?x workMateOf ?u
R_3	?p publicationAuthor ?x; ?p publicationAuthor ?y	?x coauthor ?y
R_4	?x takesCourse ?c; ?c rdf:type GraduateCourse	
R_5	?x takesCourse ?c; ?y takesCourse ?c; ?c rdf:type GraduateCourse	?x grateClassMate ?y
R_6	?x rdf:Type Student, ?x advisor ?p, ?y rdf:Type Student, ?y advisor ?p, ?p rdf:Type Professor	?x fellow ?y

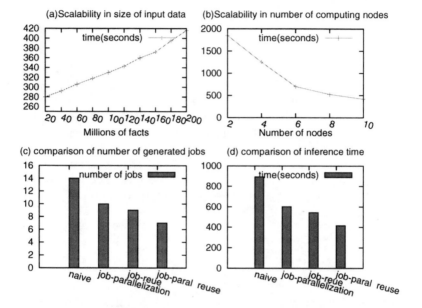

Fig. 2. Scalability of Horn-Like rule evaluation

Performance on LUBM Benchmark. From (c) of Fig. 2, a naive approach to evaluate the rule set will generate 14 MapReduce jobs. Based on naive approach, Job-Parallelization and Job-Reusing methods can reduce 5 and 4 jobs respectively. If using both methods, we can reduce the number of jobs to 7. The reduction of job numbers corresponds to the reduction of elapsed time is shown in (d) of Fig. 2.

6   Related Work

Existing works about scalable semantic inference concentrate on RDFS and OWL ter Horst semantic rule set. Urbani, J .et first introduces MapReduce-based RDFS rule inference in [9]. They exploit the fact that all the RDFS rules require a join between one schema triple and one instance triple, and load all the schema triples into the memory of each computing node in advance, to make their algorithms scalable. To tackle the rules that contain "owl:sameAs", "owl:someValuesFrom" and "owl:allValuesFrom", they then design special algorithms in [5]. Liu [11] also pay some attention to scalable RDFS reasoning on MapReduce, but they focus on annotated RDFS reasoning. Some challenges, such as unnecessary derivation and fixpoint calculation in scalable annotated RDFS reasoning are proposed, and but the authors do not describe their solutions in detail. These optimizations for RDFS and OWL rule inference are not applicable for domain- and application-specific rule inference.

7 Conclusion and Future Work

In contrast to standard rules e.g. RDFS and OWL ter Horst rules, Horn-Like rule can be used to express domain- and application-specific rule naturally, and is widely used in semantic-based applications. In this paper, we introduce a scalable Horn-Like rule execution mechanism based on MapReduce. Besides, we present optimization algorithms such as job-reusing and job parallelization to improve the performance of rule evaluation. In future work, we intend to perform an extensive experimental evaluation to verify the performance for different input data set morphologies. We also plan to extend this approach to inference semantic data incrementally.

References

1. Kalyanpur, A., et al.: Structured data and inference in DeepQA. IBM Journal of Research and Development 10, 1–10 (2012)
2. Baclawski, K., Kokar, M.M., Waldinger, R., Kogut, P.A.: Consistency checking of semantic web ontologies. In: Horrocks, I., Hendler, J. (eds.) ISWC 2002. LNCS, vol. 2342, pp. 454–459. Springer, Heidelberg (2002)
3. Ma, Y., Liu, L., Lu, K., Jin, B., Liu, X.: A Graph Derivation Based Approach for Measuring and Comparing Structural Semantics of Ontologies. IEEE Transactions on Knowledge and Data Engineering 26, 1039–1052 (2013)
4. Motik, B., Sattler, U.: A comparison of reasoning techniques for querying large description logic ABoxes. In: Proceedings of the 13th international conference on Logic for Programming. Artificial Intelligence, and Reasoning, pp. 227–241 (2006)
5. Urbani, J., Kotoulas, S., Maassen, J., van Harmelen, F., Bal, H.: WebPIE: a Webscale parallel inference engine using MapReduce. J. of Web Semantics. 10, 59–75 (2012)
6. Horn, A.: On sentences which are true of direct unions of algebras. Journal of Symbolic Logic 16, 14–21 (1951)
7. Dean, J., Ghemawat, S.: Mapreduce: Simplied data processing on large clusters. In: Proceedings of the USENIX Symposium on Operating Systems Design & Implementation (OSDI), pp. 137–147 (2004)
8. Hitzler, P., Krotzsch, M., Parsia, B., Patel, P.F., Rudolph, S.: OWL 2 Web Ontology Language Primer, W3C recommendation (2012)
9. Urbani, J., Kotoulas, S., Oren, E., van Harmelen, F.: Scalable distributed reasoning using mapReduce. In: Bernstein, A., Karger, D.R., Heath, T., Feigenbaum, L., Maynard, D., Motta, E., Thirunarayan, K. (eds.) ISWC 2009. LNCS, vol. 5823, pp. 634–649. Springer, Heidelberg (2009)
10. Horrocks, I., et al.: SWRL: A semantic web rule language combining OWL and RuleML. W3C Member submission (2004)
11. Liu, C., Qi, G.: Toward scalable reasoning over annotated RDF data using mapReduce. In: Krötzsch, M., Straccia, U. (eds.) RR 2012. LNCS, vol. 7497, pp. 238–241. Springer, Heidelberg (2012)

Model Checking (k, d)-Markov Chain with ipLTL

Lianyi Zhang, Qingdi Meng, and Guiming Luo

School of Software, Tsinghua University
100084, Beijing, China
lyzhang117@gmail.com

Abstract. In this paper a hierarchical Markov model is proposed for the temporal analysis of periodic stochastic systems. For analyzing the system behavior, an interval linear temporal logic, ipLTL, is presented, which is an LTL with linear inequalities on the probability mass functions (pmfs) as an atomic proposition. We prove that the proposed model converges to a steady state which enables us to develop an algorithm to determine the bound of the system execution time and check the specification written in ipLTL. Some properties of the manufacturing systems are analyzed and verified to illustrate the efficiency of our methods.

Keywords: Markov chain, model checking, linear temporal logic, convergency, probability mass function, linear inequalities.

1 Introduction

Markov chains are good models for analyzing the behaviors of embedded and distributed systems that are stochastic and memoryless. Many interesting researches about probabilistic logic and associated model checkers have been developed. In particular, PRISM[1] is a synthesized tool that can model check Discrete Time Markov Chain (DTMC), Continuous Time Markov Chain (CTMC) and Markov decision procedure for specifications written in PCTL*[2] or CSL[3]. An advantage of using Markov chains is that we can directly model a system by estimating the probabilistic mass functions (the distributions) over time. In a regular Markov chain, over a large time, there exists a unique limiting distribution. iLTLChecker[4,5] is a model checker, which can check regular Markov chain for the properties of the probabilistic mass functions. Different from regular Markov chains, periodic Markov chains will cycle, moving through different distributions, which have no unique limiting distributions (steady states). We develop an algorithm for model checking the behavior of systems modeled by periodic Markov chains of order k, of which in each transition, behaves nondeterministically either as a periodic Markov chain or as a periodic Markov chain of order k. We prove that the constructed model converges to a steady state from the initial states.

Under the convergence property, we develop an algorithm for checking Markov chains with period d and order k against the temporal properties in ipLTL, which

R. Buchmann et al. (Eds.): KSEM 2014, LNAI 8793, pp. 278–289, 2014.
© Springer International Publishing Switzerland 2014

can express the expected reward of the system. A predicate about an expected reward on the interval execution of the periodic system can be expressed, as instance, the expected number of messages in the buffer in a periodic network with time delay transitions, or the expected profit and loss of a manufacturing system with interval time delay control.

The main contribution of this paper are: (1) construct a periodic Markov chain of order k, and prove that our model is converged which captures the long-run behavior of systems; (2) present ipLTL, which specify the temporal logic about the predicate of pmfs; (3) carry on the bounded feasibility checking through a variant Büchi automata. The rest of this paper is organized as follows: The proposed (k, d)-Markov chain is presented and the long-run convergence properties of our Markov model is demonstrated in Section 2. In Section 3, we firstly present the interval probability temporal logic and its operators, secondly provide the underlying foundation needed to check the temporal properties of the period stochastic system, and finally give the model checking of feasible checking. We do some experiments on a manufacturing system in Section 4, and conclude the paper in Section 5.

2   Extended Markov Chain Model

In this section, we first introduce the extended Markov chain model with period d and order k $((k, d)$-Markov chain), and then demonstrate the long-run convergence properties of our Markov model.

2.1   Extended Markov Chain Model

In this part, based on discrete-time Markov chain (DTMC), periodic Markov chain and Markov chain of order k, we presented the extended Markov chain model with period d and order k $((k, d)$-Markov chain).

Let S be a countable set of states and let $\mathcal{P} = (\Omega, \mathcal{F}, \mathrm{P})$ be a probability space, where Ω is a sample description space, $\mathcal{F} \subseteq 2^\Omega$ is a Borel σ field of events, and $\mathrm{P} : \mathcal{F} \to [0, 1]$ is a probability measure. A *Discrete Time Markov Chain*([6]) is a function $\boldsymbol{X} : N \times \Omega \to S$ satisfying the following memoryless property:

$$\mathrm{P}[X(t) = s_t | X(t - 1) = s_{t-1}, ..., X(0) = s_0]$$
$$= \mathrm{P}[X(t) = s_t | X(t - 1) = s_{t-1}],$$

where $P[X(t) = s]$ stands for $P(\zeta \in \Omega : X(t, \zeta) = s)$.

We represent the STD $\mathcal{T}_X(S, \mathbf{M})$ for a Markov process $\boldsymbol{X}$ where $S = s_1, ..., s_n$ is a set of states and $\mathbf{M} \in \mathbb{R}^{n \times n}$ is a probability transition matrix such that $\mathbf{M}_{ij} = Pr(X_t = s_i | X_{t-1} = s_j)$. To simplify the explanation, we define a *probability mass function* $\mathbf{x} : \mathbb{N} \to \mathbb{R}^{n \times 1}$ such that $\mathbf{x}(t) = [Pr(X_t = s_1) ... Pr(X_t = s_n)]^T$. A discrete-time Markov chain is a deterministic model that once the initial probability vector $\mathbf{x(0)}$ is given, the rest $\mathbf{x}(t), t \geq 1$ are determined, concisely as $\mathbf{x}(t) = \mathbf{M} \cdot \mathbf{x}(t - 1) = \mathbf{M}^t \cdot \mathbf{x}(0)$.

For a Markov chain $X = (X_0, X_1, X_2, ...)$ with countable state space S and transition probability matrix $\mathbf{M}$, the period [6] of state $s \in S$ is

$$d(s) = gcd\{n \in \mathbb{N}_+ : \mathbf{M}^n(s, s) > 0\}$$

Thus, starting in s, the chain can return to s only at multiples of the period d, and d is the largest such integer. State s is *aperiodic* if $d(x) = 1$ and periodic if $d(s) > 1$. Suppose that $s \in S$. If $\mathbf{M}(s, s) > 0$, then x is aperiodic. A Markov chain is aperiodic if every state is aperiodic. An irreducible Markov chain only needs one aperiodic state to imply all states are aperiodic.

A *periodic Markov chain model*[7] is a tuple $\mathcal{D}_X = (S, \mathbf{M})$, where S and $\mathbf{M}$ has the same definition of the discrete Markov model $\mathcal{T}_X$, moreover $\mathbf{M}$ is a periodic probabilistic matrix with period number d.

Consider that there exists a Markov chain with period d. Given an arbitrary initial probabilistic mass function, we have $\lim_{t \to \infty} x(t + d) = \lim_{t \to \infty} x(t)$.

A *Markov chain of order k* (or a Markov chain with memory k)[6], where k is finite, is a process satisfying

$$Pr(X_n = s_n | X_{n-1} = s_{n-1}, ..., X_0 = s_0)$$
$$= Pr(X_n = s_n | X_{n-1} = s_{n-1}, ..., X_{n-k} = s_{n-k}) \text{ for } n > k$$

In other words, the future state depends on the past k states. When applied memory k affects the pmfs, we have $\mathbf{x}(t) = f(\mathbf{x}(t - 1), \mathbf{x}(t - 2), ..., \mathbf{x}(t - k))$, where f is an appropriate function. In the following model, we consider f as a linear combination of past k pmfs.

Based on the above models, we now present the Markov chain extended with period d and order k ((k, d)-Markov chain).

Definition 1 ((k, d)-Markov chain). *Consider that there exist accidental order k (or memory k) transitions on a d-period Markov chain. Formally, the generated structure (k, d)-Markov chain ((k, d)-MC) is denoted by a tuple $\mathcal{A}_{\mathcal{D}_X} = (\mathcal{D}_X, \overrightarrow{\alpha})$ where*

- $\mathcal{D}_X = (S, \mathbf{M})$ *is a Markov chain with period d;*
- $\overrightarrow{\alpha}$ *is a k dimension positive coefficient vector satisfying that $\sum_{i=0}^{k} \alpha_i = 1$, where α_i is the ith element of $\overrightarrow{\alpha}$ which represents the memory contribution of the last i^{th} pmf to the current pmf.*

From the above structure $\mathcal{A}_{\mathcal{D}_X} = (\mathcal{D}_X, \overrightarrow{\alpha})$ for a periodic Markov chain X with accidental k memory transitions, we can construct a nondeterministic model for the evolution of pmfs extended with d periodicity and k memory, in which there exist two kinds of transitions:

- one is the transition on pmf of periodic Markov chain applying $\mathbf{x}(t) = \mathbf{M} \cdot \mathbf{x}(t - 1)$;

– the other is accidental k memory transition. The simple k memory transition is a linear combination of past k pmfs that hold the following constraint
$$\mathbf{x}(t) = \sum_{i=1}^{|\overrightarrow{\alpha}|} \alpha_i \mathbf{x}(t - i)$$

Given a (k, d)-MC $\mathcal{A}_{\mathcal{D}_X}$, we use pmf $\mathbf{x}(t)$ of Markov chain to describe system state at time instant t. At start of the system, we have one possible initial pmf $\mathbf{x}(0)$. Then the system passes at least an interval of $|\overrightarrow{\alpha}| - 1$ time instants during which evolve under periodic Markov chain as $\mathbf{x}(t) = \mathbf{M}\mathbf{x}(t - 1)$. After that, the system may make a memory transition at t_{I_1}. We get a periodic interval $I_0 = [t_{I_0}, t_{I_1}]$ from the initial state to the memory state. From the above definitions, we can use (k, d)-MC to model a discrete stochastic linear system with probabilistic distribution.

Definition 2 (Paths constraint of $\mathcal{A}_{\mathcal{D}_X}$). *Given a (k, d)-MC $\mathcal{A}_{\mathcal{D}_X}$, where $\mathcal{D}_X = (S, \mathbf{M})$ and $\overrightarrow{\alpha}$ is k-memory coefficient vector, one path of $\mathcal{A}_{\mathcal{D}_X}$ is*

$$\sigma = \mathbf{x}(0) \xrightarrow{\tau_0} \mathbf{x}(1) \xrightarrow{\tau_1} \dots$$

where τ_i is the transition from $\mathbf{x}(i)$ to $\mathbf{x}(i + 1)$. We constrain that if τ_i is a memory transition for $i \geq (k - 1)$, then $\forall j \in (i - k, i - 1) \Rightarrow \tau_j$ is a transition for which the behaviors are expressed as the periodic Markov chain $\mathcal{D}_X$.

Definition 3 (Interval trace of $\mathcal{A}_{\mathcal{D}_X}$). *Given a path σ of $\mathcal{A}_{\mathcal{D}_X}$, an interval trace of path is*

$$\sigma_I = \sigma_I^0 \xrightarrow{\tau_{I_1}} \sigma_I^1 \xrightarrow{\tau_{I_2}} \dots$$

where τ_{I_i} is the ith memory transition and $I_i \subseteq \mathbb{N}$ is time-interval between the ith and $(i + 1)^{th}$ memory transition during which there exist $|I_i|$ transitions on the periodic Markov chain.

After giving the definitions to model a discrete stochastic linear system with (k, d)-Markov chain, in this part, we define the reward of (k, d)-Markov chain. By introducing a reward Markov model, we can augment the Markov process with reward (or cost) information about which a predicate can be expressed by linear inequality.

A *Markov reward model*([8]) is a Markov chain extended with a reward function that associates reward values with states. The reward associated with a state can be regarded as an earned value when a process visits that state. We represent a Markov reward model as a triple $(S, \mathbf{M}, r)$, where $(S, \mathbf{M})$ is a STD of a DTMC and $r : S \to \mathbb{R}$ is a reward function. In this paper, we consider only the constant rewards which can be represented by a row vector $\mathbf{r} \in \mathbb{R}^{1 \times |S|}$ as follows:

$$\mathbf{r} = r(s_i) \text{ for } i = 1, ..., |S|.$$

The expected reward at time t can be written as

$$E[r](t) = \sum_{s_i \in S} r(s_i) \cdot \mathbf{P}[X(t) = s_i] = \mathbf{r} \cdot \mathbf{x}(t).$$

By introducing a reward Markov model, we can augment the Markov process with reward (or cost) information about which a predicate can be expressed by linear inequality. Simply, we can define an *average unit time accumulated expected reward* in an interval. So, we define the atomic proposition (*ap*) as

$$ap ::= \frac{1}{|I_i| + 1} \sum_{j=0}^{|I_i|} \mathbf{r} \cdot \mathbf{x}(t_{I_i} + j) < b, \qquad (1)$$

where I_i is the ith interval, t_{I_i} is the beginning time instant of the interval, and $|I_i|$ is the interval length.

2.2   Long-Run Convergence

In this part, we demonstrate the long-run convergence properties of our extended Markov chain for the temporal analysis of the period stochastic system with memory. After a large time, the successive possible pmfs become periodic repeated vectors. Given a (k, d)-MC $\mathcal{A}_{\mathcal{D}_X}$, we use pmfs $\mathbf{x}$ of Markov chain to describe system states. At the memory time instant, system transition matrix is transient aperiodic. Pmfs converge by a linear memory transition.

Theorem 1. *Given a (k, d)-MC $\mathcal{A}_{\mathcal{D}_X} = (\mathcal{D}_X, \vec{\alpha})$ such that $\mathcal{D}_X = (S, \mathbf{M})$ if*

- *$|\vec{\alpha}| > 1$,*
- *The memory transition is fired infinitely along the path,*

then $\mathbf{x}(\infty) = \boldsymbol{\pi}$ such that $\boldsymbol{\pi} = \mathbf{M}\boldsymbol{\pi}$.

Proof. As constrained by the valid path, when a time delay transition is taken,

$$\mathbf{x}(t) = \sum_{i=1}^{|\vec{\alpha}|} \alpha_i \mathbf{x}(t - i)$$

$$= \sum_{i=1}^{|\vec{\alpha}|} \alpha_i \mathbf{M}^{(d+1-i)} \mathbf{x}(t - 1)$$

$$= (\sum_{j=0}^{d-1} \beta_j \mathbf{M}^j) \cdot \mathbf{x}(t - 1)$$

where $\beta_j = \sum \alpha_i$ and $(i - 1) \mod d \equiv j \ (1 \leq i \leq |\vec{\alpha}|)$.

Since there may be infinite delay transitions along the path, and multiplication of polynomials of singular matrix $\mathbf{M}$ is commutative, so

$$\lim_{t \to \infty} \mathbf{x}(t) = \lim_{t \to \infty, g \to \infty} (\sum_{i=0}^{d-1} \beta_i \mathbf{M}^i)^g \cdot (\mathbf{M}^{t-g} \mathbf{x}(0)) = \lim_{t \to \infty, g \to \infty} (\vec{\beta} \vec{\mathbf{M}}^T)^g (\mathbf{M}^{t-g} \mathbf{x}(0))$$

where $\mathbf{M}_D = \vec{\beta} \vec{\mathbf{M}}^T$ is an matrix computed by the inner product of $\vec{\beta}$ ($\beta_0 > 0$) and $\vec{\mathbf{M}} = (\mathbf{I}, \mathbf{M}, ..., \mathbf{M}^{d-1})$. Since β_0 is multiplied by $\mathbf{I}$, diagonal elements of $\mathbf{M}_D$

are positive, and then $\mathbf{M}_D$ is aperiodic. If $\mathbf{M}$ is irreducible, then as we do not remove existing transitions at STD structure, the result is that $\mathbf{M}_D$ is aperiodic and irreducible if the original matrix $\mathbf{M}$ is irreducible. Then there exists unique final distribution probabilistic vector $\boldsymbol{\pi}$ from arbitrarily initial pmf $\mathbf{x}(0)$, such that

$$\lim_{t \to \infty} \mathbf{x}(t) = \lim_{g \to \infty} (\vec{\beta} \vec{\mathbf{M}}^T)^g \mathbf{x}(t') = \boldsymbol{\pi}$$

where $\mathbf{x}(t') = \mathbf{M}^{t-g} \mathbf{x}(0)$.

We have proved that, in the long run, $\mathbf{x}(\infty)$ converges to unique $\boldsymbol{\pi}$. However, in terms of being affected by the periodic multiplication component $\mathbf{x}(t')$, we care more about whether the successive $\mathbf{x}(t)$ converges at the same time in the same interval, i.e., $\mathbf{x}$ having the same convergence rate in the same period. A n-dimension square matrix $\mathbf{P}$ is doubly stochastic [9] if both its row vectors and column vectors are stochastic distribution vectors.

Lemma 1. *Assume that* $\mathbf{P}$ *is a* $n \times n$ *doubly stochastic matrix, as* $t \to \infty$, $\mathbf{P}^t$ *is converging to* $[\frac{1}{n}]_{n \times n}$.

Theorem 2. *Given initial d-periodic pmfs* $\mathbf{X}_0 = (\mathbf{x}(0), ..., \mathbf{x}(d-1))^T$, *in the long-run convergence, the same periodic pmfs* $\mathbf{X}_j = (\mathbf{x}(0+d \cdot j), ..., \mathbf{x}(d-1+d \cdot j))^T$ *converge at the same rate.*

Proof. Define $\mathbf{X}_j$ as the column vector of the periodic pmf at the jth period after j times long-run converge. Then $\mathbf{X}_{j+1} = \mathbf{P} \cdot \mathbf{X}_j$, where $\mathbf{P}$ is a $d \times d$ doubly stochastic matrix. $\mathbf{P}_1$ is the first row of $\mathbf{P}$ as $(\beta_0, \beta_1, ..., \beta_{d-1})$, and $\mathbf{P}_2$ as $(\beta_{d-1}, \beta_0, ..., \beta_{d-2})$, and so on. Then we have the following: $\lim_{j \to \infty} \mathbf{X}_j = \mathbf{P} \cdot \mathbf{X}_j = \mathbf{P}^j \cdot \mathbf{X}_0 = [\frac{1}{d}]_{d \times d} \cdot \mathbf{X}_0 = [\boldsymbol{\pi} \; ... \; \boldsymbol{\pi}]^T$ from which we can conclude that pmfs of same period converge at the same rate.

We use pmfs $\mathbf{x}$ of the (k, d)-MC $\mathcal{A}_{\mathcal{D}_X}$ to describe system states. After a large time, the successive possible pmfs become periodic repeated vectors. At the memory time instant, system transition matrix is transient aperiodic. Pmfs converge by a linear memory transition. After the moment, an interval begins and continues until the next encountered memory transition instant, and generate successive intervals as $I_1, I_2, ...$ that is infinite. From the above definitions, we can use (k, d)-Markov chain to model a discrete stochastic linear system with probabilistic distribution, accidental memory and long run stability.

3 Verification of (k, d)-Markov Chain

In this section, we firstly present the interval probability temporal logic and its operators, secondly provide the underlying foundation needed to check the temporal properties of the period stochastic system, and finally give the feasible checking.

3.1   IPLTL: Interval Probability Linear Temporal Logic

We now describe the syntax and the semantics of ipLTL. ipLTL has the same logic and temporal operators as linear temporal logic [10]. With ipLTL, we can specify the average of the accumulated expected rewards about the interval trace I_i, in which the interval length is denoted by $|I_i|$ for the ith interval.

Syntax. We use ipLTL as the specification logic where the atomic propositions of the ipLTL are linear inequalities about the interval accumulated reward of pmf $\mathbf{x}(t)$. The syntax of ipLTL formula ψ is as follows:

$$\psi ::= \; \top \mid \bot \mid (\psi) \mid ap \mid \neg\psi \mid \psi \vee \psi \mid \psi \wedge \psi \mid \mathbf{X}_I\psi \mid \psi\mathbf{U}_I\psi \mid \psi\mathbf{R}_I\psi$$

where $\mathbf{r} \in \mathbb{R}^n$, $b \in \mathbb{R}$, ap defined as (1). As usual, $\rightarrow$, $\mathbf{G}_I$ and $\mathbf{F}_I$ are defined as $\psi \rightarrow \phi \equiv \neg\psi \vee \phi$, $\mathbf{G}_I\psi \equiv \bot\mathbf{R}_I\psi$, $\mathbf{F}_I\psi \equiv \top\mathbf{U}_I\psi$.

Given the (k,d)-Markov chain $\mathcal{A}_{\mathcal{D}_X} = (\mathcal{D}_X, \overrightarrow{\alpha})$, the average accumulated expected reward that σ_I is in interval I_i is given by

$$E[r](I_i) = \frac{1}{|I_i|+1} \sum_{j=0}^{|I_i|} \mathbf{r} \cdot \mathbf{x}(t_{I_i} + j)$$

$$= \frac{1}{|I_i|+1} \mathbf{r}\mathbf{M}^{\sum_{n=0}^{i-1} |I_n|}(\mathbf{M_D})^i \cdot \sum_{j=0}^{|I_i|} \mathbf{M}^j\mathbf{x}(0)$$

where $\mathbf{M}$ is the period Markov Chain transition matrix and $\mathbf{M}_D$ is the polynomial of periodic matrix $\mathbf{M}$.

Semantics. A binary satisfaction relation $\models$ over tuples of an interval trace σ_I of (k,d)-MC $\mathcal{A}_{\mathcal{D}_X}$ and an ipLTL formula is recursively defined as follows:

$$\sigma_I^0 \models \top, \sigma_I^0 \not\models \bot, \sigma_I^0 \models ap \; \Leftrightarrow \; \frac{1}{|I|+1} \sum_{j=0}^{|I|} \mathbf{r} \cdot \mathbf{x}(t_I + j) < b$$

$$\sigma_I^0 \models \neg\psi \Leftrightarrow \sigma_I^0 \not\models \psi$$

$$\sigma_I^0 \models \psi \vee \phi \Leftrightarrow \sigma_I^0 \models \psi \text{ or } \sigma_I^0 \models \phi$$

$$\sigma_I^0 \models \psi \wedge \phi \Leftrightarrow \sigma_I^0 \models \psi \text{ and } \sigma_I^0 \models \phi$$

$$\sigma_I^0 \models \mathbf{X}_I\psi \Leftrightarrow \sigma_I^1 \models \psi$$

$$\sigma_I^0 \models \psi\mathbf{U}_I\phi \Leftrightarrow \text{there is } j \geq 0 \text{ such that } \sigma_I^j \models \phi \text{ and } \sigma_I^i \models \psi \text{ for } i = 0,...,j-1$$

A binary satisfaction relation $\models$ over tuples of a (k,d)-MC $\mathcal{A}_{\mathcal{D}_X}$ and an ipLTL formula is define as follows:

$$\mathcal{A}_{\mathcal{D}_X} \models \psi \text{ iff for all initial pmf } \mathbf{x}(0) = \mathbf{M}^d\mathbf{x}(0), \; \sigma_I^0 \models \psi,$$

where $\mathcal{A}_{\mathcal{D}_X}$ may take either the period transition matrix or convergence transition matrix at every time instant under valid path semantics. Although both the initial pmf and the evolved interval trace are infinite, we can still construct a model checking method for the above satisfaction relation.

3.2 Feasible Checking

In this part, we provide the underlying foundation needed to check the temporal properties of the period stochastic system, and then do the feasible checking.

We have proved that the compositional Markov process has a stationary final pmf $\mathbf{x}(\infty)$, when infinite time interval delays apply to Markov matrix $\mathbf{M}_D$. Furthermore, since $\mathbf{x}(t)$ is a pmf, $0 \leq \mathbf{x}_i(t) \leq 1$ for all t. This constraint on the initial pmf and the final stationary pmf leads to bounding functions within which left-hand side of the inequalities, $\frac{1}{|I_i|+1} \sum_{j=0}^{|I_i|} \mathbf{r} \cdot \mathbf{x}(t_{I_i} + j) < b$, will remain for $t_{I_i} \geq 0$. So, for the set of inequalities of a given specification, there is a number N after which the truth values of the inequalities become constant for every initial pmf $x(0)$. This can guarantee the termination of the model checking procedure.

Theorem 3. *Let $\mathbf{M}_D$ be a matrix that is diagonalizable with eigenvalues $|\lambda_i| < 1$ for $i = 2...n$ and $\lambda_1 = 1$ [11,12], and let $\mathbf{x}(t) \in [0, 1]^{n \times 1}$ be the pmf at the ith interval. Then for all inequalities $\frac{1}{|I_i|+1} \sum_{j=0}^{|I_i|} \mathbf{r} \cdot \mathbf{x}(t_{I_i} + j) < b$ of a given ipLTL formula ψ, if $\frac{1}{|I_i|+1} \sum_{j=0}^{|I_i|} \mathbf{r} \cdot \mathbf{x}(\infty) \neq b$, then there is an integer N such that for any integer $N' \geq N$,*

$$\frac{1}{|I_{N'}| + 1} \sum_{j=0}^{|I_{N'}|} \mathbf{r} \cdot \mathbf{x}(t_{I_{N'}} + j) < b \ \textit{iff} \ \frac{1}{|I_N| + 1} \sum_{j=0}^{|I_N|} \mathbf{r} \cdot \mathbf{x}(t_{I_N} + j) < b.$$

In order to check the model $\mathcal{A}_{\mathcal{D}_X}$ against a specification of ipLTL formula ψ, we first build a labeled generalized Büchi automaton [13] by the expansion algorithm. An LGBA is a tuple $\mathcal{B} = (Q, \Sigma, L, \Delta, Q_0, \mathcal{F})$, where Q is a set of states, Σ is an alphabet consisting of all atomic propositions AP, $\Delta : Q \rightarrow 2^Q$ is a transition relation, $Q_0 \subseteq Q$ is a set of initial states, $\mathcal{F} = \{F_i \subseteq Q\}$ is a set of sets of accepting states, and $L : Q \rightarrow 2^{AP}$ is a labelling function that returns a set of inequalities that a state should satisfy. Let $w = a_0 a_1...$ be an ω-word over the alphabet Σ. $\rho = q_0 q_1...$ is a run of $\mathcal{B}$ on the word if $q_0 \in Q_0$, and for each $i \geq 0$, $q_{i+1} \in \Delta(q_i)$ and $a_i \in L(q_i)$. $\mathcal{B}$ accepts exactly those runs in which $inf(\rho) \cap F_i \neq \emptyset$. We say a run accepts an interval trace σ_I^τ if $\mathbf{x}(t_{I_\tau})$ satisfies all inequalities of $L(q_\tau)$ for $\tau \geq 0$. An LGBA $\mathcal{B}$ accepts an interval trace σ_I^0 if there is an accepting run of $\mathcal{B}$ that accepts σ_I^0.

We define a set of states $F_+ = \{q_i$: a run of $\mathcal{B}$ $\rho = q_0...q_i...q_k...$ accepts σ_I^∞, and q_k is cycled by a accepting run of $\mathcal{B}\}$. Note that any state in F_+ is reachable to an accepting run of $\mathcal{B}$. So, we check only those search paths of length N that end with a state in F_+.

Theorem 4. *Let $\mathcal{B}$ be an LGBA $(Q, \Sigma, L, \Delta, Q_0, \mathcal{F})$ for a specification ψ, let $\mathcal{A}_{\mathcal{D}_X}$ be with period matrix $\mathbf{M}$ and matrix $\mathbf{M}_D$, and let a set of inequalities $Ineq(\rho)$ for a run ρ be $Ineq(\rho) = \{\frac{1}{|I_i|+1}\mathbf{r}\cdot\mathbf{M}^{\sum_{n=1}^{i-1}|I_n|}(\mathbf{M_D})^i\cdot\sum_{j=0}^{|I_i|}\mathbf{M}^j\mathbf{x} : \frac{1}{|I_i|+1}\mathbf{r}\cdot$ $\sum_{j=0}^{|I_i|}\mathbf{M}^j\mathbf{x} \in L(q_i), i \in [0, N]\}$ where N is the bound computed in Theorem 3. An interval trace σ_I^0 is accepted by $\mathcal{B}$ iff there is a run $\rho = q_0q_1...q_\omega$ of $\mathcal{B}$ with $q_N \in F_+$ and $\mathbf{x}(0)$ is a feasible point of $Ineq(\rho)$.*

Proof. $\rightarrow$: Since $\mathcal{B}$ accepts σ_I^0, there is an accepting run ρ of $\mathcal{B}$ that accepts σ_I^0. In other words, $\mathbf{x}(t_{I_k})$ satisfies all inequalities in $L(q_k)$ for $k \geq 0$. Since $\frac{1}{|I_i|+1}\mathbf{r}\cdot\sum_{j=0}^{|I_i|}\mathbf{M}^j\mathbf{x}(t_{I_i}) < b \equiv \frac{1}{|I_i|+1}\mathbf{r}\cdot\mathbf{M}^{\sum_{n=0}^{i-1}|I_n|}(\mathbf{M_D})^i\cdot\sum_{j=0}^{|I_i|}\mathbf{M}^j\mathbf{x}(0) < b$, $\mathbf{x}(0)$ is a feasible point of $Ineq(\rho)$.

Let ρ^N be the suffix of ρ after the first N states. Since ρ^N is an accepting run of $\mathcal{B}$ with infinite length over a finite set of states, there is a loop in the run that includes at least one state of F_i for all $F_i \in \mathcal{F}$. Since some states of the loop are cycled by an accepting run of $\mathcal{B}$, and since q_N is reachable to those states, $q_N \in F_+$.

$\leftarrow$: Since $\mathbf{x}(0)$ satisfies all inequalities of $Ineq(\rho)$, ρ accepts σ_I^0, and since $q_N \in F_+$, ρ is an accepting run of $\mathcal{B}$. Hence, $\mathcal{B}$ accepts σ_I^0.

Let $\mathcal{B}_{\neg\psi}$ is an LGBA of negative ipLTL formula $\neg\psi$. Since $\mathcal{B}_{\neg\psi}$ accepts exactly those runs that satisfy $\neg\psi$, if there is a feasible solution for $Ineq(\rho)$ of a run ρ of $\mathcal{B}_{\neg\psi}$ with $q_N \in F_+$ then $\sigma_I^0 \models \neg\psi$. Hence, $\mathbf{x}(0)$ is a counterexample of the original specification ψ. The feasibility can be checked by solving a linear programming problem with artificial variables. However, since we check the feasibility by linear programming [14], we have a problem of inequality. For example, $\neg(a > b)$ will be translated as $-a < -b$ instead of $-a \leq -b$. However, for continuous values, precise equality is not meaningful. Instead, we can check properties such that $|a - b| < \epsilon$ or $|a - b| > \epsilon$. Note that $\neg(a < b)$ is converted to the inequality $-a < -b$ while finding the negation normal form of ψ.

4   Experiment

We consider a manufacturing system as the experiment object. The manufacturing system has three successive procedures, with the two steps of *Produce, Test* in each procedure. In each Test step, the product has a probability that fails the test and turn back to the prior procedure Test step. For example, in the procedure 1,if the *Test_1* fail, then it will turn back to the *Test_3*, else it will proceed to the *Product_2*. As production resources may be limited and profit must be maximized at each step, we want to check whether some interval resource limits and profit demands could be satisfied in the long run for the whole system.

The manufacture system can be simplified as periodic DTMC model is shown in Fig. 1. We assume memory effect vector $\overrightarrow{\beta}$ is (0.5, 0.3, 0.2). Reward vector represent cost of production resource a,b and unit profit at each state as $\mathbf{r}_a$: $(10, 5, 5, 1, 1, 1), \mathbf{r}_b$: $(1, 5, 1, 10, 5, 1)$, and $\mathbf{r}_{pro}$: $(1.0, -1.0, -1.0, 1.0, -1.0, 1.0)$,

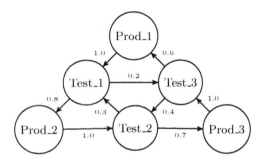

Fig. 1. Periodic Markov chain for the manufacture system

respectively. We assume an interval between delays lasts for 5 time instant and $|I| = 4$. We define the proposition that

- $P_1 : \mathbf{G}_I(\sum_{j=0}^{|I|} \mathbf{r}_a \cdot \mathbf{x}(t_I + j) < 17)$, always in an interval, accumulated cost of resource a is less than 17;
- $P_2 : \mathbf{F}_I(\sum_{j=0}^{|I|} \mathbf{r}_b \cdot \mathbf{x}(t_I + j) < 19.45)$, eventually in an interval, accumulated cost of resource b is less than 19.45;
- $P_3 : \mathbf{F}_I\mathbf{G}_I(\sum_{j=0}^{|I|} \mathbf{r}_c \cdot \mathbf{x}(t_I + j) < -0.9)$, eventually always in an interval, accumulated profit is less than -0.90;
- $P_4 : \neg(\sum_{j=0}^{|I|} \mathbf{r}_a \cdot \mathbf{x}(t_I + j) < 17)\mathbf{U}_I(\sum_{j=0}^{|I|} \mathbf{r}_{pro} \cdot \mathbf{x}(t_I + j) < -0.9)$, accumulated cost of resource a is not less than 17 until profit becomes less than -0.9;
- $P_5 : \mathbf{G}_I[(\sum_{j=0}^{|I|} (\mathbf{r}_b - \mathbf{r}_a) \cdot \mathbf{x}(t_I + j) < 0) \rightarrow \mathbf{F}_I(\sum_{j=0}^{|I|} \mathbf{r}_{pro} \cdot \mathbf{x}(t_I + k) < -0.9)]$, always if accumulated cost of a is more than b, then eventually profit becomes less than -0.9.

We implement feasibility checking methods, construct Büchi automata by the expansion algorithm, and get inequalities for negative ipLTL formula as described above. By the simplex method, we get the resulting initial pmfs that satisfy the inequalities and are the counterexamples of temporal properties to those of the original formula.

- P_1 is not satisfied and a counterexample initial pmf is (0.028229, 0.494350, 0.000001, 0.000001, 0.054476, 0.422944);
- P_2 is not satisfied and a counterexample initial pmf is (0.114460, 0.176970, 0.191294, 0.144981, 0.220887, 0.151408);
- P_3 is not satisfied and a counterexample initial pmf is (0.287844, 0.084431, 0.000001, 0.000001, 0.555488, 0.072236);
- P_4 is not satisfied and a counterexample initial pmf is (0.000000, 0.526011, 0.013627, 0.010328, 0.000001, 0.450031);
- P_5 is satisfied.

Fig.2 shows how the accumulated expected reward evolves along the interval path of three different reward vectors as $\mathbf{r}_a$ and $\mathbf{r}_b - \mathbf{r}_a$ when the initial pmf of the counterexample in $P1$. By the interval delay effect, we can conclude that the original periodic DTMC has the property of long-run convergence. After about

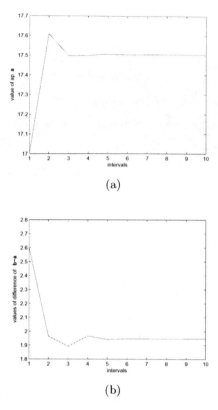

(a)

(b)

Fig. 2. Values of interval accumulated reward of (a) $\mathbf{r}_a$ and (b) $(\mathbf{r}_b - \mathbf{r}_a)$

6 intervals (30 time instants), the reward vector becomes stable and is the truth value of the atomic proposition constant.

5   Conclusions

An extended Markov model has been presented for the temporal analysis of periodic stochastic systems with memory. The model is constructed based on a d-period Markov chain with memory k, which is known as the (k, d)-Markov Chain ((k, d)-MC). We checked the (k, d)-MC for the properties written in ipLTL, shorted for interval probabilistic linear temporal logic. Using ipLTL, we can express the predicate about the expected interval behaviors of the system evolution expressed by the probabilistic mass functions. Since the interval pmfs of the (k, d)-MC converge, we can find a bound for the system execution time, after which the predicate about the interval pmfs (the inequality) becomes constant. Hence, feasibility checking is enabled.

Manufacturing systems are practical systems. We model checked some properties about profit and loss of one such system to illustrate the use of our method. We are currently conducting additional research, and we plan to use our model checking method in other large-scale applications. In this paper, we only focused on the same interval. In future work, we want to investigate the properties of different intervals by introducing the partial order reduction [15] technique.

Acknowledgment. This work is supported by the Funds NSFC61171121, NSFC60973049, and the Science Foundation of Chinese Ministry of Education-China Mobile 2012.

References

1. Kwiatkowska, M., Norman, G., Parker, D.: Prism: Probabilistic symbolic model checker. In: Field, T., Harrison, P.G., Bradley, J., Harder, U. (eds.) TOOLS 2002. LNCS, vol. 2324, pp. 200–204. Springer, Heidelberg (2002)
2. Aziz, A., Singhal, V., Balarin, F., Brayton, R.K., Sangiovanni-Vincentelli, A.L.: It usually works: The temporal logic of stochastic systems. In: Wolper, P. (ed.) CAV 1995. LNCS, vol. 939, pp. 155–165. Springer, Heidelberg (1995)
3. Bianco, A., De Alfaro, L.: Model checking of probabilistic and nondeterministic systems. In: Thiagarajan, P.S. (ed.) FSTTCS 1995. LNCS, vol. 1026, pp. 499–513. Springer, Heidelberg (1995)
4. Kwon, Y., Agha, G.: Linear inequality ltl (iltl): A model checker for discrete time markov chains. In: Davies, J., Schulte, W., Barnett, M. (eds.) ICFEM 2004. LNCS, vol. 3308, pp. 194–208. Springer, Heidelberg (2004)
5. Kwon, Y., Agha, G.: Verifying the evolution of probability distributions governed by a dtmc. IEEE Transactions on Software Engineering 37(1), 126–141 (2011)
6. Papoulis, A., Pillai, S.U.: Probability, random variables, and stochastic processes. Tata McGraw-Hill Education (2002)
7. Delcher, A.L., Harmon, D., Kasif, S., White, O., Salzberg, S.L.: Improved microbial gene identification with glimmer. Nucleic Acids Research 27(23), 4636–4641 (1999)
8. Pinsky, M., Karlin, S.: An introduction to stochastic modeling. Academic press (2010)
9. Sinkhorn, R.: A relationship between arbitrary positive matrices and doubly stochastic matrices. The Annals of Mathematical Statistics 35(2), 876–879 (1964)
10. Gerth, R., Peled, D., Vardi, M.Y., Wolper, P.: Simple on-the-fly automatic verification of linear temporal logic. In: Proceedings of the Fifteenth IFIP WG6. 1 International Symposium on Protocol Specification, Testing and Verification. IFIP (1995)
11. Strang, G.: Introduction to linear algebra. SIAM (2003)
12. Lay, D.C.: Linear algebra and its applications. Univ. of Maryland-College Park (2003)
13. Büchi, J.R.: On a decision method in restricted second order arithmetic. In: The Collected Works of J. Richard Büchi, pp. 425–435. Springer (1990)
14. Luenberger, D.G.: Linear and nonlinear programming. Springer (2003)
15. Alur, R., Brayton, R.K., Henzinger, T.A., Qadeer, S., Rajamani, S.K.: Partial-order reduction in symbolic state space exploration. In: Grumberg, O. (ed.) CAV 1997. LNCS, vol. 1254, pp. 340–351. Springer, Heidelberg (1997)

Argument Ranking with Categoriser Function*

Fuan Pu, Jian Luo, Yulai Zhang, and Guiming Luo

School of Software, Tsinghua University, Beijing, China
{pfa12,j-luo10,zhangyl08}@mails.tsinghua.edu.cn,
gluo@mail.tsinghua.edu.cn

Abstract. Recently, ranking-based semantics is proposed to rank-order arguments from the most acceptable to the weakest one(s), which provides a graded assessment to arguments. In general, the ranking on arguments is derived from the strength values of the arguments. Categoriser function is a common approach that assigns a strength value to a tree of arguments. When it encounters an argument system with cycles, then the categoriser strength is the solution of the non-linear equations. However, there is no detail about the existence and uniqueness of the solution, and how to find the solution (if exists). In this paper, we will cope with these issues via fixed point technique. In addition, we define the categoriser-based ranking semantics in light of categoriser strength, and investigate some general properties of it. Finally, the semantics is shown to satisfy some of the axioms that a ranking-based semantics should satisfy.

Keywords: abstract argumentation, ranking semantics, graded assessment, categoriser function, fixed point technique.

1 Introduction

The field of computational models of argumentation [1] aims at reflecting on how human argumentation utilizes incomplete and inconsistent knowledge to construct and analyze arguments about the conflicting options and opinions.

The most popularly used framework to talk about general issues of argumentation is that of abstract argumentation [2], which provides a unifying and powerful tool for the study of many formal systems developed for common-sense reasoning. In the past nearly 20 years, several different kinds of semantics for abstract argumentation system have been proposed that highlight various aspects of argumentation [3, 4, 5]. Those semantics partition the set of arguments into two classes: extensions and non-extensions. Each extension is a set of arguments, which is able to "survive together" and represents a coherent point of view. In order to reason with a semantics one has to take either a credulous or skeptical perspective. In other words, an argument is ultimately *accepted* with respect to a semantics if it belongs to every extension; an argument is *rejected* if it dose not belong to any extension; and an argument is *undecided* if it is in some extensions and not in others.

---

* This work is supported by the Funds NSFC61171121, NSFC60973049, and the Science Foundation of Chinese Ministry of Education-China Mobile 2012.

R. Buchmann et al. (Eds.): KSEM 2014, LNAI 8793, pp. 290–301, 2014.

However, those semantics may exhibit a variety of problematic aspects such as emptiness, non-existence, multiplicity [6] when encountering cycles, and are not suitable for practical applications in some scenarios. Considering an argument system whose grounded extension is empty, for example, if one must make a choice, then the grounded semantics is unavailable since all arguments are unacceptable in this case.

Recently, [7] introduces a new family of semantics, which provide a graded assessment to arguments, i.e., it ranks arguments from the most acceptable to the weakest one(s). In fact, this line of thinking has been mentioned in [8], in which two approaches, generic local valuation and global valuation, are proposed to evaluate the strength of an argument, and then a preordering (ranking) on arguments is induced by those strength values. In particular, the authors show that the approach for local valuation generalizes the categoriser function [9], and enables to handle cycles, then the strength valuation is the solution of second-degree equations. However, there is no detail about the following questions: Does there exist a solution for these equations? If it exists, is it unique or multiple and how to find them?

In this paper, we expect to tackle these issues by fixed-point technique. In addition, a ranking-based semantics, called categoriser-based ranking semantics, is defined in the light of categoriser valuation, and some of its properties are investigated. Lastly, we prove that the semantics satisfies some of the axioms, proposed by [7], which a ranking-based semantics should satisfy. The remainder of this paper is structured as follows. In Section 2, we briefly recall some backgrounds on abstract argumentation and the ranking-based semantics for argumentation frameworks. In Section 3, we employ the fixed-point technique to analyze the categoriser strength valuation for argumentation system, and the categoriser-based ranking semantics is defined. We relate the semantic with [7] in Section 4 and conclude in Section 5.

2   Preliminaries

2.1   Abstract Argumentation Framework

Abstract argumentation frameworks [2] convey a very simple view on argumentation since they do not presuppose any internal structure of an argument. Here, the interactions among arguments are attack relations, which express conflicts between them.

Definition 1 (Abstract Argumentation Framework). *An argumentation framework is a pair AF = $\langle \mathcal{X}, \mathcal{R} \rangle$ where $\mathcal{X}$ is a finite set of arguments and $\mathcal{R} \subseteq \mathcal{X} \times \mathcal{X}$ is a binary relation on $\mathcal{X}$, also called attack relation. $(a,b) \in \mathcal{R}$ means that a attacks b, or a is a (direct) attacker of b. Often, we write $(a,b) \in \mathcal{R}$ as $a\mathcal{R}b$.*

We denote by $\mathcal{R}^-(x)$ (respectively, $\mathcal{R}^+(x)$) the subset of $\mathcal{X}$ containing those arguments that attack (respectively, are attacked by) the argument $x \in \mathcal{X}$, extending this notation in the natural way to sets of arguments, so that for $S \subseteq \mathcal{X}$, $\mathcal{R}^-(S) \triangleq \{x \in \mathcal{X} : \exists y \in S \text{ such that } x\mathcal{R}y\}$ and $\mathcal{R}^+(S) \triangleq \{x \in \mathcal{X} : \exists y \in S \text{ such that } y\mathcal{R}x\}$.

A set $S \subseteq \mathcal{X}$ is *conflict-free* iff $S \cap \mathcal{R}^-(S) = \emptyset$. Let $\mathfrak{F} : 2^{\mathcal{X}} \mapsto 2^{\mathcal{X}}$ be the *characteristic function* of an argument system such that $\mathfrak{F}(S) = \{x \in \mathcal{X} : \mathcal{R}^-(x) \subseteq \mathcal{R}^+(S)\}$. We define the *defenders* of an argument x, denoted by $\mathcal{D}(x)$, are the attackers of the elements of $\mathcal{R}^-(x)$. Formally, $\mathcal{D}(x) = \{y \in \mathcal{X} : y \in \mathcal{R}^-\left(\mathcal{R}^-(x)\right)\}$.

To define the solutions of an argument system, we mean selecting a set of arguments that satisfy some acceptable criteria. Let $S \subseteq \mathcal{X}$ be conflict-free, then, S is **admissible** iff $S \subseteq \mathfrak{F}(S)$; S is a **preferred extension** iff it is a maximal (w.r.t. $\subseteq$) admissible set; S is a **complete extension** iff $S = \mathfrak{F}(S)$; S is a **grounded extension** iff it is the minimal (w.r.t. $\subseteq$) complete extension (or, alternatively, it is the least fixed point $\mathfrak{F}$); S is a **stable extension** iff $\mathcal{R}^+(S) = \mathcal{X} \backslash S$.

Example 1. *Let us consider the abstract argumentation framework illustrated in Figure 1, in which vertices represent arguments and direct arcs correspond to attacks (i.e. elements of $\mathcal{R}$). For this example, $\{x_1, x_3\}$ is the preferred, complete and grounded extension, however, there exist no stable extensions at all.*

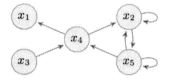

Fig. 1. A simple example of argumentation framework

From the above example, it is shown that an abstract argumentation framework can be represented as a digraph, known as attack graph. One of the often-used ways is to represent a digraph as 0-1 matrix for computational purposes. For an argumentation system $AF = \langle \mathcal{X}, \mathcal{R} \rangle$ with $\mathcal{X} = \{x_1, x_2, \cdots, x_n\}$, we define the **attack matrix** of AF as the $n \times n$ matrix $\mathbf{D} = [d_{ij}]$ such that $d_{ij} = 1$ if $x_j \mathcal{R} x_i$; otherwise, 0.[1] For instance, the attack matrix of the argument system in Example 1 is

$$\mathbf{D} = \begin{bmatrix} 0 & 0 & 0 & 1 & 0 \\ 0 & 1 & 0 & 1 & 1 \\ 0 & 0 & 0 & 0 & 0 \\ 0 & 0 & 1 & 0 & 1 \\ 0 & 1 & 0 & 0 & 1 \end{bmatrix}$$

Moreover, we denote the i-th row of $\mathbf{D}$ by $\mathbf{D}_{i*}$, which can indicate some information about the direct attackers of argument x_i, e.g., the sum of $\mathbf{D}_{i*}$ shows the number of attackers of x_i.

2.2   Ranking-Based Semantics for Argument System

In order to provide a graded assessment to arguments, [7] proposes ranking-based semantics which rank-order the set of arguments from the most acceptable to the weakest one(s). This novel approach is distinct from the already existing semantics which assign an absolute status (*accepted*, *rejected* and *undecided*) for each argument. It compares pairs of arguments in the light of their respective sets of attackers, and states which arguments is more acceptable than another.

---

[1] In fact, the attack matrix of an argumentation framework is the transpose of the adjacency matrix of its corresponding attack graph.

Before proceeding let us first formally characterize what we mean by the statements "ranking" in light of linear orderings [10].

Definition 2 (Ranking). *Let* $\mathcal{T}$ *be some set. A ranking* $\succeq$ *on* $\mathcal{T}$ *is a binary relation on* $\mathcal{T}$ *such that:*

- $\succeq$ *is total (i.e. for all* $x, y \in \mathcal{T}$, $x \succeq y$ *or* $y \succeq x$);
- $\succeq$ *is transitive (i.e. for all* $x, y, z \in \mathcal{T}$, *if* $x \succeq y$ *and* $y \succeq z$, *then* $x \succeq z$).

Let $\mathfrak{R}(\mathcal{T})$ *be the set of all rankings on* $\mathcal{T}$.

In this paper, we give $x \succeq y$ the meaning that x is at least as acceptable as y. This may be more intuitive than that of [7], in which the meaning of $x \succeq y$ is just the opposite. Formally, $x \simeq y$ if and only if $x \succeq y$ and $y \succeq x$, which means x and y are equally acceptable. Moreover, $x \succ y$, means x is strictly more acceptable than y, if and only if $x \succeq y$ but not $y \succeq x$.

Definition 3 (Ranking-based Semantics). *Let* $\mathbb{G}_{\mathcal{X}}$ *be the set of all argument systems with finite argument set* $\mathcal{X}$. *A ranking-based semantics is a function* $\Gamma : \mathbb{G}_{\mathcal{X}} \mapsto \mathfrak{R}(\mathcal{X})$.

In other words, for a given argumentation framework $AF = \langle \mathcal{X}, \mathcal{R} \rangle$, the ranking-based semantics Γ will transform $\mathcal{X}$ into a ranking $\succeq_{\Gamma}^{AF} \in \mathfrak{R}(\mathcal{X})$.

Generally, the ranking on arguments is induced by the strength values of the arguments. One of the most common approaches is categoriser function [9], which assigns a strength value to each argument. We will discuss it in the next section.

3   Argument Ranking with Categoriser Function

3.1   Categoriser Function for Strength Valuation

"Categoriser" function is originally used for "deductive" arguments, where an argument is structured as a pair $\langle \Phi, \phi \rangle$, where Φ is a set of formulae, called **premise**, ϕ is a formula, called **claim**, and Φ entails ϕ. The attack relation considered here is canonical undercut and cycles are not allowed. The notion of an "argument tree" captures a precise and complete representation of attackers and defenders of a given argument, root of the tree. Then, categoriser function assigns a value to a tree of arguments. This value represents the relative strength of an argument (root of the tree) given all its attackers and defenders. The categoriser function, denoted by C, is defined as

$$C(x_i) = \begin{cases} 1 & \text{if } \mathcal{R}^-(x_i) = \emptyset \\ \frac{1}{1+C(x_1')+\cdots+C(x_n')} & \text{if } \mathcal{R}^-(x_i) \neq \emptyset \text{ with } \mathcal{R}^-(x_i) = \{x_1', \cdots, x_n'\} \end{cases} \quad (1)$$

Intuitively, the larger the number of defeaters of an argument, the lower its value. The larger the number of defenders of an argument, the larger its value.

Note that, in the work of [9], categoriser function solely handles acyclic graphs. However, Cayrol et al. reveal that the categoriser function is an instance of their generic local valuation [8], thus making it possible to cope with cycles. In this case, the strength values are the solution of non-linear equations. Specifically, let $\langle \mathcal{X}, \mathcal{R} \rangle$ be an argument system with $\mathcal{X} = \{x_1, x_2, \cdots, x_n\}$, and its attack matrix be $\mathbf{D}$, and suppose the strength values of all arguments be column vector v, of which the i-th component, denoted by v_i or $v(x_i)$, represents the strength value of x_i, then the strength values are the solution of the following n equations:

$$v_i = 1/(1 + \mathbf{D}_{i*} \cdot v), \quad i = 1, 2, \cdots, n \tag{2}$$

Remark 1. (2) exactly expresses the categoriser functions in (1) irrespective of whether $\mathcal{R}^-(x_i)$ is empty or not, as the item $\mathbf{D}_{i*} \cdot v$ exactly indicates the sum of the strength values of all attackers of x_i.

Remark 2. We merely consider all strength values as nonnegative real numbers, i.e., $v_i \geq 0$ for all i. Combining with (2), we can easily know $v_i \in [0, 1]$ (actually $v_i \in (0, 1]$). This means that if the solution of (2) exists, it must be in $[0, 1]^n$.

3.2 Fixed Point Schema for Categoriser Equations

In [8], the authors show a simple example to evaluate arguments in a isolated cycle with categoriser valuation by solving second degree equations. For a complex argumentation system, however, no details are available about these questions: Do these equations always exist solutions in the reals and how many real solutions exist? If the real solutions exist, how should we find them? In this subsection, we will address these questions through fixed-point techniques.

Firstly, let us transform the equations into the fixed-point form [11]:

$$v = F(v) = [f_1(v), f_2(v), \cdots, f_n(v)]^T \tag{3}$$

where function F maps $[0, 1]^n$ into $[0, 1]^n$, and the function f_i from $[0, 1]^n$ to $[0, 1]$, called the coordinate function of F, is defined by the categoriser function, i.e.,

$$f_i(v) = 1/(1 + \mathbf{D}_{i*} \cdot v) \tag{4}$$

Intuitively, $F(0) = 1$, where the bold 0 (respectively, 1) is an appropriately dimensioned column vector of all 0's (respectively 1's). Sometimes, we also write $f_i(v)$ as the function of v and $\mathbf{D}_{i*}$, i.e.,

$$f_i(v) = f(v, \mathbf{D}_{i*}) \tag{5}$$

Clearly, the function $f(v, \mathbf{D}_{i*})$ is a non-increasing function with respect to v and $\mathbf{D}_{i*}$. Note that $f(v, 0) = 1$ for any $v \in [0, 1]^n$.

We convert the original problem into a fixed point problem. Then, finding the solutions of the categoriser equations is equivalent to finding the fixed-points of F. In other words, a fixed-point of F is a solution of (2). Now, let us give the following theorem, which shows that the solution of categoriser equations always exists.

Theorem 1 (Existence of categoriser valuation). *For any argumentation framework* $AF = \langle \mathcal{X}, \mathcal{R} \rangle$ *with* $\mathcal{X} = \{x_1, x_2, \cdots, x_n\}$, *the categoriser valuation defined in (2) has at least one solution in* $[0, 1]^n$.

Proof (Sketch). We prove the equivalence result that function F has at least one fixed point. The proof uses Brouwer's fixed point theorem [12, Thm 2.14, pp. 24] and the observation that $[0, 1]^n$ is homeomorphic to a closed ball (closed, bounded, connected and without holes) and function F is continuous on it. $\square$

The previous theorem is of utmost importance if we are to widely use categoriser valuation, since one would be turned away from an argumentation system that is not capable of assigning meaningful strength values (real solution) to arguments in all case.

Next we focus on the existence of a unique valuation. Assigning multiple solutions to an argumentation framework may be more interesting from a theoretical perspective, but we look forward to the kind of users of this framework to expect a unique valuation. One intuitive application of a unique categoriser valuation is that it may help removing ambiguity on argument ranking. We will show that for every argumentation system there always exists a unique categoriser valuation, and the valuation can be calculated by fixed point iteration.

Theorem 2 (Uniqueness of categoriser valuation). *Let $AF = \langle \mathcal{X}, \mathcal{R} \rangle$ be an argument system with $\mathcal{X} = \{x_1, x_2, \cdots, x_n\}$. Then, the categoriser equations defined in (2) has a unique solution $v^* \in [0, 1]^n$, which is the limit of the sequence of $\{v^{(k)}\}_{k=0}^{\infty}$, defined from an arbitrarily selected $v^{(0)}$ in $[0, 1]^n$ and generated by*

$$v^{(k)} = F(v^{(k-1)}), \quad \text{for each } k \geq 1 \tag{6}$$

Proof. Let $u^{(0)} = 0$, $u^{(1)} = F(u^{(0)}) = 1$ and $u^{(k)} = F(u^{(k-1)})$ for each $k \geq 2$. Then, we can easily know that

$$u^{(0)} \leq u^{(2)} \leq u^{(1)} \tag{7}$$

and that there exists $0 < \varphi < 1$ such that

$$\varphi u^{(1)} \leq u^{(2)} \tag{8}$$

Since F is non-increasing (i.e., for any $u, v \in [0, 1]^n$, if $u \leq v$ then $F(u) \geq F(v)$), by applying F on (7) and by induction, it is easy to see that

$$0 = u^{(0)} \leq u^{(2)} \leq \cdots \leq u^{(2k)} \leq \cdots \leq u^{(2k+1)} \leq \cdots \leq u^{(3)} \leq u^{(1)} = 1 \tag{9}$$

On the other hand, from (8) and (9), we find $\varphi u^{(2k-1)} \leq u^{(2k)}$ for each $k \geq 1$. Letting $\pi_k = \sup\{\pi : \pi u^{(2k-1)} \leq u^{(2k)}\}$, then $\pi_k u^{(2k-1)} \leq u^{(2k)}$ and $0 < \varphi \leq \pi_1 \leq \cdots \leq \pi_k \leq \cdots \leq 1$. In the following, we prove that $\lim_{k \to \infty} \pi_k = 1$.

Note that $f_i(\pi u) = \frac{1}{\pi + f_i(u)(1-\pi)} f_i(u)$ for all $i \in \{1, 2, \cdots, n\}$, then there exists $0 < \alpha < 1$ and a continuous function $\psi(\pi) = \frac{1}{\pi + \alpha(1-\pi)}$ such that

$$F(\pi u) \leq \psi(\pi) F(u), \quad \forall \pi \in [\varphi, 1), u \in [\varphi, 1]^n \tag{10}$$

Then, by (9), (10) and the non-increasing property of F, we have

$$u^{(2k+1)} = F(u^{(2k)}) \leq F(\pi_k u^{(2k-1)}) \leq \psi(\pi_k) u^{(2k)} \leq \psi(\pi_k) u^{(2k+2)} \tag{11}$$

Algorithm 1. Fixed-point iteration for categoriser valuation

Input: $\mathbf{D}_{n \times n}$: attack matrix; ϵ: prescribed tolerance;
Output: $\boldsymbol{v}^{(k)}$: the approximate solution of the categoriser equations

1 **begin**
2 $\quad$ $k \longleftarrow 0;\ \boldsymbol{v}^{(0)} \longleftarrow \mathbf{1};$
3 $\quad$ **repeat**
4 $\quad\quad$ $k \longleftarrow k + 1;$
5 $\quad\quad$ $v_i^{(k)} = f(\boldsymbol{v}^{(k-1)}, \mathbf{D}_{i*})$ for each $i \in \{1, 2, \cdots, n\};$
6 $\quad$ **until** $\|\boldsymbol{v}^{(k)} - \boldsymbol{v}^{(k-1)}\| \leqslant \epsilon;$
7 $\quad$ **return** $\boldsymbol{v}^{(k)};$
8 **end**

---

which implies that $\pi_{k+1} \geq \frac{1}{\psi(\pi_k)} = \pi_k + \alpha(1 - \pi_k)$. So,

$$1 - \pi_{k+1} \leq (1 - \alpha)(1 - \pi_k) \leq \cdots \leq (1 - \alpha)^k(1 - \pi_1) \leq (1 - \alpha)^k(1 - \varphi) \quad (12)$$

As $0 < \alpha < 1$, thus by (12) we have

$$\lim_{k \to \infty}(1 - \pi_{k+1}) = 0 \quad \Rightarrow \quad \lim_{k \to \infty} \pi_k = 1 \quad (13)$$

Therefore, by (9) we get, for any integer $p \geq 1$

$$0 \leq \boldsymbol{u}^{(2k+2p)} - \boldsymbol{u}^{(2k)} \leq \boldsymbol{u}^{(2k+1)} - \boldsymbol{u}^{(2k)} \leq (1 - \pi_k)\boldsymbol{u}^{(2k+1)} \leq (1 - \pi_k)\boldsymbol{u}^{(1)} \quad (14)$$

Since $[0, 1]^n$ is normal, both $\{\boldsymbol{u}^{(2k+1)}\}_{k=0}^{\infty}$ and $\{\boldsymbol{u}^{(2k)}\}_{k=1}^{\infty}$ are convergence sequences. By (13) and (14), thus, there exists $\boldsymbol{u}^* \in [0, 1]^n$ such that

$$\lim_{k \to \infty} \boldsymbol{u}^{(2k+1)} = \lim_{k \to \infty} \boldsymbol{u}^{(2k)} = \boldsymbol{u}^* \quad (15)$$

Hence $\boldsymbol{u}^{(2k)} \leq \boldsymbol{u}^* \leq \boldsymbol{u}^{(2k-1)}$ and $\boldsymbol{u}^{(2k)} \leq F(\boldsymbol{u}^*) \leq \boldsymbol{u}^{(2k+1)}$. Letting $k \to \infty$ and combining with (15), it follows $F(\boldsymbol{u}^*) = \boldsymbol{u}^*$, i.e., $\boldsymbol{u}^*$ is a fixed point of F.

Now, for any $\boldsymbol{v}^{(0)} \in [0, 1]^n$ and for any $k \geq 1$, by induction, we have $\boldsymbol{u}^{(2k)} \leq \boldsymbol{v}^{(2k)} \leq \boldsymbol{u}^{(2k-1)}$ and $\boldsymbol{u}^{(2k)} \leq \boldsymbol{v}^{(2k+1)} \leq \boldsymbol{u}^{(2k+1)}$. Then $\boldsymbol{v}^{(k)} \to \boldsymbol{v}^* = \boldsymbol{u}^*$ as $k \to \infty$. In particular, let $\boldsymbol{v}^{(0)} = \boldsymbol{w}^*$, where $\boldsymbol{w}^*$ is any fixed point of F in $[0, 1]^n$, then $\boldsymbol{v}^{(k)} = \boldsymbol{w}^*$ for all $k \geq 1$, and we get $\boldsymbol{w}^* = \boldsymbol{u}^*$. So, F has a unique fixed point in $[0, 1]^n$. $\qquad\square$

The proof of this theorem mainly refers to [13, Lmm 2.1]. An approximate calculation of the unique categoriser valuation $\boldsymbol{v}^*$ is done by using Algorithm 1. In this paper, we set the initial strength values $\boldsymbol{v}^{(0)} = \mathbf{1}$ since we assume that each argument is not attacked at the beginning and has the maximum strength value 1. The iteration terminates when the change of the sequence $\{\boldsymbol{v}^{(k)}\}_{k=0}^{\infty}$ is under a given tolerance ϵ. As the proof of uniqueness suggests, the estimation of convergence rate of this algorithm is

$$\|\boldsymbol{v}^{(2k)} - \boldsymbol{v}^*\| \leq \|\boldsymbol{v}^{(2k)} - \boldsymbol{u}^{(2k)}\| + \|\boldsymbol{v}^* - \boldsymbol{u}^{(2k)}\| \leq 2\|\boldsymbol{u}^{(2k+1)} - \boldsymbol{u}^{(2k)}\| \quad (16)$$

By (12) and (14), we have $\|\boldsymbol{v}^{(2k)} - \boldsymbol{v}^*\| \leq 2(1 - \alpha)^{k-1}(1 - \varphi)\|\boldsymbol{u}^{(1)}\|$. Similar argument gives that $\|\boldsymbol{v}^{(2k+1)} - \boldsymbol{v}^*\| \leq 2(1 - \alpha)^{k-1}(1 - \varphi)\|\boldsymbol{u}^{(1)}\|$.

Remark 3. In Algorithm 1, we can see that at each iterative step the strength value of any argument x_i is simultaneously recomputed in the light of its direct attackers (represented by $\mathbf{D}_{i*}$) and the strength values in the previous step (i.e., $v^{(k-1)}$). This exactly embodies the idea of "local approach" (i.e., the value of an argument only depends on the values of its direct attackers) in [8].

3.3   Categoriser-Based Ranking Semantics

Now, we have shown that for the categoriser equations there always exists a unique solution for any argumentation framework. The solution assigns a numerical value to each argument, which can be interpreted as the strength of the argument. The greater the strength value, the more acceptable the argument. Thus, we can induce a ranking on arguments from the unique solution.

Definition 4 (Categoriser-based ranking semantics). *Let* $AF = \langle \mathcal{X}, \mathcal{R} \rangle$ *be an argumentation framework, and* v^* *be the unique solution of* (2). *The categoriser-based ranking semantics is a ranking-based semantic and transforms AF into the ranking* $\succeq$ *such that* $\forall x_i, x_j \in \mathcal{X}$, $x_i \succeq x_j$ *if and only if* $v^*(x_i) \geq v^*(x_j)$.

Obviously, the categoriser-based ranking semantics satisfies that for any $x \in \mathcal{X}$, $v^*(x_i) = 1$ if $\mathcal{R}^-(x_i) = \emptyset$; else $v^*(x_i) < 1$, i.e., non-attacked arguments are more acceptable than attacked ones. Non-attacked arguments are supported by extension-based semantics. They are part of any extension under complete, preferred, stable and grounded semantics. Therefore, it is naturel to believe that an argument which has no attackers is ranked higher than another argument which has attackers.

In addition, we give other properties of the categoriser-based ranking semantics:

Proposition 1. *Let* $x_i, x_j \in \mathcal{X}$. *The categoriser-based ranking semantics satisfies:*

[P1]: *If* $\mathcal{R}^-(x_i) = \mathcal{R}^-(x_j)$, *then* $x_i \simeq x_j$.
[P2]: *If* $\mathcal{R}^-(x_i) \subseteq \mathcal{R}^-(x_j)$, *then* $x_i \succeq x_j$.

Proof. By (6), the categoriser strength of any argument x_i can be written as

$$v^*(x_i) = \lim_{k \to \infty} v^{(k)}(x_i) = \lim_{k \to \infty} f_i(v^{(k-1)}) = \lim_{k \to \infty} f(v^{(k-1)}, \mathbf{D}_{i*}) \qquad (17)$$

For [P1], $\mathcal{R}^-(x_i) = \mathcal{R}^-(x_j)$ implies that $\mathbf{D}_{i*} = \mathbf{D}_{j*}$, which implies $f(v^{(k-1)}, \mathbf{D}_{i*}) = f(v^{(k-1)}, \mathbf{D}_{j*})$. By (17), we have $v^*(x_i) = v^*(x_j)$, i.e., $x_i \simeq x_j$. For [P2], $\mathcal{R}^-(x_i) \subseteq \mathcal{R}^-(x_j)$ means that $\mathbf{D}_{i*} \leq \mathbf{D}_{j*}$. Since $f(v, \mathbf{D}_{i*})$ is a non-increasing function of $\mathbf{D}_{i*}$, we have $f(v^{(k-1)}, \mathbf{D}_{i*}) \geq f(v^{(k-1)}, \mathbf{D}_{j*})$. Thus, $v^*(x_i) \geq v^*(x_j)$, i.e., $x_i \succeq x_j$. $\square$

This proposition states that two arguments with the same direct attackers have the same ranking, and an argument, whose direct attackers pertain to the set of direct attackers of another argument, is at least as more acceptable than the argument.

Let us show an example of how the semantics works:

Example 2. *Consider again the argument system in Fig. 1. Let* $\epsilon = 10^{-3}$ *and* $v^{(0)} = \mathbf{e}$. *Then, the valuation sequence* $\{v^{(k)}\}_{k=0}^{\infty}$, *calculated by Algorithm 1, is shown in Fig. 2.*

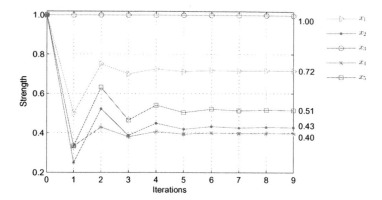

Fig. 2. Categoriser valuation sequence of Example 1

When $k = 0$, all arguments have the maximum strength value 1 as we presuppose each argument is not attacked at the beginning.

When $k = 1$, then the strength value of each argument merely depends on the number of its direct attackers since the strength values of all argument from the previous step are 1. Thus, x_3 has the maximum strength value 1 since it has no attacker, followed by x_1 with one attacker, and followed by x_4 and x_5 with two attackers, and followed by x_2 with three. From another perspective, since $\mathcal{R}^-(x_3) \subset \mathcal{R}^-(x_1) \subset \mathcal{R}^-(x_5) \subset \mathcal{R}^-(x_2)$, then by Proposition 1 we have $x_3 \succ x_1 \succ x_5 \succ x_2$.

When $k = 2$, after a new round of calculation, the strength value of each argument is recomputed. But, since $\mathcal{R}^-(x_3) \subset \mathcal{R}^-(x_1) \subset \mathcal{R}^-(x_5) \subset \mathcal{R}^-(x_2)$ always holds, the ranking among them will not be changed. Note that the ranking on x_2 and x_4 is altered as the sum of the strength values of the attackers of x_2 is greater than that of x_4.

......

After finitely many iterations, the valuation sequence gradually tends to be stable and converge to an approximative solution $\boldsymbol{v}^* = [0.72, 0.43, 1.00, 0.40, 0.51]^T$ within a tolerable range. Actually, the valuation sequence reflects how argument strength values change with iterations. Note that x_1 has a maximum strength value 1, since it is not attacked, and all other arguments have the strength values less than 1 since they are attacked by at least one argument. With the solution $\boldsymbol{v}^*$, the categoriser-based ranking semantics gives: $x_3 \succ x_1 \succ x_5 \succ x_2 \succ x_4$.

4 Relating with Ranking Axioms

In [7], the authors set up a set of axiom (postulates) that ranking-based semantics should satisfy. In this section, we will formally show that the categoriser-based ranking semantics meets some of these postulates.

The first axiom is that a ranking on a set of arguments does not rely on their identity but only on the attack relations among them. In other words, if two argumentation system are isomorphic then they should have the same ranking semantics. The isomorphisms between argumentation frameworks $AF_1 = \langle \mathcal{X}_1, \mathcal{R}_1 \rangle$ and $AF_2 = \langle \mathcal{X}_2, \mathcal{R}_2 \rangle$ is a bijective function $\tau \colon \mathcal{X}_1 \mapsto \mathcal{X}_2$ such that for all $x, y \in \mathcal{X}_1$, $x\mathcal{R}_1 y$ if and only if $\tau(x)\mathcal{R}_2\,\tau(y)$. Now we define the first axiom, called *abstraction*, as follows:

Axiom 1 (Abstraction (Ab)). *A ranking-based semantics* Γ *satisfies* abstraction *iff for any two argumentation framework* $AF_1 = \langle \mathcal{X}_1, \mathcal{R}_1 \rangle$ *and* $AF_2 = \langle \mathcal{X}_2, \mathcal{R}_2 \rangle$, *for any isomorphism* τ *from* AF_1 *to* AF_2, *we have* $\forall x, y \in \mathcal{X}_1, x \succeq_\Gamma^{AF_1} y$ *iff* $\tau(x) \succeq_\Gamma^{AF_2} \tau(y)$.

The second axiom states the question that whether an argument x is at least as acceptable as an argument y should be independent of any argument z that is not connected to x or y, i.e., there is no path from x or y to z (neglecting the direction of the edges). Let $\mathcal{C}(AF)$ be the set of weakly connected components of AF. Each weakly connected component of AF is a maximal subgraph of AF in which any two arguments are mutually connected by a path (neglecting the direction of the edges).

Axiom 2 (Independence (In)). *A ranked-based semantics* Γ *satisfies* independence *iff for any AF and for any* $AF_c = \langle \mathcal{X}_c, \mathcal{R}_c \rangle \in 2^{\mathcal{C}(AF)}, \forall x, y \in \mathcal{X}_c, x \succeq_\Gamma^{AF} y$ *iff* $x \succeq_\Gamma^{AF_c} y$.

The third axiom, called *void precedence*, encodes the idea that non-attacked arguments are more acceptable than attacked ones.

Axiom 3 (Void Precedence (VP)). *A ranked-based semantics* Γ *meets* void precedence *iff for any* $AF = \langle \mathcal{X}, \mathcal{R} \rangle, \forall x, y \in \mathcal{X}$, *if* $\mathcal{R}^-(x) = \emptyset$ *and* $\mathcal{R}^-(y) \neq \emptyset$ *then* $x \succeq_\Gamma^{AF} y$.

The fourth axiom states that having attacked attackers is more acceptable than non-attacked attackers, i.e., being defended is better than not.

Axiom 4 (Defense precedence (DP)). *A ranked-based semantics* Γ *satisfies* defense pre-cedence *iff for every* $AF = \langle \mathcal{X}, \mathcal{R} \rangle, \forall x, y \in \mathcal{X}$, *if* $|\mathcal{R}^-(x)| = |\mathcal{R}^-(y)|, \mathcal{D}(x) \neq \emptyset$ *and* $\mathcal{D}(y) = \emptyset$ *then* $x \succeq_\Gamma^{AF} y$.

The next axiom says that an argument x should be at least as acceptable as argument y, when the direct attackers of y are at least as numerous and well-ranked as those of x. This involves the concept of *group comparison*: Let $\succeq_\Gamma$ be a ranking on a set of arguments $\mathcal{X}$. For any $S_1, S_2 \subseteq \mathcal{X}$, $S_1 \succeq_\Gamma S_2$ iff there exists an injective mapping δ from S_2 to S_1 such that $\forall x \in S_2, \delta(x) \succeq_\Gamma x$. Moreover, $S_1 \succ_\Gamma S_2$ is a *strict group comparison* iff (1) $S_1 \succeq_\Gamma S_2$; (2) $|S_2| < |S_1|$ or $\exists x \in S_2, \delta(x) \succ_\Gamma x$.

Axiom 5 (Counter-Transitivity (CT)). *A ranked-based semantics* Γ *satisfies* counter-transitivity *iff for every* $AF = \langle \mathcal{X}, \mathcal{R} \rangle, \forall x, y \in \mathcal{X}$, *if* $\mathcal{R}^-(y) \succeq_\Gamma^{AF} \mathcal{R}^-(x)$ *then* $x \succeq_\Gamma^{AF} y$.

Axiom 6 (Strict Counter-Transitivity (SCT)). *A ranked-based semantics* Γ *satisfies* strict **(CT)** *iff for any* $AF = \langle \mathcal{X}, \mathcal{R} \rangle, \forall x, y \in \mathcal{X}$, *if* $\mathcal{R}^-(y) \succ_\Gamma^{AF} \mathcal{R}^-(x)$ *then* $x \succ_\Gamma^{AF} y$.

The following two axioms represent two opinions: give precedence to cardinality over quality (i.e. two weakened attackers is worse for the target than one strong attacker), or vice versa. In some situations, both choices are reasonable.

Axiom 7 (Cardinality Precedence (CP)). *A ranked-based semantics* Γ *satisfies* cardinality precedence *iff for arbitrary argumentation framework* $AF = \langle \mathcal{X}, \mathcal{R} \rangle, \forall x, y \in \mathcal{X}$, *if* $|\mathcal{R}^-(x)| < |\mathcal{R}^-(y)|$ *then* $x \succ_\Gamma^{AF} y$.

Axiom 8 (Quality Precedence (QP)). *A ranked-based semantics* Γ *satisfies* quality precedence *iff for arbitrary argumentation framework* $AF = \langle \mathcal{X}, \mathcal{R} \rangle, \forall x, y \in \mathcal{X}$, *if* $\exists y' \in \mathcal{R}^-(y)$ *such that* $\forall x' \in \mathcal{R}^-(x), y' \succ_\Gamma^{AF} x'$, *then* $x \succ_\Gamma^{AF} y$.

The last axiom focuses on the way arguments are defended. The main idea is that an argument which is defended against more attackers is more acceptable than an argument which is defended against a smaller number of attacks. There are two types of defense: simple and distributed. The defense of an argument x is *simple* iff each defender of x attacks exactly one attacker of x, formally, $\forall y \in \mathcal{D}(x)$ such that $|\mathcal{R}^+(y) \cap \mathcal{R}^-(x)| = 1$. The defense of an argument x is *distributed* iff every attacker of x is attacked by at least one argument, i.e., $\forall y \in \mathcal{R}^-(x)$ such that $|\mathcal{R}^-(y)| \geq 1$.

Axiom 9 (Distributed-Defense Precedence (DDP)). *A ranked-based semantics Γ satisfies* distributed-defense precedence *iff for any AF* $= \langle \mathcal{X}, \mathcal{R} \rangle$, $\forall x, y \in \mathcal{X}$ *such that* $|\mathcal{R}^-(x)| = |\mathcal{R}^-(y)|$ *and* $|\mathcal{D}(x)| = |\mathcal{D}(y)|$, *if the defense of x is simple and distributed and the defense of y is simple but not distributed then* $x \succ_\Gamma^{AF} y$.

In addition, [7] provides some relationships between these axioms: if a ranking-based semantics Γ satisfies (SCT) then it satisfies (VP); if Γ satisfies both (CT) and (SCT), then it satisfies (DP); Γ can not satisfy both (CP) and (QP). Now, we show which axioms are or are not compatible with the categoriser-based ranking semantics.

Theorem 3. *The categoriser-based ranking semantics satisfies* (Ab), (In), (VP), (DP), (CT) *and* (SCT), *and does not satisfy* (CP), (QP) *and* (DDP).

From the definition of the categoriser function, it can be easily seen that categoriser-based ranking semantics satisfies (Ab) and (In). To some extent, Proposition 1 is a special case of (CT). In particular, when $\mathcal{R}^-(x_i) \subset \mathcal{R}^-(x_j)$ then the semantics gives $x_i \succ x_j$, which is a special case of (SCT). The (VP) and (DP) can be implied from (CT) and (SCT). Now, let us give a counter example to show that the semantics does not satisfy (CP) and (QP):

Example 3. *Consider the argument system in Fig. 3, in which* $\mathcal{R}^-(x) = \{x_1, x_2, x_3\}$ *and* $\mathcal{R}^-(y) = \{y_1, y_2, y_3, y_4\}$. *Clearly,* $|\mathcal{R}^-(x)| < |\mathcal{R}^-(y)|$. *However, the categoriser-based ranking semantics gives* $y \succ x$ *(since* $v^*(x) = 0.40$ *and* $v^*(y) = 0.43$*), which conflicts with* (CP). *Note that* $y_1 \succ x_i$ *for all* $i \in \{1, 2, 3\}$ *(since* $y_1 = 0.60$ *and* $x_i = 0.50$*). From* (QP), $x \succ y$ *should hold, but it is not true for the semantics.*

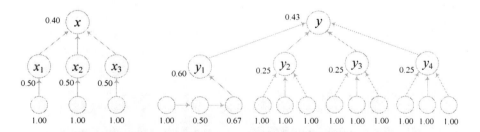

Fig. 3. A counter-example of axiom (CP) and (QP)

The main reason of the counter situation in the above example is that these two axioms represent two extreme: one treats all attackers equally, and one merely focuses on some attacker (with highest rank with respect to the set of attackers of the argument) of an argument. In categoriser valuation, however, the value of an argument (represented by $f(\boldsymbol{v}, \mathbf{D}_{i*}))$ depends on both the number and quality (i.e., the strength values) of its attackers, the attackers of its attackers, etc.

Another reason that (CP) is not satisfied by the categoriser valuation is that (CP) concentrates too much on quite local topological aspects of an argumentation framework, but ignores the global topology [14]. However, the categoriser valuation is a global approach since the strength value of an argument depends on the strength values of its attackers, which is a recursive definition. For the same reason, the categoriser-based ranking semantics does not satisfy (DDP).

5 Conclusion

In this paper, we firstly investigated the existence and uniqueness of the categoriser strength valuation via fixed-point technique. On this basis, we then defined a new ranking-based semantics, called categoriser-based ranking semantics, for abstract argumentation framework. We analyzed some general properties of the semantics, and prove that it satisfies some of the postulates that a ranking-based semantics should satisfy. Our ongoing work is about a deeper analysis of the approach and its relationships to other approaches.

References

[1] Rahwan, I., Simari, G.R.: Argumentation in artificial intelligence. Springer (2009)
[2] Dung, P.M.: On the acceptability of arguments and its fundamental role in nonmonotonic reasoning, logic programming and n-person games. Journal of Artificial Intelligence 77(2), 321–357 (1995)
[3] Baroni, P., Giacomin, M., Guida, G.: SCC-recursiveness: A general schema for argumentation semantics. Artificial Intelligence 168(1), 162–210 (2005)
[4] Dung, P.M., Mancarella, P., Toni, F.: A dialectic procedure for sceptical, assumption-based argumentation. COMMA 144, 145–156 (2006)
[5] Baroni, P., Cerutti, F., Giacomin, M., Simari, G.R.: Computational Models of Argument, vol. 216. Ios Press (2010)
[6] Bench-Capon, T.J., Dunne, P.E.: Argumentation in artificial intelligence. Artificial Intelligence 171(10), 619–641 (2007)
[7] Amgoud, L., Ben-Naim, J.: Ranking-based semantics for argumentation frameworks. In: Liu, W., Subrahmanian, V.S., Wijsen, J. (eds.) SUM 2013. LNCS (LNAI), vol. 8078, pp. 134–147. Springer, Heidelberg (2013)
[8] Cayrol, C., Lagasquie-Schiex, M.C.: Graduality in argumentation. J. Artif. Intell. Res (JAIR) 23, 245–297 (2005)
[9] Besnard, P., Hunter, A.: A logic-based theory of deductive arguments. Artificial Intelligence 128(1), 203–235 (2001)
[10] Altman, A., Tennenholtz, M.: Axiomatic foundations for ranking systems. J. Artif. Intell. Res (JAIR) 31, 473–495 (2008)
[11] Burden, R.L., Faires, J.D.: Numerical analysis. Brooks/Cole, USA (2001)
[12] Teschl, G.: Nonlinear functional analysis. Lecture notes in Math, Vienna Univ., Austria (2001)
[13] Li, K., Liang, J., Xiao, T.J.: Positive fixed points for nonlinear operators. Computers & Mathematics with Applications 50(10), 1569–1578 (2005)
[14] Thimm, M., Kern-Isberner, G.: Stratified labelings for abstract argumentation. arXiv preprint arXiv:1308.0807 (2013)

Investigating Collaboration Dynamics
in Different Ontology Development Environments

Marco Rospocher[1], Tania Tudorache[2], and Mark A. Musen[2]

[1] Fondazione Bruno Kessler—IRST,
Via Sommarive 18, Trento, I-38123, Italy
rospocher@fbk.eu
[2] Stanford Center for Biomedical Informatics Research,
Stanford University, 1265 Welch Road, Stanford, CA 94305, USA
{tudorache,musen}@stanford.edu

Abstract. Understanding the processes and dynamics behind the collaborative construction of ontologies will enable the development of quality ontologies in distributed settings. In this paper, we investigate the collaborative processes behind ontology development with two Web-based modeling tools, WebProtégé and MoKi. We performed a quantitative analysis of user activity logs from both tools. This analysis sheds light on (i) the way people edit an ontology in collaborative settings, and (ii) the role of discussion activities in collaborative ontology development. To explore whether the ontology tool influences the collaboration processes, we conducted five investigations using the collaborative data from both tools and we found that users tend to collaborate in similar ways, even if the tools and their collaboration support differ. We believe these findings are valuable because they advance our understanding of collaboration processes in ontology development, and they can serve as a guide for developers of collaborative tools.

1 Introduction

In the last few years, researchers have investigated the processes and dynamics behind the collaborative development of ontologies by teams of users (e.g., [1],[2],[3],[4]). These studies – most of them facilitated by state of the art ontology editing environments, such as WebProtégé [5] and MoKi [6], that trace the activities performed by users on the ontology – provide useful insights on ontology development related aspects, especially for tool engineers and ontology project managers. The former benefit from the outcome of these analyses as they have the opportunity to optimize their tools to make the work of the users more straightforward and effective. The latter gain from these studies metrics and tools to assess and monitor the development status and the quality of the ontology under their responsibility.

This paper contributes to this research stream by conducting some explorative investigations on (i) the way people edit an ontology in collaborative settings, and (ii) the role of discussion activities in collaborative ontology development. What sets apart our contribution from previous works is that for the first time the collaborative modeling processes of ontologies developed with two different ontology development frameworks are compared: WebProtégé – a generic Web-based ontology editor – and MoKi – a

R. Buchmann et al. (Eds.): KSEM 2014, LNAI 8793, pp. 302–313, 2014.

Wiki-based tool for modeling ontological (and procedural) knowledge. In addition to *editing* activities, i.e., changes users perform on the actual formal model under development, we also consider *discussion* activities, i.e., user contributions such as notes, comments, or threaded discussions that do not directly correspond to a change of an ontology entity, but that may influence its formalization. For the purpose of this work, we consider collaboration to be characterized by two objective measures: (a) *the number of changes* and (b) *the number of notes* performed on ontology entities by different users.

The research question we are trying to answer is the following: *Do the editing patterns and the level of collaboration vary across different projects developed with different ontology development tools?* To answer this question, we performed five investigations on collaborative projects developed with the two tools:

I1: Is the editing process localized, meaning that after editing an ontology entity, do users tend to edit a *semantically related* entity?

I2: Is the formalization of an ontology entity truly collaborative, meaning that an entity is the result of editing by two or more users?

I3: Are *discussed* ontology entities actually discussed by two or more users?

I4: Are highly discussed ontology entities, especially those discussed by two or more users, also highly edited?

I5: Do users tend to edit more than to discuss?

Two important aspects of our study are worth noting: First, instead of setting up some "artificial" ontology development experiments, the investigation was performed on five real ontology development projects, where the ontology modeling took place independently from our analysis. Even though the ontologies considered for the two tools inevitably vary in size, domain, as well as in the number of users participating in their development, we could still find similarities in the collaboration processes that span across the different tools and ontologies. Second, although we reckon that some ontology development related activities took place outside the tools (e.g., conceptualization, face-to-face discussions), our analysis is based *exclusively* on the logs of the activities performed by users while using the tools, as only for these activities we can rely on truly objective data that we can measure and compare without interfering with the ontology development projects.

The paper is organized as follows: We discuss relevant related work (Section 2), then we introduce the tools and the ontology development projects considered in our study (Section 3). We present the results for the five investigations (Section 4), discussing them as well as some further findings and limitations (Section 5). We conclude with some final remarks and outlook for future work (Section 6).

2 Related Work

Some works have previously investigated the collaborative process behind the development of ontologies. Randall et al. [7] present an ethnographic study to identify how the ontology development process unfolded in practice during the development of a cell-type ontology. Leenheer et al. [8] propose a set of indicators that they apply to understand the social arrangements in community-based ontology evolution. Falconer et al. [2] investigate the implicit roles of authors in collaborative ontology modeling, and analyze the relationship between ontology changes and how users communicate. Di Francescomarino et al. [3] present the results of a controlled experiment assessing the effectiveness and the impact on the modeling process of wiki-like collaborative features. Strohmaier, et al. [4] introduce a set of measures, that they applied to several ontology development projects, to analyze the hidden social dynamics behind the collaborative ontology development process.

With respect to all these works, our contribution is novel under several aspects. First, for the first time ontology developments performed with two different collaborative tools are considered in the same study. Second, new investigations are performed (e.g., I3, I4, and I5), taking into account both editing activities and discussion activities, two distinct yet complementary aspects in the collaborative development of an ontology.

3   Material and Methods

First, we introduce the tools (Section 3.1) and the ontology development projects (Section 3.2) that we used in performing the research investigations proposed in Section 1.

3.1   Tools

WebProtégé [1] [5] is a collaborative ontology authoring tool for the Web. The user interface of WebProtégé is designed as a portal with tabs, and each tab contains one or more portlets that provide different functionalities, such as browsing the class tree or editing the properties of the ontology entities. The forms mechanism in WebProtégé supports the creation of Web forms, through which domain experts can easily fill the content of an ontology using templates [9,10,11].

WebProtégé provides extensive collaboration support. All ontology changes are tracked in a structured change log as instances of the Changes and Annotation ontology (ChAO) [12]. Users may create notes on a particular ontology entity, or engage in threaded discussions either at the level of an entity or the ontology itself. Users may watch a particular entity or a branch, and they will receive email notifications when changes or new discussions affect the watched entities.

We use in this paper the change logs generated by a custom configuration of WebProtégé, called iCAT [9,10,11] that medical experts are using to author two important medical ontologies, ICD and ICTM, used in our study (see Section 3.2).

MoKi (the Modelling Wiki) [2] [6] is a collaborative wiki-based tool for modeling ontological and procedural knowledge in an integrated manner[3] via a Web-browser. Each entity in the ontology (class, property, individual) has a wiki page associated to it, containing both unstructured and structured content. Each page has a multi-mode access to the content to support users with different skills and competencies.

---

[1] WebProtégé can be accessed at: http://webprotege.stanford.edu
[2] MoKi can be tested online at: https://moki.fbk.eu/moki/tryitout2.0/
[3] Though MoKi allows to model both ontological and procedural knowledge, here we will limit our description only to the features for building ontologies.

To support user collaboration, several mechanisms are also in place: (i) discussions by means of talk pages,[4] where users can document modeling choices, and debate possible modeling options in order to converge to a shared formalization; (ii) watchlists allow users to be aware (and notified with messages and email alerts) of changes performed on pages (and, thus, ontology entities) they are monitoring; (iii) a notification mechanism allows users to be updated of changes on the ontology that are relevant for them; finally, (iv) recent activity awareness features, like browsing the last changes performed, newly created ontology entities, specific user contributions, recent/new discussions, most active users, etc.

Relevant Similarities and Differences between WebProtégé and MoKi. We focus here only on the comparison of the features and functionalities of the two systems that are relevant for the conducted study.

Both tools are Web-based, so users do not need to install any specific tool on their devices. WebProtégé and MoKi both provide functionalities for supporting the editing and discussion of ontology entities. Furthermore, in the projects considered for this study, users edit the ontology in both tools by mainly using forms that guide them in providing content. The collaboration features, such as notifications, watchlists, and change history, are also similar.

A major difference concerns the granularity and the modality of the editing and discussion activities performed by users: in WebProtégé, each atomic change a user performs – e.g., a change on the property value of some entity, or the annotation of a specific element – is automatically stored, without requiring any explicit saving action from the user; in MoKi, users are presented with the page or discussion page of an ontology entity, they can perform several modifications in the content or discussion of the entity, but all these changes are recorded in the log as a single editing or discussion activity only when the user explicitly applies the changes by clicking the *Save* page button. This difference should be considered when interpreting the results we obtained.

The navigation and hierarchy awareness in the tools is different: In WebProtégé, the class hierarchy is always visible while users edit the ontology, while in MoKi, users work with wiki pages that display one entity at a time, and the class hierarchy must be accessed separately through the sidebar menu.

Discussion awareness is offered in both tools in different ways: in WebProtégé, callout icons are shown in the class hierarchy widget next to the name of classes having discussions, while in MoKi dedicated functionalities to review recent/new discussions are provided through the sidebar.

Motivation for Using WebProtégé and MoKi in Our Study. In this research, we proposed to study the collaboration processes as they take place in real-world projects, which requires that the ontology editing tools record detailed change and discussion logs. Although other collaborative ontology editing tools exist, we are not aware of any of them that make such detailed logs available for any kind of projects. Both WebProtégé and MoKi provide access to these kind of logs and have been used

---

[4] A talk page is associated to each page describing an ontology entity; discussions on the ontology as a whole are associated to the MoKi Main Page.

in several real-world projects, which makes them suitable for the analysis conducted in this study.

3.2 Ontologies

We next describe the five real-world ontology development projects considered in our study. The first two projects are using WebProtégé, while the remaining three are using MoKi for the development. Note that each project team is composed of users coming from different organizations, and with different background and experience. All users were trained on how to use the modeling tool to build ontologies.

The 11th Revision of the International Classification of Diseases (ICD) is a project led by the World Health Organization (WHO) with the goal of updating ICD to reflect the scientific progress in the medical field.[5] ICD contains a taxonomy and descriptions of diseases, and is used in all United Nations countries for creating mortality and morbidity statistics, for insurance claims, and for policy making. To keep up to date with the scientific progress, WHO publishes new revisions of ICD every decade or more. The current revision in use is ICD-10. ICD-11 is the revision currently under development and is an OWL ontology authored using a custom installation of WebProtégé, called iCAT [9,10,11]. ICD-11 has currently over 50,000 classes, 225,000 individuals and 220 properties, and users have added more than 50,000 notes. More than 200 medical experts from around the world are using WebProtégé since 2009, out of which more than 100 have performed active changes for the period considered. WebProtégé tracks every change users make in the system, and stores the metadata associated to each change. We use the structured change logs from WebProtégé in the analyses performed in this paper.

The International Classification of Traditional Medicine (ICTM) is a collaborative project led by the WHO that aimed to produce an international standard terminology and classification for diagnoses and interventions in Traditional Medicine.[6] The classification tries to unify knowledge from the traditional medicine practices from China, Japan and Korea, for which country-specific classifications already exist. The ICTM classification is represented as an OWL ontology and its content is authored in 4 languages: English, Chinese, Japanese and Korean. More than 20 domain experts from the three countries develop ICTM using a customized version of the system used for ICD-11, called iCAT-TM, which possesses all collaboration features of the original iCAT system. ICTM has over 1,500 classes, 18,000 individuals and 200 properties. As for ICD, we use the structured change logs from WebProtégé to perform our analyses.

Organic Agriculture (OA) and Viticulture (Vit) ontologies are developed within the context of the Organic.Lingua European project.[7]

---

[5] http://www.who.int/classifications/icd/en/
[6] http://tinyurl.com/ictmbulletin
[7] http://www.organic-lingua.eu

The OA ontology[8] is designed to classify educational material in a multi-lingual web-portal containing organic agriculture and agro-ecology resources. Ten domain experts are developing the multi-lingual OWL ontology (currently containing 15 languages) by inspecting and reusing existing taxonomies and categorizations. In the analyses conducted in this paper, we use a five months period of activity logs for the OA ontology development, corresponding to the modeling phase in which the experts used the (multi-lingual) MoKi.

The Vit ontology describes concepts related to the science, production, and study of grapes. In particular, it covers grapevine descriptors, like the color of the berry skin, leaf degree resistance, time of bloom, and so on. Three domain experts are currently developing the Vit OWL ontology, which references standards like the IPGRI Descriptors for Grapevine.[9] As the domain experts have so far not used the discussion functionalities in the development of the Vit ontology, we are only using this ontology for investigations I1 and I2. In the analyses conducted in this paper, we use the MoKi activity logs of the whole development period (three months) of the Vit ontology.

Motivation and Emotion (ME) ontology covers the motivational and emotional aspects of the learning process in the pedagogical field. A team of two domain experts (a pedagogist and psychologist employed in a publishing house specialized in educational books) and a knowledge engineer developed this OWL ontology from scratch with MoKi within a short time period. The ME ontology deals with concepts such as motivation and its aspects, emotion, as well as, the educational material and the interventions to be used for facing motivational or emotional difficulties. The activity logs of the whole development period of the ME ontology are considered in our analyses.

4   Results

In this section, we describe the results of the five investigations that we introduced in Section 1. We used the change and discussion logs for the five collaborative projects (Section 3.2) recorded by WebProtégé and MoKi, respectively.

4.1   I1: Is the Editing Process Localized?

In our first investigation (I1), we tested whether users, after editing a class, tend to edit another class *closely* or *semantically* related to the previous one. Let x be the class edited at a certain time, and let y, with $x \neq y$, be the next class edited by the same user. We considered the following cases: (i) x and y are direct subclasses of a common class (*siblings*), (ii) y is a direct subclass of x (*child*), (iii) x is a direct subclass of y (*parent*), (iv) y is a subclass of x but not a direct one (*descendant*), (v) x is a subclass of y but not a direct one (*ancestor*), and (vi) none of the previous case holds (*none*). We counted the number of these occurrences, normalizing the values over the total number of cases.[10]

---

[8] The first version of the OA ontology was produced in the context of the Organic.Edunet (http://www.organic-edunet.eu) European project.

[9] http://tinyurl.com/grapevinevit

[10] For the sake of investigating I1, any sequence of consecutive editing activities on the same class are considered as a single operation.

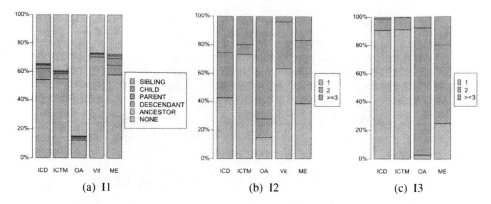

Fig. 1. (a) I1: Normalized distribution of the semantic relation (segments from bottom to top: sibling, child, parent, descendant, ancestor, none) between a class and the next edited class by the same user; (b) I2: Normalized distribution of the entities having editing activities by only one user (lower segment), two users (middle segment), and three or more users (upper segment); (c) I3: Normalized distribution of the entities having discussion activities by only one user (lower segment), two users (middle segment), and three or more users (upper segment)

Figure 1a shows the results obtained for the various datasets. With the exception of OA, we observe that most of the times (ranging between 60% and 73% for the different ontologies considered) users edited as the next entity a sibling ($\geq$ 50%), a child or a parent of the previous class. This may suggest that users tend to work *locally* on the ontology, and in doing so may be facilitated by the functionality offered by both tools to navigate the class hierarchy of the ontology. Though similar in both tools, it is worth mentioning that this functionality bears some differences: among them, in WebProtégé the class hierarchy is always shown to the user, while in MoKi users have to explicitly access that functionality in a separate page. The development of the OA ontology shows a quite different behaviour: 75% of the times users edited a class not closely related to the previous one. However, this is mainly due to the multilingual focus of the ontology, and the way users "self-organized" to built it: as among their modelling activities users have to add also language related information to the classes (e.g., title, description, synonyms), they preferred to accomplish this task by editing the classes according to the alphabetical order provided them by the MoKi list functionality.

4.2   I2: Is the Editing Truly Collaborative?

In our second investigation (I2), we examined how many distinct users usually edit an ontology entity, whether a class, individual, or property. We classified ontology entities in three categories: those edited (i) by only one user, (ii) by two distinct users, and (iii) by three or more distinct users.

Figure 1b shows the normalized distribution of the entities in these three categories. With the exception of OA, we observe that most of the ontology entities (ranging between 75% and 96% for the different ontologies considered) are edited by at most 2 users: it is unlikely that more than two users edit the same entity. Indeed, in a couple of cases most of the ontology entities are edited only by a single user (ICTM: 73%,

Vit: 63%). Again, OA represents an exception. Approximately 72% of the entities in the OA ontology are edited by at least three users: a close-up look actually shows that 65% of the entities are actually edited by at least five distinct users. This is again related to the multilingual focus of the ontology, as users from different nationalities are actually adding title, description, and synonyms in different language to each ontology entity. It has also to be observed that OA has a rather low entities / user ratio (~ 40) compared to other ontologies such ICD (~ 382) or Vit (~ 160), and therefore multiple users editing activities on the same entity are more likely to occur.

4.3 I3: Are Discussions Truly Collaborative?

In our third investigation (I3),[11] we explored how many distinct users usually discuss an ontology entity. Similarly to I2, we classified ontology entities in three categories: those discussed (i) by only one user, (ii) by two distinct users, and (iii) by three or more distinct users.

Figure 1c shows the normalized distribution of the entities in these three categories. In ICD and ICTM, most of the ontology entities ($\sim 91\%$) are discussed by a single users, while in OA and ME most of the ontology entities (resp., 97% and 75%) are discussed by at least 2 users (resp., with an 8% and 20% of ontology entities discussed by at least three users). On one side, this may be due to the difference in size of the ontology and the number of users involved in the project, especially when considering the entity / user ratio (ICD: ~ 382 and ICTM: ~ 171; OA: ~ 40 and ME: ~ 33). On the other side, it may be due to the different discussion-awareness support provided by the tools used to develop the ontology. In MoKi, users awareness of the latest discussion activities, newly started discussion, and mostly discussed entities are easily accessible either through the sidebar or are provided as notifications when logging to the tool, while in WebProtégé getting the same information is less straightforward.

4.4 I4: Are Highly Discussed Entities Also Highly Edited?

In our forth investigation (I4), we explored if there is a correlation between the discussion activities and the editing activities on an entity. We classified the ontology entities in two groups: those having at least two distinct users discussing each of them, and those having zero or at most one user discussing them. We then computed the following metrics on these groups: (i) the average/median of the number of distinct users editing an entity; and, (ii) the average/median of the number of editing activities on an entity.

Figure 2 shows the results of computing the metrics on the ontologies considered. All ontologies show greater values for all the metrics when computed on the set of entities having at least two users discussing them (blue/right bars of each dataset) than the other set (red/left bars of each dataset).[12] In particular, the average and median of the number of editing activities per ontology entity is greater on those entities having at least two users discussing them. Indeed, if we consider (A) the distribution of the number of editing activities for the entities having at least two users discussing them,

---

[11] We will exclude Vit from I3, I4 and I5 as discussions were not used.

[12] With one only exception being the median value of the number of distinct users editing an entity in ME, equals on both sets.

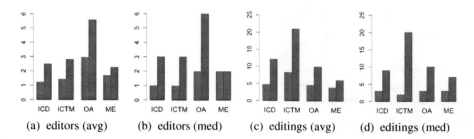

Fig. 2. Comparison of metrics computed on entities having at least two users discussing them (blue/right bars of each dataset) and entities having one or no user discussing them (red/left bar): (a) average and (b) median of the number of distinct users editing an entity; and, (c) average and (d) median of the number of editing activities on an entity

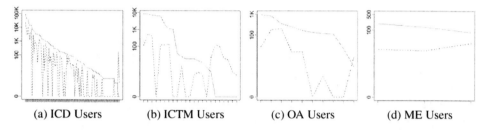

Fig. 3. Comparison between the number of editing and discussion activities of each user (logarithmic scale): each tick on the x axis correspond to a user, the solid/blue line shows the number of the user editing activities, wile the dotted/black line shows the number of the user discussion activities. Users are sorted in descending order according to the number of editing activities.

and (B) the distribution of the number of editing activities for the entities having at most one user discussing them, we observe by computing the unpaired Wilcoxon test that (A) is statistically significantly greater than (B) in all the ontology developments considered (p-value ≤ 0.05).

4.5   I5: Do Users Edit More Than Discuss?

In our fifth investigation (I5), we examined whether users tend to perform more editing activities than discussion activities. For each user, we counted the number of editing activities and discussion activities performed.

Figure 3 shows for each ontology the number of editing activities (solid/blue line) and the number of discussion activities (dotted/black line) performed by each user (each tick on the x axis). In most of the cases (ICD = 98%, ICTM = 69%, OA = 90%, ME = 100%), users performed much more editing activities than discussion activities, although it is worth mentioning that there are few cases in which the contrary holds (e.g., some users of ICTM, ICD, OA). For each ontology, we then computed the ratio between the average number of editing activities per user and the average number of discussion activities per user. The ratio is greater than 4 for all the ontologies considered (ICD = 4.64, ICTM = 23.67, OA = 6.44, ME = 7.82), thus confirming that users tend to perform more editing activities than discussion activities.

5   Discussion and Limitations

The results of our investigations suggest that the editing patterns and the collaboration processes are similar across different ontology development environments and different ontologies. However, as we discovered, there are also some differences that we discuss below. We consider this work to be the first step in trying to understand how a generic collaborative ontology development process works, and we plan to deepen our research in our future work.

Our first investigation (I1) examined whether the **changes are local in nature**. The results show that in four out of five of the ontologies considered users mostly edited entities closely related to the previously edited ones. As we observed in Section 4.1, one explanation for this is the availability, both in MoKi and in WebProtégé, of a functionality to navigate the class hierarchy tree. Instead, in OA users performed their editing activities according to the alphabetical sorting of the entities provided by the MoKi listing functionality. This may suggest that available tool functionalities actually impact the way people perform their editing activities, and this is one subject for our further investigation.

The **distribution of pages** edited by one, two, or three and more users (i.e., the metric used in I2) may provide useful insights also to ontology project managers: entities having a very few number of editors, in particular those with only one editor, may require additional attention as their formalization may reflect the point of view of only a single user, something that, especially in community driven ontology developments, may not be advisable.

In our third investigation (I3), we examined whether **discussions are truly collaborative**. The results show that the usage of notes and discussions are different in the ontologies developed with WebProtégé and MoKi: more than 90% of notes in ICD and ICTM, developed with WebProtégé, are single-user, rather than being threads of discussions involving multiple users, while in the Vit and ME, developed with MoKi, only 3% and 18% respectively, were single-user. This difference can be explained by a few factors. First, the sizes of the ICD and ICTM ontologies are much larger, e.g., more than 5,000 classes in ICD have discussions involving multiple users, which is much higher than the total number of classes in either Vit or ME. Given the larger size of the ICD and ICTM ontologies, the ratio users / entity is much smaller in WebProtégé ontologies compared to the MoKi ontologies. Second, the features available in MoKi, which presents the latest discussion activity when the user logs in, may favour a higher user awareness of the activity in the system, which may have an influence on the reaction of the users to current discussions and changes. We plan to investigate this interesting research topic as part of our future work. Third, users of ICD and ICTM have used the notes mechanism not only for discussing the ontology entities, but mostly for taking notes or providing additional documentation on the entities, for example, for providing external references to scientific literature, or documenting outside provenance for a piece of the content.

In our forth investigation (I4), we observed that the average **number of editing activities per entity** is greater for those ontology entities having at least two users discussing them. Although we cannot draw a strong implication between these two facts – the more the discussion activities, the more the editing activities – it is very interesting that this behavior is common to all the ontology developments that we investigated, independently on the tool used. This may suggest that, in order to favor the increase of users editing activities, tools could encourage and facilitate the use of discussion support functionalities.

In general, the results of our fifth investigation (I5) show that users perform much **more editing activities than discussion activities**; however, it is interesting to observe that there are a few users that contribute to the development of the ontology mostly by means of discussions. Although we need further analyses to investigate this aspect in more detail, it may indicate the existence of different types of users, for example, users who prefer to pursue the formalization of an ontology entity by first sharing thoughts and opinions with other users, or users that mainly review and comment work performed by other members.

In the study we conducted, we used five real-world ontologies developed with Web-Protégé and MoKi. We focused our analysis only on the features that were common to both tools and we used objective measures that both tools provided (detailed changes and discussions logs). Still, to further confirm the generality of the claims we derived from our results, additional tools should be considered in the analysis.

The ontologies we analyzed in our study vary in size and in the number of users participating in their development, spanning from very large ontologies developed by a large team of people (e.g., ICD) to smaller ontologies built by few users (e.g., Vit). To further confirm the results of our study, in our future work, we will consider additional ontology projects, such as the development of some large ontologies with MoKi, or the modeling of a small focused ontology with WebProtégé.

To conclude, we recall that our analysis is based exclusively on the logs of the activities performed by users while using the tools. Although we acknowledge that a lot of discussions and decisions are made outside the ontology editing environments, these were the only activities for which we could rely on truly objective data. In our future work, we plan to complement our analysis with additional experimental study techniques that may also cover activities taking place outside the modelling tool.

6   Conclusions

We investigated in this paper the collaborative process behind the development of five real-world ontologies modeled with WebProtégé and MoKi. Our results shed light on several aspects related to the collaborative development of ontologies, including the way users edit the ontology and the role of discussion activities. Among the findings of our study, we observed that: (i) users tend to edit ontology entities closely related (i.e., a sibling, a parent, or a child) to the previously edited one; (ii) any ontology entity is edited/discussed by few users (generally, at most two); (iii) the more an ontology entity is discussed, the more likely that entity is highly edited as well; and, (iv) users tend to edit more than to discuss. The results of our analysis raise some aspects that ontology development tool engineers could further investigate in order to improve their tools: from offering different ontology browsing functionalities, to better support discussion activities by enhancing user discussion awareness.

To further validate our findings, we plan to extend our study to consider additional ontology development projects. We will also consider to include additional modeling tools, although we are not aware of any other ontology development environment offering the detailed logs of user editing and discussion activities needed for our study. We will also conduct a more in-depth investigation of the influence that the user interface and the availability of certain features may have on the dynamics of the collaboration.

Acknowledgements. The paper was written while the first author was Visiting Faculty at the Stanford Center for Biomedical Informatics Research and he would like to thank the Center for its hospitality. We thank Natasha Noy and Chiara Di Francescomarino for providing feedback on the paper. This work was supported in part by grants GM086587 and GM103316 from the US National Institutes of Health.

References

1. Schober, D., Malone, J., Stevens, R.: Observations in collaborative ontology editing using collaborative protégé. In: Workshop on Collaborative Construction, Management and Linking of Structured Knowledge (2009)
2. Falconer, S., Tudorache, T., Noy, N.F.: An analysis of collaborative patterns in large-scale ontology development projects. In: Sixth International Conference on Knowledge Capture (K-CAP), vol. 11, pp. 25–32 (2011)
3. Di Francescomarino, C., Ghidini, C., Rospocher, M.: Evaluating wiki-enhanced ontology authoring. In: ten Teije, A., Völker, J., Handschuh, S., Stuckenschmidt, H., d'Acquin, M., Nikolov, A., Aussenac-Gilles, N., Hernandez, N. (eds.) EKAW 2012. LNCS (LNAI), vol. 7603, pp. 292–301. Springer, Heidelberg (2012)
4. Strohmaier, M., Walk, S., Pschko, J., Lamprecht, D., Tudorache, T., Nyulas, C., Musen, M.A., Noy, N.F.: How ontologies are made: Studying the hidden social dynamics behind collaborative ontology engineering projects. Web Semantics: Science, Services and Agents on the World Wide Web 20 (2013)
5. Tudorache, T., Nyulas, C., Noy, N.F., Musen, M.A.: Webprotégé: A collaborative ontology editor and knowledge acquisition tool for the web. Semantic Web 4(1), 89–99 (2013)
6. Ghidini, C., Rospocher, M., Serafini, L.: Modeling in a wiki with moki: Reference architecture, implementation, and usages. International Journal on Advances in Life Sciences 4(4), 111–124 (2012)
7. Randall, D., Steven, R., Sharrock, W., Procter, R.N., Lin, Y., Poschen, M.: What about sea urchins? collaborative ontology building among bio-informaticians. In: 5th International Conference on e-Social Science (June 2009)
8. Leenheer, P.D., Debruyne, C., Peeters, J.: Towards social performance indicators for community-based ontology evolution. In: Workshop on Collaborative Construction, Management and Linking of Structured Knowledge (2009)
9. Tudorache, T., Falconer, S., Nyulas, C., Noy, N.F., Musen, M.A.: Will Semantic Web Technologies Work for the Development of ICD-11? In: Patel-Schneider, P.F., Pan, Y., Hitzler, P., Mika, P., Zhang, L., Pan, J.Z., Horrocks, I., Glimm, B. (eds.) ISWC 2010, Part II. LNCS, vol. 6497, pp. 257–272. Springer, Heidelberg (2010)
10. Tudorache, T., Falconer, S., Noy, N.F., Nyulas, C., Üstün, T.B., Storey, M.-A., Musen, M.A.: Ontology Development for the Masses: Creating ICD-11 in WebProtege. In: Cimiano, P., Pinto, H.S. (eds.) EKAW 2010. LNCS (LNAI), vol. 6317, pp. 74–89. Springer, Heidelberg (2010)
11. Tudorache, T., Nyulas, C.I., Noy, N.F., Musen, M.A.: Using semantic web in icd-11: Three years down the road. In: Alani, H., Kagal, L., Fokoue, A., Groth, P., Biemann, C., Parreira, J.X., Aroyo, L., Noy, N., Welty, C., Janowicz, K. (eds.) ISWC 2013, Part II. LNCS, vol. 8219, pp. 195–211. Springer, Heidelberg (2013)
12. Noy, N.F., Chugh, A., Liu, W., Musen, M.A.: A framework for ontology evolution in collaborative environments. In: Cruz, I., Decker, S., Allemang, D., Preist, C., Schwabe, D., Mika, P., Uschold, M., Aroyo, L.M. (eds.) ISWC 2006. LNCS, vol. 4273, pp. 544–558. Springer, Heidelberg (2006)

An Ontological Approach
for Specifying Provenance
into SPARQL Service Descriptions

Sabin C. Buraga and Claudia Gheorghiu

"Alexandru Ioan Cuza" University of Iasi, Romania
{busaco,claudia.gheorghiu}@info.uaic.ro

Abstract. We propose a conceptual model to describe the provenance information of the existing datasets available according to the Linked Data initiative. Using VoID and PROV, our solution defines useful metadata (e.g., licenses, publishers, server annotations, etc.) about public datasets. We also investigate how this novel approach could be effectively used by existing SPARQL endpoints, enabling a better discoverability and reuse of existing datasets made available via Web services. Several experiments are conducted to facilitate the access to legal datasets in order to enhance and build semantic Web applications.

Keywords: provenance, linked data, service description, ontology, SPARQL, knowledge engineering.

1 Introduction

There is a common trend in administrations and industry in the adoption and usage of ontologies and semi-structured data to make available their datasets. The essential idea of Linked Data initiative [10] is to publish, reuse and connect existing data sources by using the existing semantic Web standards. The interest and growth are most visible in various contexts and communities of practice which have as a main goal making data freely available to everyone and publishing it as RDF (Resource Description Framework) constructs by interconnecting data items from different sources via URLs, for myriad further uses [2].

In order to manipulate a multitude of available datasets, applications must overcome the problem of discoverability, but also the provenance question. Software needs (formal) mechanisms to enable the automatic identification of the process of creation and the origins of these sources. Additionally, the varying sources on the Web have different quality. For a given query, multiple answers can be obtained, where the problem of source trust arises, plus reusability and/or copyright conditions [8]. Valuable (big) data could also include: what is the original source of a dataset, where, how, and when has the data (or source data) been retrieved from the Web; who is responsible with modifying the data, etc. These aspects regards data provenance and are discussed in Section 2.

With the expansion of the data exchanged in digital formats, and with the growth of the linked data cloud, it becomes necessary to develop techniques

R. Buchmann et al. (Eds.): KSEM 2014, LNAI 8793, pp. 314–325, 2014.

for tracking and correlating the relevant aspects of a dataset provenance. We adopted the PROV data model together with VoID and SPARQL Service Description (SD) vocabularies to describe the provenance information of the data we are interested in. Our aims are to investigate how these models are used by existing SPARQL endpoints and to enable a better discoverability and reuse of existing datasets made available via Web services. Sections 3 and 4 give explanations about our approach.

Section 5 describes our extension of the SD vocabulary specifying provenance information as part of a service description. This conceptual model is able to represent useful metadata (e.g., licenses, publishers, Web server status, and many others) about each dataset. Using these enhancements, a provenance repository was created in order to conduct several practical experiments detailed in Section 6. The paper ends with conclusions and further work.

2 Expressing Provenance in Linked Data

Provenance (also referred to as lineage, pedigree, parentage, genealogy, and filiation) can represent a description – in metadata terms – of the origins of data and the process by which it originates [4,12]. Also, it could be viewed as metadata recording workflows, annotations, and notes about certain experiments [7].

Provenance is critical in contexts ranging from scientific reproducibility to journalism. Furthermore, a number of sub-disciplines make use of diverse datasets like biological data webs, pharmacological knowledge bases, semantic Web services or platforms.

We define *data provenance* as the knowledge that helps determine the derivation history of a data resource, starting from its original sources. We use the term *data resource* or *dataset* to refer to data available as RDF[1] statements: ⟨ *subject, predicate, object* ⟩ triples – for further details, consult [2].

According to [15], the important features of the provenance of a resource are: the ancestral resource from which this data evolved, and the process of transformation of this ancestral data, that helped derive it. Also, provenance may include information on the methods adopted to generate a resource, the identities of individuals and agencies responsible for creation, plus the usage licenses applied to the dataset's contents. By knowing such details, it is possible to make inferences about quality and re-using conditions of semantic resources.

Provenance also can provide useful feedback. For example, if serious syntactic and/or semantic errors are found in the data, then it might be possible to link them to specific resources of interest.

In order to have access to the distributed – and possibly large – RDF datasets, a standardized query mechanism is used: SPARQL 1.0 [14] and SPARQL 1.1 [9].

Additionally, VoID (Vocabulary of Interlinked Datasets) [1] was developed to automate discovery and selection of datasets and fill the gap between linked data publishers and consumers. Two core classes are defined to denote a dataset

---

[1] Resource Description Framework: http://www.w3.org/RDF/

(void:Dataset) and the interlinking model – void:Linkset. A *dataset* represents a collection of RDF data, published and maintained by a single provider and accessible through dereferenceable URLs or a SPARQL endpoint (Web service). A *linkset* is considered a subclass of a dataset, describing interlinking relations between datasets in the form of RDF triples. This model enables a flexible way to express connections between datasets (e.g., how many links exists among them, the kind of links, and who made these statements).

To interchange provenance knowledge on the Web, the applications need a standardized framework to represent, exchange and integrate provenance information generated in various environments – PROV [13], a lightweight ontology capable to express provenance as is. The PROV ontology offers an extensible model for provenance representation, which will be stored and exposed to final users *who* (prov:Agent, prov:Organization, prov:Publisher, prov:Creator), *how* (prov:Create, prov:Derivation, prov:Modify, etc.), *when* (prov:atTime) and *where* (prov:Location) different operations on a dataset took place.

SPARQL Service Description [16] is a method for discovering SPARQL services (endpoints). Also, it provides a vocabulary for describing them, such as various capabilities: which query features, I/O formats or entailments are supported, how default and named graphs are configured. As a result, the developer could access a list of SPARQL service features, made available via the SPARQL 1.1 Protocol for RDF – SPROT [6].

Relatively few existing approaches could be mentioned. A recent implementation of conTEXT [11] – a platform for text analysis based on linked data – illustrates the use of provenance metadata in the context of social feeds by adopting different methods for data exploration and visualization.

Another notable project is the RKB VoID Store[2] collecting the existing VoiD descriptions and providing query and browsing functionalities. Also, it offers a SPARQL endpoint over all the VoiD descriptions about different public datasets. However, the majority of indexed VoiD resources are not updated, and unfortunately refer to non-existing URIs or datasets.

Considering the context of Web science, [5] highlights the challenges in determining and creating data provenance in molecular biology datasets. Several other basic case studies are described in [2] and [7].

3   Accessing Provenance Information

First step of our study is to investigate how existing provenance information could be accessed in a proper manner.

A sizable proportion of the semantically-annotated content (in form of vocabularies, ontologies, linked data, or mashups) is generated by governmental entities. This means that many assertions were created using explicit licenses and reuse terms, some of them containing sensible information. It would be useful to record – in as much detail as possible – what were the original sources of a

---

[2] RKB VoID Store: http://void.rkbexplorer.com/

dataset, what licenses applies for these resources, how one could reuse or extend a collection, etc. [1,5,15]. However, this kind of information is not commonly captured in current practice. The lack of proper dataset annotation published by organizations, governments or individuals makes it hard for others to reuse existing datasets. One step in this direction is using discoverability and provenance vocabularies for published linked data, like VoID, SD and PROV.

An analysis of existing SPARQL endpoints that offer VoID and SD descriptions for their data is given by [3]: only about one third of all registered endpoints on DataHub[3] follow recommendations and enable a better discoverability by implementing VoID and SD.

We identified two major online catalogues that a consumer could use in order to find available semantic datastores: the previously mentioned repository (DataHub) and the RKB VoID Store.

However, [3] revealed that DataHub contains up-to-date information about available SPARQL endpoints and also a larger number of indexed services. We spotted a number of 494 SPARQL endpoints on DataHub and only 99 endpoints available on the VoID Store. Therefore, the DataHub represents our choice for obtaining and searching available SPARQL endpoints descriptions of their features and content via VoID and SPARQL 1.1 Service Descriptions (SD) vocabularies.

Using the catalogue API[4], we retrieved the list of registered VoID descriptions. Our analysis shows that only 8 of 240 found VoID files are registered on the recommended URI, according to the VoID specifications[5]. A number of 53 VoID resources are being indexed on the RKB VoID Store, while other 179 VoID URLs referred to the root domain or a `models/` folder including a `void.ttl` RDF file.

To access these descriptions, the consumer must know their location and how to retrieve and interpret them. Given the nature of distributed data publication and merging, provenance information about data can be published by any parties, according to any publication approaches or vocabularies considered by their publishers as fit for the task. Although PROV is a W3C standard, our research reveals that most endpoints have not yet updated their provenance contents and still use other schema representations – so, applications must be able to access and make sense of diverse information described by various vocabularies.

In the next section, we will employ several methods for discovering useful data provenance.

4   Provenance Discovery Strategies

Given the list of 240 endpoints identified as having a VoID description on the DataHub catalog, we picked two approaches for finding and consuming provenance metadata.

---

[3] DataHub: `http://datahub.io/`
[4] Retrieval request of DataHub VoID files: `http://datahub.io/api/2/search/resource?format=api/sparql&all_fields=1&limit=10000`
[5] VoID Vocabulary: `vocab.deri.ie/void/autodiscovery`

First possibility was to examine the list of vocabularies included in each VoID file found on DataHub (by using the `void:vocabulary` property). A quick analysis of these properties shows that most common vocabularies are DC Terms[6], FOAF[7] and SKOS[8] which do not offer enough provenance semantics that can be further used by a consumer. Therefore, poor VoID annotations can give little insight on provenance of the associated datasets.

Another strategy is querying each endpoint and identifying possible vocabularies while searching for interesting knowledge – e.g., license, copyright, or other reuse conditions indexed in the SPARQL endpoint list. For this job, we chose a list of most common open vocabularies that may be used to represent provenance: DC Terms, FOAF, SKOS, DC elements[9], OPMV[10] and Provenance Vocabulary[11].

We issued the following template query to every SPARQL endpoint:

```
PREFIX void: <http://rdfs.org/ns/void#>
SELECT DISTINCT ?p WHERE { ?s ?p ?o
   FILTER (REGEX (STR (?p), "^%%vocab")) } LIMIT 100
```

We instantiated this query for each of the 240 endpoints by substituting the placeholder – `%%vocab` – with the given vocabulary URL. As expected, this was a time consuming process. Also, the heterogeneity of the obtained results would made it difficult to interpret them.

Table 1 illustrates the results of each experiment, noting that 157 of 240 accessed VoID files were successfully retrieved (about 65%), while only 42 of these queried endpoints have responded to the given queries.

We noticed that Dublin Core is the most common vocabulary used for expressing provenance in a VoID file (available as RDF in Turtle format), as well as the most common result of issuing direct queries. Our research illustrates that only 13 of the 240 analyzed endpoints (i.e. about 9%) uses PROV for provenance metadata. We can conclude that, in the current situation, it is difficult for a consumer to reach, interpret and inference over provenance of a dataset.

5   A Proposal for Building Provenance Datasets

From a pragmatic point of view, we intend to develop a service-oriented Web system which – based on existing VoID, PROV or Dublin Core annotations and SPARQL Service Descriptions – automatically identifies and builds the complete

---

[6] Dublin Core Terms Vocabulary: `http://dublincore.org/documents/dcmi-terms/`
[7] FOAF (Friend Of A Friend) Specification: `http://xmlns.com/foaf/spec/`
[8] SKOS (Simple Knowledge Organization System) Specification:
`http://www.w3.org/TR/skos-reference/`
[9] Dublin Core Elements: `http://dublincore.org/documents/dces/`
[10] Open Provenance Model Vocabulary Specification:
`http://open-biomed.sourceforge.net/opmv/ns.html`
[11] Provenance Vocabulary Core Ontology Specification:
`http://trdf.sourceforge.net/provenance/ns.html`

Table 1. Experiment results

Vocabulary	Endpoint responses to direct queries	Vocabulary references in VoID files
DC Terms	35	8
SKOS	34	17
FOAF	35	24
DC Elements	20	8
PROV	13	9
Provenance Vocabulary	9	0
OPMV	2	1

provenance graph of a dataset. The system acts as a directory service that, given some provenance properties, check its address space to find endpoints having those properties. We planned to gather and process metadata about available datasets published by existing public SPARQL endpoints. The resulting information is stored in a unifying dataset, having the essential provenance data needed for determining quality of the existing resources, reuse terms and, also, contributors.

5.1 Creating the PROV Store Datasets

To create the dataset needed by the platform, we used existing VoID files indexed on DataHub. According to SPARQL endpoint references found in the VoID files (via `void:sparqlEndpoint`), we performed HTTP GET requests against each endpoint URL and obtained any available SD description. A typical SD document [16] contains details about the available dataset, query features, I/O formats or entailments are supported, how default and named graphs are configured. We extended these information with several metadata crucial for determining provenance – our solution is modeling provenance vocabularies available in each dataset, licenses, publishers, and contributors to the current data.

Additionally, we considered to be important for consumers to have an overview about the availability of the Web server hosting the SPARQL endpoint (it may not be operational) and its transfer rate capabilities, metadata available during the transfer of VoID resources and, also, during the provenance vocabularies enquiry on each endpoint. To offer a provenance file according to the W3C recommendations, we used the Dublin Core PROV mappings[12] and translated any Dublin Core metadata to its corresponding PROV entity.

5.2 Provenance Extensions to SPARQL Service Descriptions

In [10], a linked data(set) publisher was advised to reuse terms from well-known vocabularies wherever possible, and one should only define new terms one cannot find in existing vocabularies. Reusing existing vocabularies takes advantage of

---

[12] Dublin Core to PROV Mapping: `http://www.w3.org/TR/prov-dc/`

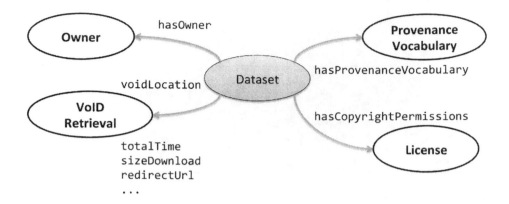

Fig. 1. SDPROV ontology: main classes and properties

the ease of bringing together diverse domains within RDF, and it makes data more reusable. By reusing vocabularies, the data is no longer isolated nor locked within a single context designed for a single use. We adhered to this advice and have made use of common ontologies such as VoID, SD, PROV, Dublin Core, and OWL-Time[13].

In addition, we defined several concepts for Web server annotations, copyright and licenses of provenance vocabularies (see Fig. 1).

The main proposed classes of our `sdprov` lightweight ontology are:

- **ProvenanceVocabulary** specifies a vocabulary that could be used to determine provenance information for a dataset;
- **Copyright** is a container for the list of exclusive rights granted to the author or creator of an dataset;
- **License** models the set of rights and permissions for the end-user – e.g., Creative Commons[14];
- **Owner** denotes the owner of a copyright for a dataset; it is a subclass of the **Agent** class defined by the PROV ontology;
- **VoIDRetrieval** is a container used to describe metadata about the process of retrieving VoID files for every registered endpoint; its properties include information about the actual retrieval URL, server response time or file size.

The corresponding properties are:

- **hasOwner** specifies that a dataset has an explicit owner; its domain is `sdprov:Owner`, while its range is a `sd:Dataset`;
- **hasProvenanceVocabulary** specifies a provenance vocabulary for a dataset;
- **hasCopyrightPermissions** is used to state the copyright permissions concerning a dataset; has as domain the `sd:Dataset` class and the `sd:License` class as range;

---

[13] Time Ontology in OWL: http://www.w3.org/TR/owl-time/
[14] Creative Commons Licenses: http://creativecommons.org/licenses/

- **voidLocation** is used in association with a **VoIDRetrieval** individual, to describe the location of the retrieved VoID file;
- **voidUrl** denotes the actual URI of the processed VoID file;
- **redirectCount** stores the number of performed HTTP redirects during the VoID retrieval; it is an important metadata, giving an overview of the total request time and performance when accessing the VoID file in its original location;
- **totalTime** is an object property expressing the total download time;
- **sizeDownload** represents the number of bytes of the original VoID file;
- **speedDownload** specifies the VoID download's measured speed; offers an estimated value of the hosting server bandwidth;
- **startTransferTime** provides the file retrieval timestamp; important for keeping track of the latest update for the file;
- **redirectUrl** – the URL from which the file was downloaded; can be used for direct access on future requests, thus improving the system's performance.

6   Practical Experiments and Results

To prove various advantages of our solution, we conducted several experiments mainly concerning queries about legislation. We deployed a virtual machine running Ubuntu Server (having 1 GB RAM and 2 GHz x86 processor).

6.1   Enhancing a Query Service Description

We chose a SPARQL Service Description (SD) representation for the UK Legislation[15] service:

```
<http://gov.tso.co.uk/legislation/sparql> a sd:Service ;
    sd:feature sd:UnionDefaultGraph ;
    # data will be available in these formats
    sd:resultFormat <http://www.w3.org/ns/formats/N-Triples>,
        <http://www.w3.org/ns/formats/RDF_XML>,
        <http://www.w3.org/ns/formats/SPARQL_Results_JSON>,
        <http://www.w3.org/ns/formats/SPARQL_Results_XML>,
        <http://www.w3.org/ns/formats/Turtle> ;
    sd:supportedLanguage sd:SPARQL10Query ;
    sd:url <http://gov.tso.co.uk/legislation/sparql> .
```

We used this representation for the dataset available at the selected endpoint. These descriptions were inserted in our RDF database – a Fuseki store[16]. The following fragment shows RDF/Turtle schema representation of the enhanced descriptions having provenance metadata – denoted by our proposed **sdprov** vocabulary – for the legislation of the UK, Ireland and Wales dataset.

---

[15] UK Legislation endpoint: http://gov.tso.co.uk/legislation/sparql
[16] Fuseki Triple Store: http://jena.apache.org/documentation/serving_data/

```
<http://gov.tso.co.uk/legislation/sparql> a sd:Service ;
    sd:feature sd:UnionDefaultGraph ;
    # result formats supported by the endpoint
    sd:resultFormat <http://www.w3.org/ns/formats/N-Triples>,
        <http://www.w3.org/ns/formats/SPARQL_Results_JSON>,
        <http://www.w3.org/ns/formats/Turtle> ;
    # default dataset metadata
    sd:defaultDataset [ a prov:Entity , sd:Dataset ;
        # additional info about copyright
        sdprov:hasCopyrightPermission
            <http://gov.uk/doc/open-government-licence/> ;
        sdprov:hasOwner <http://..../legislation/void/tna> ;
        # supplemental info about employed vocabularies
        sdprov:hasProvenanceVocabulary [ a
          # provenance knowledge is using Dublin Core and FOAF
          sdprov:ProvenanceVocabulary ; sdprov:vocabulary dc: ] ;
        sdprov:hasProvenanceVocabulary
          [ a sdprov:vocabulary foaf: ] ;
        # primary dataset has as source www.legislation.gov.uk
        prov:hadPrimarySource <http://www.legislation.gov.uk/> ;
        # primary dataset has associated the following publisher
        prov:qualifiedAssociation [ a prov:Association ;
        prov:agent <http://gov.tso.co.uk/legislation/void/tna> ;
        prov:hadRole  [ a  prov:Publisher ]
    ] ;
    prov:specializationOf
        <http://gov.tso.co.uk/legislation/void/Legislation> ;
    prov:wasAttributedTo <http://.../legislation/void/tna> ;
    prov:wasGeneratedBy [ # other PROV constructs...
        ] ;
        prov:used <http://www.legislation.gov.uk/> ] ;
    sd:defaultGraph [ a sd:Graph ;
    sd:name <http://gov.tso.co.uk/legislation/void/Legislation> ]
    ] ;
    sd:endpoint <http://gov.tso.co.uk/legislation/sparql> .
```

6.2   Support for Creating a Mashup

We considered the case of a developer who wants to create a legal mashup and provide information about available semantic legal datasets. Data without explicit license is a potential legal liability and leaves consumers unclear what the usage conditions are. Therefore, it is very important that publishers make explicit the terms under which the dataset can be used. Our proposal provides an easy to use support for this problem.

To access metadata about the dataset license, original source and publishers, the system would create the following query – the `hasCopyrightPermission` property provided by our conceptual model is used:

```
SELECT ?endpoint ?dataset ?primarySource ?copyright ?agent
WHERE {
  GRAPH ?g {
  ?endpoint a sd:Service ; sd:defaultDataset ?dataset .
  ?dataset dct:title ?title .
  # using proposed vocabulary to obtain copyright info
  ?dataset sdprov:hasCopyrightPermission ?copyright ;
           prov:hadPrimarySource ?primarySource ;
           prov:qualifiedAssociation ?association .
  ?association prov:hadRole ?role .
  ?role a prov:Publisher .
  ?association prov:agent ?agent . }
FILTER regex (str (?title), "legislation", "i")
}
```

The results are (this dataset is available under the terms of the UK Open Government License for public sector information):

```
endpoint: <http://gov.tso.co.uk/legislation/sparql>
title: "Legislation"
agent: <http://gov.tso.co.uk/legislation/void/tna>
copyright:
  <http://nationalarchives.gov.uk/doc/open-government-licence/>
primarySource: <http://www.legislation.gov.uk/>
```

By performing the query via Fuseki Web interface, we obtained the response in 844 ms. An equivalent, but limited query (without copyright or publisher query parameters) issued against the VoID Store returns – with no results for legislation datasets – in 476 ms. The VoID Store also includes a HTTP redirect for every query, which adds to the total request time an average of 870 ms, therefore the response is given to the end-user in 1346 ms.

6.3   Accessing Data Specific to a Given Provenance Vocabulary

Another experiment was implemented to get what result formats a SPARQL endpoint supports, which provenance vocabularies does it uses or which is the default dataset graph against could a developer launch further queries and retrieve data. Our system issued the following query by using our vocabulary (in this context, the `hasProvenanceVocabulary` property is considered):

```
SELECT DISTINCT ?format ?vocabURI ?graphName
WHERE {
 GRAPH ?g {
```

```
    ?endpoint a sd:Service ; sd:defaultDataset ?dataset .
    ?dataset  dct:title ?title .
    ?endpoint sd:resultFormat ?format .
    ?dataset  sd:defaultGraph ?graph .
    ?graph    sd:name ?graphName .
    ?dataset  sdprov:hasProvenanceVocabulary ?vocab .
    ?vocab    sdprov:vocabulary ?vocabURI . }
 FILTER regex (str (?title), "legislation", "i")
}
```

Several results (the dataset about UK legislation uses Dublic Core and FOAF constructs and is available in several formats like RDF/XML, N-Triples, etc.):

```
<http://www.w3.org/ns/formats/RDF_XML>,
<http://purl.org/dc/elements/1.1/>,
<http://gov.tso.co.uk/legislation/void/Legislation>
========================
<http://www.w3.org/ns/formats/RDF_XML>,
<http://xmlns.com/foaf/0.1/>,
<http://gov.tso.co.uk/legislation/void/Legislation>
========================
<http://www.w3.org/ns/formats/N-Triples>,
<http://purl.org/dc/elements/1.1/>,
<http://gov.tso.co.uk/legislation/void/Legislation>
========================
<http://www.w3.org/ns/formats/SPARQL_Results_JSON>,
<http://xmlns.com/foaf/0.1/>,
<http://gov.tso.co.uk/legislation/void/Legislation>
. . .
```

The results are obtained in 178 ms when using our SD PROV store endpoint. Searching for vocabularies available on legislation datasets on the VoID Store – which does not make use of our designed conceptual model – returns no results in 1356 ms (870 ms for the HTTP redirect performed by the endpoint and 486 ms for the actual query).

7   Conclusion

In the current Web of data, where heterogeneous information is provided by various sources of disparate qualities, a mechanism that allows consumers to automatically determine the provenance of data(sets) is needed. We introduced a lightweight ontology as a convenient extension for the SPARQL 1.1 Service Description vocabulary that allows SPARQL endpoint providers to add provenance and license metadata to their published datasets.

The paper also enhanced a VoID specification regarding the UK legislation dataset used to made experiments showing how this vocabulary extension can be practically utilized.

In future work, we will extend our results to be a foundation of a service oriented platform, able to support provenance queries for all indexed datasets. Based on the constructed provenance graph, we intend to perform complex inferences on data provenance and also make recommendations based on computed results.

References

1. Alexander, K., Cyganiak, R., Hausenblas, M., Zhao, J.: Describing Linked Datasets. In: Bizer, C., Heath, T., Berners-Lee, T., Idehen, K. (eds.) Proceedings of the Linked Data on the Web Workshop (LDOW 2009), vol. 538, CEUR Workshop Proceedings, Madrid (2009)
2. Allemang, D., Handler, J.: Semantic Web for the Working Ontologist, 2nd edn. Elsevier, Amsterdam (2011)
3. Buil-Aranda, C., Hogan, A., Umbrich, J., Vandenbussche, P.-Y.: SPARQL Web-Querying Infrastructure: Ready for Action? In: Alani, H., Kagal, L., Fokoue, A., Groth, P., Biemann, C., Parreira, J.X., Aroyo, L., Noy, N., Welty, C., Janowicz, K. (eds.) ISWC 2013, Part II. LNCS, vol. 8219, pp. 277–293. Springer, Heidelberg (2013)
4. Buneman, P., Khanna, S., Tan, W.-C.: Why and Where: A Characterization of Data Provenance. In: Van den Bussche, J., Vianu, V. (eds.) ICDT 2001. LNCS, vol. 1973, pp. 316–330. Springer, Heidelberg (2000)
5. Buneman, P., Khanna, S., Tan, W.-C.: Data Provenance: Some Basic Issues. In: Kapoor, S., Prasad, S. (eds.) FST TCS 2000. LNCS, vol. 1974, pp. 87–93. Springer, Heidelberg (2000)
6. Feigenbaum, L., Williams, G.T., Clark, K.G., Torres, E.: SPARQL 1.1 Protocol. W3C Recommendation, http://www.w3.org/TR/sparql11-protocol/
7. Greenwood, M., Goble, C., Stevens, R., Zhao, J., Addis, M., Marvin, D., Moreau, L., Oinn, T.: Provenance of e-Science Experiments – Experience from Bioinformatics. In: Proceedings of the UK OST e-Science. EPSRC (2003)
8. Groth, P., Gil, Y., Cheney, J., Miles, S.: Requirements for Provenience on the Web. International Journal of Digital Curation 7(1), 39–56 (2012)
9. Harris, S., Seaborne, A. (eds.): SPARQL 1.1 Query Language. W3C Recommendation, http://www.w3.org/TR/sparql11-query/
10. Heath, T., Bizer, C.: Linked Data: Evolving the Web into a Global Data Space. Morgan & Claypool, California (2011)
11. Khalili, A., Auer, S., Ngonga Ngomo, A.-C.: conTEXT – Lightweight Text Analytics Using Linked Data. In: Presutti, V., d'Amato, C., Gandon, F., d'Aquin, M., Staab, S., Tordai, A. (eds.) ESWC 2014. LNCS, vol. 8465, pp. 628–643. Springer, Heidelberg (2014)
12. Lanter, D.: Design of a Lineage-Based Meta-Data Base for GIS. Cartography and Geographic Information Systems 18(4), 255–261 (1991)
13. Lebo, T., Sahoo, S., McGuinness, G. (eds.): PROV-O – The PROV Ontology. W3C Recommendation (2013), http://www.w3.org/TR/prov-o/
14. Prud'hommeaux, E., Seaborne, A. (eds.): SPARQL Query Language for RDF. W3C Recommendation, http://www.w3.org/TR/rdf-sparql-query/
15. Simmhan, Y., Plale, B., Gannon, D.: A Survey of Data Provenance in e-Science. SIGMOD Record 34(3), 31–36 (2005)
16. Williams, G. (ed.): SPARQL 1.1 Service Description. W3C Recommendation, http://www.w3.org/TR/sparql11-service-description/

Complex Networks' Analysis
Using an Ontology-Based Approach: Initial Steps

Alex Becheru and Costin Bădică

University of Craiova,
Blvd. Decebal nr. 107, Craiova 200440, Romania
becheru@gmail.com,
costin.badica@software.ucv.ro

Abstract. This paper presents a new ontology that enables the knowledge-based analysis of complex networks. The purpose of our research was to develop a new approach for the knowledge-based analysis of complex networks based on various network attributes and metrics. Our approach is both easy to use and easy to understand by a human. It facilitates the automated classification of different types of networks. For the creation of this ontology we applied an already known methodology from the scientific literature. The ontology was also enriched with our own developed methods. We applied our ontology to the analysis scenarios of complex networks obtained from real world problems, thus supporting its generality, as well as its usability across domains.

Keywords: Complex Networks, Ontology, Graph Types.

1 Introduction

Our current understanding of the surrounding environment, either geographical or biological, shows us that nature is formed out of complex interconnecting systems. Networks created by these systems support phenomena that are far from being deterministic trough traditional methods. Each element influences the network, while the network puts its mark on every element. Now we can say with certainty that the *butterfly effect* imagined by Edward Lorenz is truly possible.

In order to understand complex interconnected systems a new field of research emerged – *Network Science* (NS) or *Complex Networks Analysis* (CNA). The heart of this new research field leverages on *Graph Theory* and *Computer Science*. NS investigates non-trivial features of graph problems that usually are not addressed by lattice theory or random graphs. The understanding of such non-trivial features is of high interest, as they frequently occur in real world problems. The complexity of real world networks comes from the modeling and evaluation of overlapping and interdependent phenomena, that are neither purely regular nor purely random. Also complexity may come with the sheer size of the network itself.

NS defines some basic types of graphs by trying to understand and model the phenomena that led to their creation. Knowing the type of a complex network is very important as it gives very powerful insights into the model or phenomenon that it represents. For example an organization modeled as a *Watts-Strogatz* [1] type of complex network

R. Buchmann et al. (Eds.): KSEM 2014, LNAI 8793, pp. 326–337, 2014.

supports very well the exchange of information, as it incorporates the *Small World* phenomena [2]. But unfortunately this type of graph does not support the phenomenon of *Homophily* [3], that is also found in many real world networks.

As far as our knowledge expands, we are not aware of any other work that defines an automated process of determining the type of a complex network based on its attributes. Moreover, our proposed knowledge-based approach has the following features:

- It is easy to use, understand and share by humans.
- It is able to propose new classes of complex networks, in addition to those that can be found in the research literature.
- It is reusable and expandable, i.e. it captures in a reusable way existing classes of networks, as well as it allows the addition of new classes as the knowledge in the field will expand.

The analysis techniques considered in our approach are inspired by real world applications. We are able to exemplify the applicability of our knowledge-based method to real world problems from different application domains.

Our paper is structured as follows. The second section addresses background information about ontologies and *Network Science*. The next sections presents the ontology created by us together with the methodology used. The fourth section exemplifies the use of our ontology in real world scenarios. We finish by iterating our final remarks and further work.

2   Overview

2.1   Ontologies

Due to the vast applicability of CNA in diverse domains of research there can be widely varying viewpoints and assumptions towards phenomena that are basically the same. As people interested in CNA might have different research backgrounds and the network models can be built from different perspectives, different sets of overlapping jargons may be used for describing the resulting models. This leads to the lack of shared understanding of the resulting models, thus hindering: communication, interoperability, and reusability [4].

The need for a unifying framework for organizing and describing the knowledge in the field arises in this context. Ontologies have been used successfully in diverse cases: biology [5], public transport [6], medicine [7], linguistics [8]; to define a model that makes explicit the basic conceptualisation of a knowledge domain. T. R. Gruber which is one of the most known experts in creating ontologies defines the term ontology as a "specification of a conceptualization" [9]. According to Guarino [10], an ontology "is a logical theory accounting for the intended meaning of a formal vocabulary". The building blocks of an ontology are the concepts and the relations among them. The definitions, properties and constraints of concepts are defined as logical axioms.

We decided to use an ontology model for capturing the knowledge domain of the types of complex networks as it suits well with our needs. Moreover, ontology creation is a well established practice in the academic world. Finally, the capabilities of consistency checking and automated classification of ontology reasoning endow our approach with increased reliability [11].

2.2   Network Science

A graph (or network in NS) is composed of *nodes and links*, with each link connecting two nodes. The graph can be weighted, i.e. a label (usually a number) is associated with each link, or unweighted. Links can be unidirectional (or directed), i.e. the source and target nodes are specified, or bidirectional (or undirected). Self-looping graphs permit a node to directly connect with itself. Nodes can be seen as an abstraction of: people, entities, concepts etc. A link between two nodes models a binary relation between those nodes and has a domain-specific interpretation: physical interconnection, shared similarity, interaction, etc. In this paper we address bidirectional, unweighted and non-self-looping graphs. They are the most common types of graphs found in the literature. The other types of graphs can be converted to this type without losing valuable information.

There are two important papers standing as the building blocks of *Network Science*. Paul Erdös and Alfréd Rényi wrote about random graphs in 1959 [12]. In 1973, Mark Granovetter discovered the "strength of weak ties" [13]. A graph usually consists of a number of subgraphs, nodes inside these subgraphs are tightly connected among them and loosely (weak ties) connected with other subgraphs. One may think that those weak ties are not relevant, but without their presence the graph of subgraphs would not exist. CNA emerged at the beginning of the 1990's as a result of the progress in applied computational sciences. But the most important factor was the access to data describing real world networks. The emergence of the World Wide Web, as well as the explosion of the interest in detailed mapping across many sciences, especially in biology and economics, opened a multitude of research paths.

Stanley Milgram [2] and Watts et al. [1] discovered and defined the *small world phenomenon*. Otherwise called *six degrees of separation*, this phenomenon is found in many real world large networks, where contrary to the size of the network the average path length between two nodes has a very low value (6 or less). Barabasi et al. [14] showed that real world networks have a *scale free degree distribution*, also called Pareto or Zipf distribution. This means that very few nodes have high *Degree* while the majority has almost the same very low *Degree*. An explanation for the appearance of the *scale free distribution of degree* is the *preferential attachment* [15] of nodes, a node has a greater probability to be linked with nodes that have high *Degree* than with nodes with low *Degree*. Another phenomenon that is of great interest for NS is *Homophily*, described as the tendency of individuals (nodes in our case) to associate and bond with similar others [3].

NS can be used in many application domains. For example, internet companies like *Google* and *Facebook* are practically built on complex networks. In medicine, the spread of diseases is now studied with the help of CNA [16]. Security forces map the networks of acquaintances of wanted individuals, maps which could lead to alternative ways to reach them. The famous Saddam Hussein was captured using methods from NS [17]. Large oil companies use a branch of CNA known as *Organisational Network Analysis* to enhance the flow of information exchange within the companies [18]. CNA was even used to determine the best tennis players respective to different scenarios [19], e.g. best tennis player on the grass surface.

NS proposes a 3-layer approach for analyzing graphs. Each layer employs its specific measurements or otherwise called metrics. The top layer is concerned with the description of the entire graph. Metrics like the *average path length* between two nodes are employed here. By going down one layer, community detection becomes the emphasis of CNA. The last layer targets each node's properties. Using this layered approach, NS ensures that all the levels of abstraction are properly considered. In this paper we shall use metrics from the first two layers, as we are primarily interested in general graph properties.

3   Creating the Ontology

3.1   Overview

As far as we know, this is the first attempt to build a *Complex Networks* ontology, Moreover, we could not find any other ontologies that are somehow related with our purpose. Therefore, we had to start from scratch and create a new ontology. We used the following steps to define our ontology [20]:

1. Determine the domain and scope of the ontology.
2. Enumerate important terms in the ontology.
3. Define the classes and their relations to each other (hierarchy).
4. Define the properties of classes.
5. Define restrictions on the properties.
6. Create instances.

3.2   Vision on the Use of the Ontology

Our vision is to use the ontology in various application scenarios of CNA. For a given problem, CNA starts by formulating a set of specific questions regarding the network under analysis. Usually such question can be answered by performing a thorough analysis of the network using the available CNA computational tools[1]. The analysis returns a relevant set of metrics and their respective values. Based on those metrics, the network can be classified into a certain known class. This classification enables the user to obtain additional insight on the properties of the underlying network.

The ontology should be able to classify networks in different categories, depending on their support of information exchange. The basic competency questions that arise are:

- What type of graphs are studied with this ontology? Graphs can be: directed/ undirected, self-looping/non-self looping or weighted/unweighted;
- Which are the types of complex networks (complenets)? Which of them are relevant and widespread enough to be introduced in the ontology?
- Based on the determined types of complenets, can one of them be a subtype of another type? Can a complenet be at the same time of two or more types?
- Which are the canonical (present in the scientific literature) characteristics that define each complenet type?

---

[1] Like for example Gephi: https://gephi.org/

- How can we quantify complenets' characteristics? Metrics are used with this purpose, but could we include phenomena or other methods of expressing those characteristics?
- Which are the metrics and phenomena that can express the characteristics of a network? Considering the fact that different metrics and phenomena may determine the same characteristics, which will be used in the ontology and which will be considered to bring redundant information?
- As CNA is a 3 layer stratified analysis, each layer having its specific metrics, how will the metrics be represented in the ontology?
- Besides canonical ways of determining the type of a complenet, are there any other reliable ways to determine the type of a complenet? Which are the types of complenets that can be determined with alternative and reliable methods? How can these methods be used to enrich the ontology?
- Due to the fact that exemplification of the ontology's power is imperious, can we introduce powerful examples in the ontology? Well known and used networks could be used with this purpose, but which should we chose?

For the purpose of explaining our vision we shall imagine an application domain in which links between nodes model the exchange of information. Our purpose is to find insights on how the information spreads in the network. In this particular case, the following questions should be answered after the type of the complenet was determined:

- Has information good chances to spread in the entire network?
- How much time (estimated in the number of steps) it takes until the information has spread in the vast majority of the network?
- Can we compare two different networks and answer which supports better the information exchange?
- If information can suffer from alteration at each intermediary step, which networks reduce information alteration?

The proposed ontology will help us to easily classify new complex networks based on their metrics, while the resulting classification will be easy to understand and use. For all the questions iterated above you will find the answers in the next section as we shall describe the ontology.

3.3 Defining the Ontology

Our ontology is expressed using OWL 2 Web Ontology and it was created with the help of Protégé [21] open source software. The ontology contains: classes, object properties, data properties and individuals. NS metrics shall be presented while explaining the ontology construction. The ontology is available via a public *GitHub* repository[2].

Classes. OWL classes represent concepts and are interpreted as sets of individuals. For example, analyzing the class hierarchy from Figure 1, note that the top level class is called *Thing* and it has 4 subclasses (we shall call them *Top-level Classes*) as follows:

---

[2] Github repository: becheru.github.io/Ontology

General Graph metrics is a *Top-level Class* that represents value partitions of multiple graph metrics. These metrics are: *Clustering Coefficient* [22], *Density* and *Modularity*; Due to the lack of paper space we shall present only the *Modularity* metric, all the other metrics follow almost the same schema. *Modularity* was designed to measure the strength of group divisions (also called groups, clusters or communities) in a network [23]. The possible values of *Modularity* lie in the range $[-1/2, 1)$. It is positive if the number of edges within groups exceeds the number expected on the basis of chance. *Modularity's* values are represented by the following disjoint subclasses: *Negative Modularity* and *Positive Modularity*; Also *Positive Modularity* contains subclasses that divide the $(0, 1)$ range of possible *Modularity* values in 5 equal partitions: *VeryLowGraphModularity* $(0, 0.2]$, *LowGraphModularity* $(0.2, 0.4]$, *MediumGraphModularity* $(0.4, 0.6]$, *HighGraphModularity* $(0.6, 0.8]$ and *VeryHighGraphModularity* $(0.8, 1)$; The subclasses of the *Clustering Coefficient* and *Density* metrics follow the same schema as the subclasses of the *Positive Modularity*.

Graph Structure is a *Top-level Class* that contains elements which define the overall structure of a complex network. These elements are represented by the following subclasses: *Betweenness Concentration, Distribution of Degree, Number of Connected Components* and *Structure Type*;

Betweenness Concentration is based on the *Betweenness* metric. A node with high *Betweenness* appears more often in paths (shortest paths) between nodes in the network [24]. If we normalize *Betweenness* such that each node has a value between 0 and 1 with the total *Betweenness* being 1, we can determine which is the value of the highest *Betweenness* node of the graph. *Betweenness Concentration* has 3 disjoint subclasses that represent the possible value it can take. If a node has a *betweenness* value of more than 0.6, then we should chose the *Single Node Betweenness Concentration* subclass. If this is the case, then it is a sign that the graph might have a *Single Hub* structure type. If the above case is not true but the sum of multiple nodes *betweenness* surpasses the threshold of 0.6, then the right subclass to chose is *Multiple Node Betweenness Concentration*. This is a sign of a presence of a *Multiple Hub* graph structure type. If the *betweenness* value of every node is equal to zero than the right subclass to chose is *No Node Betweenness Concentration*.

The *Distribution of Degree* is obtained by making a histogram of all nodes' *Degree*. A node's *Degree* is equal to the number of edges that connect with it (undirected graph). Three possible distributions are present in our ontology, each is represented by a disjoint subclass: *Uniform Distribution, Poisson Distribution* and *Exponential Distribution*. The *Exponential Distribution* contains a subclass depicting the *Pareto (Zipf or Power Law)* distribution [25]. The *Pareto* distribution is a skewed, heavy-tailed distribution that is usually found in real world networks.

With the *Number of Connected Components* class we try to determine if the graph is connected or if it it contains multiple subgraphs that are not connected among them. This class contains two disjoint subclasses that represent the possible values of the class: *Multiple Weakly Connected Components* and *One Weakly Connected Component*;

The *Structure Type* is used to determine if the overall structure of the graph is of the following types: *Single Hub*, *Multiple Hub* and *Core & Periphery* [26–28]; These types of structures are depicted by disjoint subclasses of the *Structure Type* class. See Figure 2 for a graphical exemplification of these types of structures.

The class **Graph Types** *Top-level Class* contains the classes that do the detection of different complex networks' types (defined classes). Also here we can find several examples of complex networks, they can be found under the *Named Graphs* subclass. The types of complex networks that can be found are the following: *ScatteredFragments*, *Random*, *WattsStrogatz* [1], *OverConnected*, *GoodExchange*, *Barabasi-Albert*; The *GoodExchange* type of network is based on the findings of Noah Friedkin [29], good exchange of information is possible only if the *Average Path Length* is smaller than 3. The *Barabasi-Albert* type is subdivide in other 3 subtypes: *SingleHub*, *MultipleHub* and *CorePeriphery*. Also in the ontology you can find alternative methods of determining the above mentioned complex networks types. They are represented by defined subclasses that have the *Alternative* prefix, e.g. *AlternativeBarabasiAlbert*. These alternative methods are developed based on our experience in the CNA field of study.

The *Named Graphs* subclass contains real world complex networks and well known graphs in the scientific literature. For example the *Ancient World* network describes the groups of humans in the ancient world. These groups are locally (in the same geographical zone) connected, but they form a global network that contains several components that have no means of communicating among them. The Aboriginal people in Australia were cut off for thousand of years from the rest of the world. The *DoplhinSocialNetwork* represents a network of socialisation between dolphins near the coast of New Zealand [30]. The *Erdos-Renyi* graph is an example of well known graph that is present in our ontology [12].

The **Phenomena** *Top-level Class* describes social phenomena that could be present in graphs. Our ontology describes only two phenomena as subclasses of this class: *Homophily* and *SmallWorldPhenomemon*;

Object Properties. OWL Properties represent relationships, binary relations on individuals. Object properties are relationships between two individuals. The domain of our object properties is the *Graph* class. Our ontology consists of 3 main transitive object properties: *hasPhenomena*, *hasMetric* and *hasStructure*; The *hasStructure* contains other 4 functional sub-properties: *hasBetweennesConcentration*, *hasDistributionOfDegree*, *hasNumberOfConnectedComponents* and *hasStructureType*;

Data Properties. They describe relationships between an individual and data values. The ontology has two data properties, but others can be added with ease: *hasGraphAveragePathLength* and *hasGraphNumberOfNodes*. The *hasGraphAveragePathLength* depicts the length of the average path length in a graph as a number. The *hasGraphNumberOfNodes* represents the number of nodes in a network.

Individuals. Individuals, represent objects in the domain in which we are interested. We have created two individuals for exemplification:

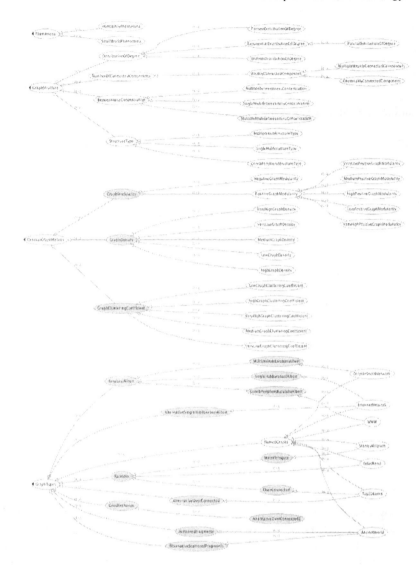

Fig. 1. Figure depicting the full class diagram of the inferred model. Defined classes are marked with the orange background. Primitive classes are transparent.

Fig. 2. Figure depicting different types of complex network structures. From left to right: *Single Hub*, *Multiple Hub*, *Core&Periphery*.

- *ExOverConnected*: this individual is of class type *Top20Banks* with two data properties: *hasGraphNumberOfNodes*=20000 and *hasGraphAveragePathLength* with
  value of 1.4.
- *ExGoodExchange* : this individual has only a data type property. The *hasGraphAveragePathLength* is equal to 2.7. Also it has a *OneWeaklyConnectedComponent*,
  meaning that the graph is connected.

In Figure 1 you can see the inferred class hierarchy. For example the Class *Dol-
phinSocialNetwork* is a subclass of the class *NamedGraphs*, this was given explicitly.
But it is also a subclass of the class *MultipleHubBarabasiAlbert* which was inferred
through our ontology. To generalise this example, every subclass of *NamedGraphs* is
also a subclass of other classes determined by inference.

3.4   Class Refinements and Constraints

For the serialization we used the *Manchester Owl syntax*, as this is easily readable. We
chose to present two classes, one primitive and one defined.

The *GoodExchange* class is a defined class, thus it used to classify network types. As
you can see below the necessary and sufficient conditions are under the *EquivalentTo*
branch. An individual has to be a *graph* made by *one weakly connected component* and
it has to have the *average path length* smaller or equal to 3. The *GoodExchange* class is
a child class of the class *GraphTypes*.

Example of a defined class/complex network type

```
Class: <#GoodExchange>
EquivalentTo:
    <#GraphTypes>
    and (<#hasNumberOfConnectedComponnets> some
            <#OneWeaklyConnectedComponnet>)
    and (<#hasGraphAveragePathLength> some xsd:double[<= "3"])
SubClassOf:
    <#GraphTypes>
```

The *ErdosRenyi* primitive class has its necessary conditions, those that characterize
it, under the *SubClassOf* branch. Because we present the inferred method the *Random*
class appears under this branch, although we did no specify it in the asserted model.
This means that the reasoner automatically detected the *ErdosRenyi* as being of *Random*
type. Each object property also presents a closure axiom, i.e. those restrictions that have
only in their definition. *Disjoint* classes with the *Erdos-Renyi* class are clearly shown.

Example of a primitive class that describes a network

```
Class: <#ErdosRenyi>
SubClassOf:
    <#hasDistributionOfDegree> only
        <#UniformDistributionOfDegree>,
    <#hasDistributionOfDegree> some
        <#UniformDistributionOfDegree>,
    <#hasNumberOfConnectedComponnets> some
        <#OneWeaklyConnectedComponnet>,
    <#hasNumberOfConnectedComponnets> only
        <#OneWeaklyConnectedComponnet>,
    <#NamedGraphs>,
    <#Random>
```

```
DisjointWith:
    <#AlternativeScatteredFragments>, <#SingleHubBarabasiAlbert>,
    <#WWW>, <#DolphinSocialNetwork>, <#BarabasiAlbert>,
    <#EthernetNetwork>, <#AncientWorld>,
    <#MultipleHubBarabasiAlbert>, <#Core&PeripheryBarabasiAlbert>
```

4   Ontology Exemplification

In this section we present a real world application example of our proposed approach involving the complex networks' ontology. We focus on human resources in business management [31]. Note however that our proposal is general enough to be employed for a large number of other application domains. Basically, every system or phenomenon that can be modeled as a complex network can be the subject of the analysis process of our proposed ontology-based approach.

Let us imagine the following case. In a company the CEO was changed due to lack of results. The company has different branches and is present in several countries. Although the company has different branches all of them develop products for the automotive world. Some branches produce highly competitive products while others are loosing money. The new CEO is interested in bringing know-how from the profitable branches to those that are loosing them. Also he suspects that collaboration inside the company is not properly working therefor he commands a study on the exchange of information inside the company.

As email is considered the standard way to communicate important things within the organization, an email exchange graph was created. Each employee is represented by a node, a link is present between nodes if an email was exchanged by those two nodes. With the help of CNA instruments the CEO found that following traits of the graph: *Pareto distribution of degree, One Weakly Connected Component* and that the structure is of type *Multiple Hub*. Next by using our ontology the CEO determined the network is a *Barabasi-Albert MultipleHub* network. By studying the properties of this network type the CEO found that there are some very few people (*Hubs*) responsible for the exchange of information within the company. Through these key employees the majority of know-how is exchanged. Also some of these *Hubs* are not even connected among them. The feedback obtained is also supported by prior knowledge of the CEO, as he observed that some key people in the company rarely met. By knowing the type of network the CEO determined that ideas and know-how get lost as these key employees filter them due to the overwhelming amount of information that they have to handle.

As proven by Valdis Krebs the best type of network for balancing innovation with communication is the *Core Periphery* structure [28]. By comparing the current network type with *Barabasi Albert Core Periphery* , both present in our ontology, the CEO can determine which are the differences. Here the *Core Periphery* structure makes the difference between the two types of networks. Now the CEO knows where to focus its attention on improving the current organizational graph. For example he could schedule regular meetings with the *Hubs*. The work of Robert L. Cross et. al. could help the CEO determine what are the steps necessary to enhance the exchange of knowledge within the organisation [18].

5 Concluding Remarks

Although our ontology is in a primitive state we already prove it to be useful in a real life scenarios. Also the ontology is ready to be used as it is, and can be used in a wide variety of cases. Wherever a complex network is built, the ontology should be able to point to the correct insights on that network. Use across different fields of research was envisioned from the beginning thus we built an easy to use and understand ontology. Also we have supplied all the necessary means by which the ontology could be reached and used by putting it in a shared version controlled repository on *GitHub*.

We acknowledge that further work is needed with this ontology. For the future we plan to enrich it with domain specific classes that will guide professionals in better understanding real life events. Also we want to incorporate the necessary steps needed to be fulfilled in order to exchange the type of a network. Thus the ontology shall be also a guide of practice not only a type detection of complex networks. More examples of uses of the ontology in the real world need to be provided, together with SPARQL[3] examples.

Acknowledgments. This work was supported by the strategic grant POS-DRU/159/1.5/2/133255. Project ID 133225(2014), co-financed by the European Social Fund within the Sectorial Operational Program Human Resources Development 2007-2013.

This work was conducted using the Protégé resource, which is supported by grant GM10331601 from the National Institute of General Medical Sciences of the United States National Institutes of Health.

References

1. Watts, D.J., Strogatz, S.H.: Collective dynamics of 'small-world' networks. Nature 393, 440–442 (1998)
2. Milgram, S.: The small world problem. Psychology Today 2, 60–67 (1967)
3. McPherson, M., Smith-Lovin, L., Cook, J.M.: Birds of a feather: Homophily in social networks. Annual Review of Sociology, 415–444 (2001)
4. Uschold, M., Gruninger, M.: Ontologies: Principles, methods and applications. The knowledge Engineering Review 11, 93–136 (1996)
5. Ashburner, M., Ball, C.A., Blake, J.A., Botstein, D., Butler, H., Cherry, J.M., Davis, A.P., Dolinski, K., Dwight, S.S., Eppig, J.T., et al.: Gene ontology: Tool for the unification of biology. Nature Genetics 25, 25–29 (2000)
6. Wang, J., Ding, Z., Jiang, C.: An ontology-based public transport query system. In: First International Conference on Semantics, Knowledge and Grid, SKG 2005, pp. 62–62. IEEE (2005)
7. Luciano, J.S., Andersson, B., Batchelor, C., Bodenreider, O., Clark, T., Denney, C.K., Domarew, C., Gambet, T., Harland, L., Jentzsch, A., et al.: The translational medicine ontology and knowledge base: Driving personalized medicine by bridging the gap between bench and bedside. J. Biomed. Semantics 2, S1 (2011)
8. Hristea, F., Colhon, M.: Feeding syntactic versus semantic knowledge to a knowledge-lean unsupervised word sense disambiguation algorithm with an underlying naïve bayes model. Fundamenta Informaticae 119, 61–86 (2012)

---

[3] Query language for RDF files http://www.w3.org/TR/rdf-sparql-query/

9. Gruber, T.R.: Toward principles for the design of ontologies used for knowledge sharing? International Journal of Human-computer Studies 43, 907–928 (1995)

10. Guarino, N.: Formal ontology in information systems: Proceedings of the first international conference (FOIS 1998), Trento, Italy, June 6-8, vol. 46. IOS Press (1998)

11. Mian, P.G., Falbo, R.D.A.: Supporting ontology development with oded. Journal of the Brazilian Computer Society 9, 57–76 (2003)

12. Erdős, P., Rényi, A.: On random graphs. Publicationes Mathematicae Debrecen 6, 290–297 (1959)

13. Granovetter, M.: The strength of weak ties. American Journal of Sociology 78, 1 (1973)

14. Barabási, A.L., et al.: Scale-free networks: A decade and beyond. Science 325, 412 (2009)

15. Newman, M.E.: Clustering and preferential attachment in growing networks. Physical Review E 64, 025102 (2001)

16. Barabási, A.L., Gulbahce, N., Loscalzo, J.: Network medicine: A network-based approach to human disease. Nature Reviews Genetics 12, 56–68 (2011)

17. Wilson, C.: Searching for saddam: Why social network analysis hasn't led us to osama bin laden. Slate (February 26, 2010)

18. Cross, R.L., Singer, J., Colella, S., Thomas, R.J., Silverstone, Y.: The organizational network fieldbook: Best practices, techniques and exercises to drive organizational innovation and performance. John Wiley & Sons (2010)

19. Radicchi, F.: Who is the best player ever? a complex network analysis of the history of professional tennis. PloS One 6, e17249 (2011)

20. Noy, N.F., McGuinness, D.L., et al.: Ontology development 101: A guide to creating your first ontology (2001)

21. Horridge, M., Knublauch, H., Rector, A., Stevens, R., Wroe, C.: A practical guide to building owl ontologies using the protégé-owl plugin and co-ode tools edition 1.0. University of Manchester (2004)

22. Schank, T., Wagner, D.: Approximating clustering-coefficient and transitivity. Universität Karlsruhe, Fakultät für Informatik (2004)

23. Lambiotte, R., Delvenne, J.C., Barahona, M.: Laplacian dynamics and multiscale modular structure in networks. arXiv preprint arXiv:0812.1770 (2008)

24. Boldi, P., Vigna, S.: Axioms for centrality. arXiv preprint arXiv:1308.2140 (2013)

25. Adamic, L.A., Huberman, B.A.: Power-law distribution of the world wide web. Science 287, 2115–2115 (2000)

26. Borgatti, S.P., Everett, M.G.: Models of core/periphery structures. Social Networks 21, 375–395 (2000)

27. Hojman, D.A., Szeidl, A.: Core and periphery in networks. Journal of Economic Theory 139, 295–309 (2008)

28. Krebs, V., Holley, J.: Building smart communities through network weaving. Appalachian Center for Economic Networks (2006), http://www.acenetworks.org (retrieved)

29. Krebs, V.: Managing the 21st century organization. IHRIM Journal 11, 2–8 (2007)

30. Lusseau, D.: The emergent properties of a dolphin social network. Proceedings of the Royal Society of London 270, S186–S188 (2003)

31. Becheru, A.: Agile development methods through the eyes of organisational network analysis. In: Proceedings of the 4th International Conference on Web Intelligence, Mining and Semantics (WIMS 2014), p. 53. ACM (2014)

Review of Knowledge Management Models for Implementation within Advanced Product Quality Planning

Bogdan Chiliban[1], Lal Mohan Baral[1,2], and Claudiu Vasile Kifor[1]

[1] "Lucian Blaga" University of Sibiu, Sibiu, Romania
bogdan.chiliban@gmail.com,
claudiu.kifor@ulbsibiu.ro
[2] Ahsanullah University of Science and Technology, Dhaka, Bangladesh
lalmohan_baral@yahoo.com

Abstract. The purpose of this paper is to review different Knowledge Management Models (KMM) based on their strengths and weaknesses and to ascertain their possible implementation within the scope of the Advanced Product Quality Planning (APQP) procedure. The KMM listed are selected based on an extensive literature review. They are presented in a visual manner and analyzed based on their structure and features, versus a predetermined set of criteria. This study offers an overview and a critical discussion of the merits and faults of a number KMM and suggests possible improvement avenues.

Keywords: Knowledge Management, Knowledge Management Models, Integrated Models, APQP.

1   Introduction

Knowledge Management involves the management of the relationship between tacit knowledge, the know-how possessed by individuals and explicit knowledge, the systemically documented know-how that becomes available to everyone in the organization [1]. The ultimate goal of knowledge management should be to transfer tacit knowledge to all members of the organization, in order to radically improve the capacity of individuals to use information strategically and apply higher-order thinking to an informed decision- making process [2]. Nonaka & Takeguci describe the cycle activities: socialization, externalization, combination, and internalization through which tacit knowledge may be transferred from one individual to another [3]. Knowledge management has adopted communication and collaboration solutions to address the unique challenge of articulating, sharing, and leveraging tacit knowledge [4]. Ho [5] cited from the literature evident that the goal of knowledge management is to deliver the right knowledge to the right individuals at the right time so that they can take appropriate actions and improve performance [6,7].

R. Buchmann et al. (Eds.): KSEM 2014, LNAI 8793, pp. 338–348, 2014.
© Springer International Publishing Switzerland 2014

The fast pace of today's world requires that attention be given not only to the existing organizational knowledge, but also to the development of new knowledge to make the organization competitive [8]. Knowledge management is a part of the field of management studies but it is also closely integrated with information and communication technologies [4], [9]. This is because of the critical role that technology plays in enabling and supporting the practice of knowledge management through information systems and social support [10]. An interesting phenomenon that has been brought about by the development of communication technology (wikis, blogs, tweets) is the development of decentralized, un- hierarchical, increasingly independent knowledge communities in organizations [11]. These horizontal networks have the capability to cut across organizational boundaries and connect formally isolated professionals, facilitating knowledge generation and dissemination [12].

Advanced Product Quality Planning (APQP) is a standardized procedure developed by the big three American motor companies such as GM, Ford and Chrysler as a response to the increased competition they faced from other European and Asian car manufacturers [13]. Its main goal is to achieve customer satisfaction through the development of quality products and processes [14]. The main steps of the APQP are: i) Prepare for APQP, ii) Plan and Define Program, iii) Product Design and Development, iv) Process Design and Development, v) Product and Process Validation and vi) Feedback Assessment and Corrective Action. The complex and varied nature o all these phases and their activities require that individuals involved in them posses high levels of interconnected knowledge, in order to achieve quality products that meet customer requirements and usher organizational performance. Although the importance of knowledge and knowledge management activities has been documented [15,16], little practical penetration was achieved in the industrial field, with few organizations recognizing the need for a formal and well coordinated managerial effort in this area.

The purpose of this study is to analyze existing knowledge management models, which are already used in different organizational environments and underline how their implementation in practice may benefit the outcomes of the APQP. After analyzing the structures and characteristics of those models, they are evaluated based on a framework which strives to identify future directions for the development of an effective knowledge management model for APQP.

2   Literature on Different KM Models/Approaches

2.1   Knowledge Management Model selection

The models presented within this work were selected based on an extensive literature review and because they possess the following elements: practical implementation, similar levels of abstraction, clear presentation of knowledge management elements or activities, identification of interdependencies between elements, implication of the need for a continuous KM process, quantification o KM results, but also because they have unique and distinguishable characteristics and approaches to the KM process.

2.2   Knowledge Tower

The knowledge tower is an enterprise level knowledge management model. Oztemel & Arslankaya [17] have proposed this model supporting the need for a hierarchical structure where each component necessitates and involves the capabilities of its predecessors. The aim of the model is to guide organizations in relation to the main elements required for effective knowledge management instantiation. In the author's opinion these are: Knowledge Infrastructure, Knowledge Management Processes, Knowledge Representation, Knowledge Planning, Knowledge Management Strategies, Knowledge Organization, Knowledge Culture, Knowledge Leverage and Assessment of KM.

Oztemel and Arslankaya also defined the activities and tools to be used for proper implementation of all the above mentioned steps. They also proposed a management structure with specific activities for KM process. This structure consists of Chief Knowledge Officer, Knowledge Manager and Knowledge Worker who will be the responsible for Strategic Knowledge, Tactic Knowledge and Operational Knowledge management process respectively.

Fig. 1. Components of EKMM [17]

2.3   Knowledge Management Process Model

The processes of KM in this model involves knowledge acquisition, creation, refinement, storage, transfer, sharing, and utilization [18]. The model clearly shows that the first steps in the KM cycles are either Creation of knowledge or Acquisition of knowledge. Knowledge Creation refers to the development of new knowledge or replacing existing knowledge with new content. This step is based on the SECI model for knowledge generation proposed by [3]. The SECI modes present the process that new knowledge must undergo in order for it to be created. It presents four distinct dimensions. These are: Socialization, Externalization, Combination and Internalization by which tacit knowledge can be disseminated across the organization for improved performance. Knowledge Acquisition is a particular process step involves existing knowledge. This is located outside of the organization and must be searched for, recognized by the stakeholder and assimilated [19]. According to researchers view, the Knowledge Refinement process involves tacit, or implicit knowledge that must be explicated, codified, organized into an appropriate format and evaluated according to a set of criteria for inclusion into the organization's formal memory.

In order for knowledge to have wide organizational impact, it must be either transferred or shared. Transfer and sharing may be conceptualized as two ends of a continuum. Transfer should be focused on purposeful communication of knowledge from a sender to a known receiver [20]. Sharing is less-focused dissemination, such as through a repository, to people who are often unknown to the contributor [21]. Once knowledge is transferred to, or shared with others, it may be utilized through elaboration (the development of different interpretations), infusion (the identification of underlying issues), and thoroughness (the development of multiple understandings by different individuals or groups) [22] in order to be helpful in facilitating innovation, collective learning, individual learning, and/or collaborative problem solving [23]. Finally Organizational performance should be evaluated through the outputs like improved productivity, revenues, profits and return on investment.

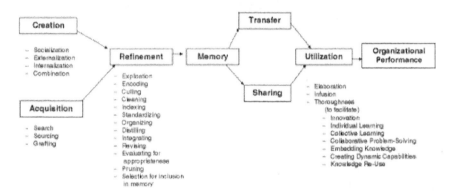

Fig. 2. Knowledge Management Process Model [18]

2.4   Knowledge Wheel

This knowledge management model presents a new way of looking at the knowledge activities as a continuous process. The idea comes from Deming's continuous improvement Plan-Do-Check-Act paradigm [24]. This model considers knowledge acquisition as the starting point of the process without differentiating between acquisition as an external source of knowledge and creation as an internal source. Knowledge is then integrated in the organization at various levels followed by storage. Stored knowledge must then be shared and transferred across the entire organization. The next step is knowledge application and as a result of the application innovation can be achieved across the organizations processes. The "turning" of the wheel is realized only in the presence of the following key factors: information and communication technology systems, organizational culture and proper management.

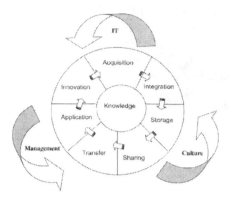

Fig. 3. The wheel of knowledge [25]

2.5   Practical Knowledge Management Model

Gilbert Probst, a member of the Geneva Knowledge Forum presented this knowledge management model [26].The Forum has identified the criteria's like Compatibility, Problem Orientation, Comprehensibility, Action Orientation and Appropriate Instruments as essential to the realization and implementation of a successful knowledge management model. The Researcher has presented the KM activities through the building blocks distributed within two cycles. The inner cycle consists of the building blocks of identification, acquisition, development, distribution, preservation, and use of knowledge. The model clearly shows that the knowledge activities present in the first cycle are not only related in a step by step fashion but also interdependent. The outer cycle consists of all these activities plus goal-setting and measurement. The model emphasizes the need for a strong correlation between organizations goals and its knowledge management activities.

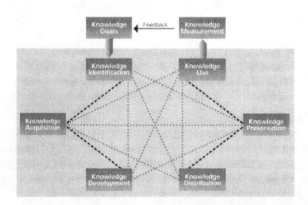

Fig. 4. The Buildings Blocks of Knowledge Management [26]

2.6   Knowledge Life Cycle

This knowledge management life cycle model emphasizes the path taken by data on its road to become knowledge [27]. Data can be gathered from both inside and outside the organization. Data is transformed into information and information is transformed into knowledge thru continuous iterations. A key element of this model is the concept of knowledge and information decay. This process is a result of the development of new ideas and new ways of thinking. Organizations must take this into consideration and as a direct result must treat the process of knowledge generation as a continuous one.

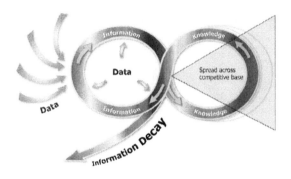

Fig. 5. Knowledge Life Cycle [27]

Another important role in the model is the need to spread knowledge across the competitive base. This seems to be elemental, but because of unwritten rules that exist in the culture of an organization more often, then not knowledge is hoarded by individuals and is not disseminated properly. Of course the main reason for this is the phrase "Knowledge is power" [28]. In today's society and ultra competitive markets, knowledge hoarding can prove to be a big competitive disadvantage for the company [29]. Knowledge must be viewed as a resource that one still possesses even after it has been given away. In fact innovation is a direct consequence of reciprocal knowledge sharing.

2.7   Integrated Knowledge Management Model for Construction Projects

This is an integrated KM approach which presents the influence of macro, meso and micro level variables on the KM process. Macro variables, such as: economic, political, cultural, social development; government policies, legal and normative documents; environmental factors, etc., have a fundamental role to play in the KM lifecycle[30]. The Meso-environment represents the systematic analysis of the sector in which the organization operates. Microenvironment factors are also very important and influence the development of a knowledge management model. These are criterion that exist at the level of the organization: local infrastructure, lifelong learning, favorable residential environment etc. The KMM also underlines the crucial strategic role that both tacit and explicit knowledge play in the elaboration of a cohesive strategy for project and organizational performance.

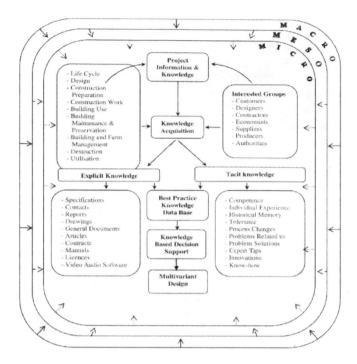

Fig. 6. Integrated Knowledge Management Model for Construction Projects [30]

3   Analysis of Knowledge Management Models

The presented models will be analyzed based on their characteristics and compatibility with the APQP procedure. The main goals of the APQP are improved product quality, reduced organizational costs and reduced time to market. All these desideratum are translated into the following criteria: continuous improvement, prioritization and parallelization of activities, empowered and leveraged workforce, clear quantification of results.

- The Knowledge Tower (KT)

The model presents in a clear and hierarchical manner the main activities and elements that need to be in place in order to achieve a successful knowledge management iteration. It postulates the idea that each comprising element needs to be fulfilled before work on the next step can commence thus creating a clear hierarchical ladder. Although it might seem like a sensible approach the lack of activity parallelization makes the model rigid and can mean increased times to the overall APQP process. The literature associated to the model also discusses the need for a formal appreciation of knowledge management roles within a company setting thus encouraging formal recognition of the impact that this resource plays on the APQP and also posts the need for a clear quantification of results at the top of the "pyramids". All these being said the model does have shortcomings when it comes to the need to actively portray its activities as a continuous process and also when it comes to the need for a coordinated knowledge dissemination effort across organizational boundaries.

- The Knowledge Management Process Model (KMPM)

The KMPM developed by King presents in a clear fashion the main activities that comprise the KM process and their connections. When it comes to its application within the APQP it offers a clear and simple guide for KM it also makes distinctions between various instances of activities (knowledge creation and acquisition) which can be parallelized and it accentuates the need for dissemination of knowledge both in the creation phase (which is based on the SECI paradigm of socialization, externalization, internalization and combination) and in the transfer and shearing phases. Following these guiding principles a leveraged and knowledgeable workforce will be achieved. Also the model clearly stipulates that any KM process is futile without taking into account its effects on organizational performance, thus having it strictly linked to organizational goals. The model as presented does lack the supporting elements that need to be in place for successful KM implementations (culture, IT) and also does not present its activities as a continuous process.

- The Knowledge Wheel (KW)

The Knowledge Wheel which is based on the Deming Cycle accentuates the need for a continuous KM process which is in complete accord with the continuous improvement paradigm utilized in organizations. Also it presents the activities their relationships and all the required supporting elements. The model presents knowledge transfer and shearing as two distinct and sequenced activities, thus putting great emphasis on distribution of tacit and explicit knowledge. However it is missing the need for a measurement activity so that the impact that knowledge has on the organization is clearly understood.

- Practical Knowledge Management Model (PKMM)

This model takes activity parallelization to a whole new level it considers that each one of its elements is in a continuous process and the exact order in which they are fulfilled is up to the individual organization. Because of this it might seem a bit complicated for first time users. It also puts great importance on a cohesive effort from a knowledge perspective and an organizational goals perspective. One of its key building blocks is the requirement for a knowledge measurement activity to cap off an iteration and justify the results obtained. The model does not present the required elements that are needed to support KM activities and makes no distinction when it comes to tacit and explicit knowledge dissemination.

- Knowledge Life Cycle (KLC)

The Knowledge Life Cycle model presents the KM process as a continuous and infinite loop thru which data is picked up and converted into information and knowledge. It concentrates on two main themes knowledge decay and dissemination. These two issues are highly important when it comes to the APQP success criteria, but the model does not concern itself with other KM activities thus being complicated for stakeholders and users to comprehend the full extent and complexity of the process. This over simplification of the KM process can be useful for focusing the efforts of experts in the field.

Table 1. Positive and Negative features of existing models

Nr.	Models	Positive features	Negative features
01	KT	Clear, easily understandable, hierarchical approach.	Rigid, with no coordinated dissemination effort.
02	KMPM	Clear concise. Focalization towards increased organizational performance. Easily understandable and un-predacious.	Does not take into consideration the necessity for continual improvement of knowledge. Does not make any reference to any means of knowledge measurement activities.
03	KW	KM process as continuous improvement. Identification of supporting elements.	Does not link knowledge to organizational objectives and does not consider the need for knowledge measurement.
04	PKMM	Views the KM process as a continuous iteration. Considers that the activities within the process do not necessarily require to be completed in a sequential manner.	Does not make any mention of the required elements that need to be in place to support a successful KM iteration. No distinction between tacit and explicit knowledge.
05	KLC	Focuses practitioners on issues of knowledge decay and dissemination.	Does not concern itself with other activities within the KM process
06	IKMMC	Takes into account not only the internal organizational factors that influence KM but also the macro and meso (external) factors that influence the process	Does not present in a clear manner the activities that make up the KM process. No clear dissemination method for tacit knowledge.

- Integrated Knowledge Management Model for Production Processes (IKMMCP)

Although this model was originally intended for the construction industry certain aspects of it fit in very well with the APQP requirements. For one the model presents in a clear manner the profound influence that the "environment" has upon any KM iteration. The idea that there are many factors that may influence a process helps practitioners and stakeholders keep things into perspective. It also makes a clear distinction of the need to process both tacit and explicit knowledge. It's main shortcoming however stems from the idea that tacit knowledge can be easily stored within a best practice repository. The model does not identify a need for clear and divers dissemination methods. In the case of the APQP due to the complexity of situations that arise a computerized decision support system is insufficient.

4   Conclusion

The paper presents an evaluation of several knowledge management models selected from the specialty literature, based on certain criteria, in order to ascertain their strengths and weaknesses and compatibility with the APQP procedure. This study may offer some guidance to find new directions to eliminate the weak points of those models by adding new ideas and methodologies. This is an initial step in the development of an integrated knowledge management model that will satisfy the need for leveraging consistent and connected knowledge across all of the APQP activities with the end goal of improving product quality, organizational performance and customer satisfaction.

References

1. Ruggles, R.: The state of the notion: Knowledge management in practice. California Management Review 40(3), 80–89 (1998)
2. Choo, C.: The Knowing Organization. Oxford University Press, Oxford (1998)
3. Nonaka, I., Takeguci, H.: The knowledge-creating company. Oxford University Press, Oxford (1995)
4. Mihalca, R., Uta, A., Andreescu, A., Intorsureanu, I.: Knowledge management in e-learning systems. Revista Informatica Economică 46(2), 60–65 (2008)
5. Ho, C.: The relationship between knowledge management enablers and performance. Industrial Management and Data Systems 109(1), 98–117 (2009)
6. O'Dell, C., Grayson, C.J.: Knowledge transfer: Discover your value proposition. Strategy & Leadership 27(2), 10–15 (1999)
7. Milton, N., Shadbolt, N., Cottman, H., Hammersley, M.: Towards a knowledge technology for knowledge management. International Journal Human-Computer Studies 51, 615–641 (1999)
8. McElroy, M.: The New Knowledge Management. Butterworth-Heinemann, New York (2003)
9. Schmidt, A.: Knowledge maturing and the continuity of context as a unifying concept for knowledge management and e-learning. In: Proceedings of I-Know, vol. 5 (2005)
10. Becerra-Fernandez, I., Sabherwa, R.: Individual, group, and organizational learning: A knowledge management perspective. In: Becerra-Fernandez, I., Leidner, D. (eds.) Knowledge Management: An Evolutionary View, pp. 13–39. M.E. Sharpe, New York (2008)
11. McAfee, A.: Enterprise 2.0: The dawn of emergent collaboration. MIT Sloan Management Review 47(3), 20–29 (2006)
12. Bate, S., Robert, G.: Knowledge management and communities of practice in the private sector: Lessons for modernising the National Health Service in England and Wales. Public. Administration 80(4), 643–663 (2002)

13. Driva, H.: The role of performance measurement during in a product design and development in a manufacturing environment. Nottingham: PhD thesis, University of Nottingham (1997)
14. Wang, K.J., Jha, V.S., Gong, D.C., Hou, T.C., Chiu, C.C.: Agent based knowledge management system with APQP: implementation of semiconductor manufacturing service industry. International Journal of Production Research 1, 01–024 (2009)
15. Dalkir, K.: Knowledge Management in Theory and Practice. Elsevier Inc., Oxford (2013)
16. Hislop, D.: Knowledge management in organizations: A critical introduction. Oxford University Press, Oxford (2013)
17. Oztemel, E., Arslankaya, S.: Enterprise knowledge management model: A knowledge. Knowledge Information Systems 31, 171–192 (2012)
18. King, W.R.: Knowledge Management and Organizational Learning. In: King, W.R. (ed.) Knowledge Management and Organizational Learning, pp. 3–13. Springer, New York (2009)
19. Huber, G.: Organizational learning: The contributing processes and the literatures. Organization Science 2(1), 88–115 (1991)
20. King, W.: Knowledge sharing. In: Schwartz, D. (ed.) The Encyclopedia of Knowledge Management, pp. 493–498. Idea Group Publishing, Hersey (2006)
21. King, W.: Knowledge transfer. In: Schwartz, D. (ed.) The Encyclopedia of Knowledge Management, pp. 538–543. Idea Group Publishing, Hersey (2006)
22. King, W., Ko, D.: Evaluating knowledge management and the learning organization: An information/knowledge value chain approach. Communications of the Association for Information Systems 5(14), 1–26 (2001)
23. King, W.: Communications and information processing as a critical success factor in the effective knowledge organization. International Journal of Business Information Systems 10(5), 31–52 (2005)
24. Deming, E.: Out of the Crisis. Massachusetts Institute of Technology, Center for Advanced Educational Services, Cambridge (1986)
25. Zhao, J., Pablos, P., Qi, Z.: Enterprise knowledge management model based on China's practice and case study. Computers in Human Behavior, 324–330 (2012)
26. Probst, G.J.B.: Practical Knowledge Management: A Model That Works. Prism, Arthur D. Little (1998), http://genevaknowledgeforum.ch/downloads/prismartikel.pdf
27. National Aeronautics and Space Administration (2008) as cited in, http://www.ryanzammit.com/pubs/Gnosis%20Kratos%20-%20Empowering%20an%20SME%20with%20its%20own%20knowledge.pdf
28. Bacon, F.: Meditationes Sacrae (1597)
29. Pillania, R.K.: Leveraging knowledge for sustainable competitiveness in SMEs. International Journa of Globalization and Small Business, 393–406 (2007)
30. Kanapeckienea, L., Kaklauskasb, A., Zavadskasc, E., Seniutd, M.: Integrated knowledge management model and system for construction projects. Engineering Applications of Artificial Intelligence, 1200–1215 (2010)

Assessing the Impact of DMAIC-Knowledge Management Methodology on Six Sigma Projects: An Evaluation through Participant's Perception

Lal Mohan Baral[1,2], Claudiu Vasile Kifor[1], and Ioan Bondrea[1]

[1] "Lucian Blaga" University of Sibiu, Sibiu, Romania
[2] Ahsanullah University of Science and Technology,
Dhaka, Bangladesh
lalmohan_baral@yahoo.com,
{claudiu.kifor,ioan.bondrea}@ulbsibiu.ro

Abstract. As an emerging concept, Knowledge Management (KM) has a great impact on organization and organizational performance. On the other hand, DMAIC is a well known methodology, used for solving organizational problems through Six Sigma projects. The main purpose of this study is to assess the impact of newly proposed DMAIC-KM integrated methodology in executing Six Sigma projects. For assessment, four Six Sigma projects have been executed using DMAIC-KM methodology within an airbag manufacturing unit. Afterward, the Opinion and Perception have been collected from the Project participant's through conducting a quantitative survey regarding the Organizational benefits achieved after executing the DMAIC-KM methodology. The results revealed that the participant's have expressed positive impression on improvement of different organizational measures. Therefore, this study would help to motivate the management of different companies to apply DMAIC-KM methodology in order to enhance their organizational performance.

Keywords: DMAIC, Knowledge Management, Six Sigma project, Impact, Organizational performance.

1   Introduction

As a quality management concept, Six Sigma has started its journey in the early 1980's from Motorola [1] and intensified its application momentum throughout the world after achieving a great savings from Six Sigma projects, applied within General Electric [2]. Karthi et al. [3] reported that the application of Six Sigma has found in more than 15 areas of various manufacturing and service sectors, though it has begun from electronics manufacturing area. As an important and well structured improvement methodology, the DMAIC (Define, Measure, Analyze, Improve and Control) has been developed for Six Sigma application [1],[4] from MAIC model, where "Define" (D) phase has been added by Jack Welch [3],[5]. Presently, the DMAIC

R. Buchmann et al. (Eds.): KSEM 2014, LNAI 8793, pp. 349–356, 2014.
© Springer International Publishing Switzerland 2014

breakthrough has become a standard quality improvement methodology for organizational performance through reducing waste, enhancing customer satisfaction, improving the procedure and the process performance [6]. A number of researchers have mentioned that there are lots of scopes to generate and disseminate knowledge during executing DMAIC breakthrough. According to their opinions, knowledge can be generated during gate review session [7], during the solutions of improvement plans [8] and during the brainstorming session among the project participants [9]. Those created knowledge can be considered as a key resource of an organization [10], and to be utilized for enhancing organizational performance through proper knowledge management procedure. KM has also shown competitive and significant impact on organization and organizational performance due to its incredible advantages in creating, sharing and leveraging the knowledge within the organization in different level through people, processes and products [7] ,[11]. For that reasons, both the DMAIC and the KM have been integrated within the Six Sigma quality management approach and applied through proposing different models. For instance, Raytheon Six Sigma and KM integrated model [12], TEKIP model [13], Six Sigma, KM and Balanced Scorecard integrated model [14], Process based knowledge creation and opportunity model [15], Knowledge flow within Six Sigma teams model [8], SECI/SIPOC continuous loop model [16], Within those models, KM is integrated either partially or scattered way with Six Sigma approach or modified DMAIC phases. Therefore, the authors have taken their newly proposed DMAIC-KM integrated methodology, in which six steps of KM activities are integrated with every phase of DMAIC in a systematic way for application to the organization [17].

The main aim of this paper is to present the impact of proposed DMAIC-KM methodology in executing Six Sigma projects. For applying DMAIC-KM methodology, authors have selected four Six Sigma projects from an airbag manufacturing unit, related to their real life process based problems. After deploying all projects, the application impact of DMAIC-KM methodology has been assessed through analysis of participant's opinion and perception, which have been gathered by means of a quantitative survey.

2   The Proposed Methodology and Application Procedure

2.1   Model Architecture

The proposed DMAIC-KM model (Fig. 1) is an integrated conceptual model consists of tasks, tools, activities, knowledge managing IT platform and project performance evaluation tools that connect the knowledge management procedure with quality management methodology. This model is developed considering the DMAIC problem solving steps (Fig.1), which requires three key factors to successfully carry out within the management procedure.

These factors are:

Factor 1: The DMAIC breakthrough should be done for executing Six Sigma projects.

Factor 2: The created knowledge should be identified in every step of DMAIC and stored while the breakthrough is performed.

Factor 3: The identified knowledge should be managed properly through six steps of KM procedure (Fig. 1) in every phase of DMAIC and available required knowledge should be reused in immediate next step to enhance the project performance.

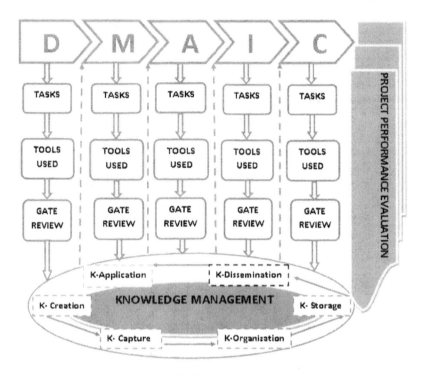

Fig. 1. Proposed DMAIC-KM model architecture [Source, 17]

During executing the Six Sigma projects using DMAIC-KM model, seven stages should be maintained. The activities and methodologies recommended for each stage are presented in the Table 1.Within aforementioned stages; stage 5 is a common stage, which should be spread over all phases of DMAIC methodology. In this stage, the created knowledge from every phase should be unveiled through six steps of knowledge management procedure (Fig.1). Here, first step (K-Creation) should be functionalized after stage 4(gate review) and the final step (K-Application) should be the input of immediate next phase of DMAIC.

During this procedure, all created knowledge will be identified according to its characteristics (tacit/explicit) with the help of a knowledge managing IT platform, and will convert the knowledge from tacit to explicit or vice versa by using Nonaka's four modes of knowledge conversion [18]. All those activities should be done in order to identify, organize and storage of created knowledge properly. The purpose of the stage 6 is to reuse the created explicit knowledge from every phase to immediate next phase of DMAIC from a knowledge storage data base within IT platform is called

"Total Recall" database. For instance, the created knowledge from Define phase should be identified, organized, converted, stored, and managed properly through KM procedure and then available required knowledge should be used for Measure phase for enhancing the execution performance. In this way every phases will be executed to complete the entire project.

Table 1. Stage wise activities and methodology

Stages	Activities	Methodology
Stage 1	Planning for DMAIC breakthrough	Six-Sigma guideline
Stage 2	Identification of Tasks for every phase of DMAIC	ISO Checklist for Six Sigma
Stage 3	Using Tools based on Tasks	Recommended Six Sigma tools for every steps
Stage 4	Gate review to organize the created knowledge	Workshop, Brainstorming and Socialization
Stage 5	Knowledge management	Six- Steps of KM methodology (Knowledge managing IT platform)
Stage 6	Knowledge reuse for next phases	Communities of practices
Stage 7	Final evaluation of project performance	Process capability calculation Survey to gather perception

2.2   Application Context and Procedure

In order to apply DMAIC-KM model, the researchers of this paper have executed a pilot project entitled **"Six Sigma project executions in textile manufacturing by using DMAIC-KM model"** within a technical textile manufacturing company situated in Sibiu, Romania, which is producing different types of safety components for automobiles, specially airbags, seatbelts, steering wheels and so on. The project was carried out between 01 October 2013 and 18 January 2014. According to the suggestions of sponsor (Factory Director) and scientific coordinator (Professor of "Lucian Blaga" University of Sibiu), four Six Sigma projects have been selected for execution under the umbrella of proposed pilot project. All four projects were selected through identifying four real life problems from airbag sewing process, which were faced by the case company during the production process. First author of this paper has coordinated the pilot project following the "collaborative action inquiry" [19] methodology and close investigation with the activities and direction of change of the object being studied [20]. The project coordinator has conducted the research with the help of four technically expert mentors from case company and another five team members for each group from the University. The researchers have followed the DMAIC break- through and KM cycle step by step as presented in the Table 1, and executed all four projects towards a logical solution of the problems. It is noteworthy to mention that the case company is data driven and well structured; moreover it was facing some real life problems during manufacturing their products. In order to solve those real life manufacturing problems as well as become knowledge based organization, the company was interested to implement DMAIC- KM methodology within the Six Sigma projects.

3   Assessment the Impact of DMAIC-KM Application

3.1   Methodology

After completing all selected projects, researchers have assessed the application impact of DMAIC-KM methodology on Six Sigma projects gathering the participant's perception. Like [20] and [21], the researchers have gathered participant's opinion from only those who were involved to execute Six Sigma projects using DMAIC-KM integrated approach through a quantitative survey for this study. The Survey questionnaires have been prepared with the concern of both the research advisor from the University and the Sponsor of the executed Six Sigma projects from case company. In order to collect the written feedback, a set of Likert scaled type questionnaires were supplied within the participants immediately after the projects have completed. The questionnaires were formulated focusing the issue like: benefits achieved by means of organizational changes from the application of DMAIC-KM methodology in executing Six Sigma projects, which has been applied within their manufacturing process. The survey officially started on 25 January 2014 and collected all the feedback within 15 of February 2014 with 100% response rate. After collecting the feedback, the data were analyzed quantitatively.

3.2   Survey Results and Discussion

In the line of quantitative analysis, the respondents were asked the questions relating to the benefits achieved through application of DMAIC-KM methodology by their organization. More specifically, the questions were asked regarding the changes of different organizational measures. The respondents scored the selected measures in between 0-10 scale, where 0=very poor, 2= below average, 4= average, 6= good, 8= very good, 10= excellent.

The evidence from the Fig. 2 revealed that two important organizational measures such as "Application of KM on process management" and "Development of knowledge based staff" are upgraded presently "very good" state from previous "average" state. "The process performance of the projects" and "Data collection efficiency" has also been upgraded from "good" to "very good" position after application of DMAIC-KM methodology as respondents have mentioned. According to the respondents rating, the performances of some important measures like "Awareness of continuous improvement", "Improvement of measure and analysis ability", "Strategic planning efficiency" and "Leadership capability" have also upgraded to "good" from the previous "average" state through application of new methodology. With respect to the benefits achieved for the organization through application of DMAIC-KM model shows that all the scales of measures have improved from previous state after introducing DMAIC-KM methodology during executing Six Sigma projects. From figure 2, it is clear that the average score for all measures lie in between "good" to "very good" at present condition, whereas previous score were in between "average" to "good". According to the respondents rating, the performances of some important measures like "Application of KM on process management", "Development of knowledge based staff", "Improvement of process performance" and "Increasing the data collection efficiency" had significantly improved. The implication is that, the application of KM tools with DMAIC breakthrough made it more effective towards improvements of organizational measures.

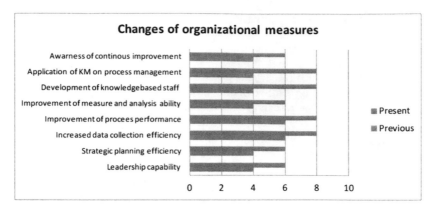

Fig. 2. Changes of organizational measures after DMAIC-KM application (n=24)

4   Conclusions and Future Research Directions

This paper has highlighted the findings from the study regarding the application impact of DMAIC-KM methodology, which has been applied in an airbag manufacturing process. The study has conducted by using a well known scientific and logical common approach (quantitative survey), which has been widely using for gathering participants opinions and perceptions. This perception based finding has unveiled that the different organizational measures have been significantly improved which have a positive impact on Six Sigma project performance. So, the results would contribute with great value not only for companies but also academic communities. For companies, it is a need for using more effective continuous improvements tools to upgrade their organizational performance. The results of this study would help to motivate those companies to use the DMAIC-KM methodology as an effective continuous improvement tool. For academics, the focus should be on searching the best practices of DMAIC-KM methodology not only for manufacturing process but also others organizations, where there is an opportunity to adopt this methodology. Although this study demonstrates positive feedback regarding the applied DMAIC-KM methodology, it has some limitations due to the fact that the questionnaire survey has been carried out within a small number of participants, which are only total 24 participants from 4 Six Sigma project teams. The outcomes may not be representative for large number of participants. This implies that the research could be conducted in engaging large number of participants to generalize the results. As a further study, we are recommending to apply the DMAIC-KM methodology within various manufacturing sectors in a wider perspective and conduct the assessment study including more participants in order to get elaborate and constructive opinions regarding the impact of DMAIC-KM application on Six Sigma projects.

Acknowledgements. This research has been carried out within Doctoral studies under the financial support of Erasmus Mundus Mobility with Asia (EMMA). We would like to express our sincere thanks to EMMA for their support. We are also thankful to factory management where the Six Sigma projects have been carried out and all project participants's both from Factory and University for their all out support to execute the projects successfully.

References

1. Mehrjerdi, Y.Z.: Six-Sigma: Methodology, tools and its future. Assembly Automation 31(1), 79–88 (2011)
2. Lee, T.Y., Wong, W.K., Yeung, K.W.: Developing a readiness self-assessment model (RSM) for Six Sigma for China enterprises. International Journal of Quality & Reliability Management 28(2), 169–194 (2011)
3. Karthi, S., Devadasan, S.R., Murugesh, R., Sreenivasa, C.G., Sivaram, N.M.: Global views on integrating Six Sigma and ISO 9001 certification. Total Quality Management & Business Excellence 23(3), 237–262 (2012)
4. Firka, D.: Six Sigma: An evolutionary analysis through case studies. The TQM Journal 22(4), 423–434 (2010)
5. Dahlgaard, J.J., Dahlgaard-Park, S.M.: Lean production, six sigma quality, TQM and company culture. The TQM Magazine 18(3), 263–281 (2006)
6. Chang, S.-I., Yen, D.C., Chou, C.C., Wu, H.C., Lee, H.P.: Applying Six Sigma to the management and improvement of production planning procedure's performance. Total Quality Management & Business Excellence 23(3), 291–308 (2012)
7. Stevens, D.E.: The Leveraging Effects of Knowledge Management Concept. In: the Deployment o Six Sigma in a Health Care Company. Ph.D. Thesis, Walden University, USA (2006)
8. Zou, X.P.T., Lee, W.B.: A study of knowledge flow in Six Sigma teams in a Chinese manufacturing enterprise. The Journal of Information and Knowledge Management Systems 40(3/4), 390–403 (2010)
9. Llorens- Montes, F.J., Molina, L.M.: Six Sigma and management theory: processes, content and effectiveness. Total Quality Management & Business Excellence 17(4), 485–506 (2006)
10. Grant, R.: Prospering in dynamically competitive environments: Organizational capability as knowledge integration. Strategic Management Journal 17, 109–122 (1996)
11. Becera-Fernandez, I., Sabherwal, R.: Individual, group and organizational learning. In: Becera-Fernandez, I., Leidner, D. (eds.) knowledge Management: An Evolutionary View, pp. 13–39. M.E. Sharpe, Armonk (2008)
12. Lanyon, S.: At Raytheon Six Sigma Works, Too, To Improve HR Management Process. Journal of Organizational Excellence 22(4) (Autumn 2003)
13. Yeung, R.: Integrating Six Sigma with Knowledge Management, Lecture slides in City University Hong Kong, (2004), http://www.sixsigma.org.hk (accessed date: October 12, 2012)
14. Paladino, R.E., Newman, K.B.: Integrating Balanced Scorecard, Six Sigma and Knowledge Management to Drive Value at Crown Castle. APQC (2004), http://www.apqc.org (accessed date: September 12, 2012)
15. Wu, C., Lin, C.: Case study of knowledge creation facilitated by Six Sigma. International Journal of Quality & Reliability Management 26(9), 911–932 (2009)
16. Herbert, N.: Merging Knowledge Creation Theory with the Six-Sigma Model for Improving Organizations: The Continuous Loop Model. International Journal of Management 28(2) (2011)
17. Kifor, C.V., Baral, L.M.: An integrated DMAIC-Knowledge Management conceptual model for Six Sigma quality management. In: 6th International Conference on Manufacturing Science and Education MSE, Sibiu, Romania, pp. 12–15 (2013)

18. Nonaka, I., Toyama, R., Konno, N.: SECI, Ba and leadership: A unified model of dynamic knowledge creation. Long Range Planning 33(1), 5–34 (2000)
19. Lewin, K.: Action Research and Minority Problems. Journal of Social Issues 2(3), 34–46 (1946)
20. Cronemyr, P.: Six Sigma Management, Action Research with some Contributions to theories and methods, Ph.D. Thesis (2007)
21. Orbak, A.Y.: Shell scrap reduction of foam production and lamination process in automotive industry. Total Quality Management and Business Excellence 23(3), 314–325 (2012)

A Knowledge-Transfer System Integrating Workflow, A Rule Base, Domain Ontologies and a Goal Tree

Nobuhito Marumo[1,*], Takashi Beppu[1], and Takahira Yamaguchi[2,*]

[1] Graduate School of Science and Technology,
Keio University, 3-14-1 Hiyoshi, Kohoku-ku,
Yokohama-shi Kanagawa-ken 223-8522, Japan
MC-NYramaty@z6.keio.jp, tb-0325@z3.keio.jp
[2] Faculty of Science and Technology, Keio University
yamaguti@ae.keio.ac.jp

Abstract. This paper discusses how to develop a knowledge-transfer system (KTS) by integrating four knowledge sources: workflow, a rule base, domain ontologies, and a goal tree with domain ontology centered structure. When novice workers acquire knowledge from experienced workers, they should not only learn a single form of knowledge, but also understand the interrelationships among these four knowledge sources. In this study, we look at a case study involving a snow control plan for highways. This study present a case in which KTS is being implemented well.

Keywords: Knowledge Management, Knowledge Transfer, Ontologies.

1 Introduction

In this paper, we discuss how to develop an Off the Job Training (Off-JT) knowledge-transfer system (KTS) to support novice workers in learning knowledge from experienced workers. Because novice workers have difficulties in understanding work assigned to them from companies, they need to keep in mind the 3W1H questions: When, What, Why and How (the Who and Where questions are not important here). To manage the 3W1H questions, a KTS should include four knowledge sources: workflow, rule base, domain ontology, and goal tree. When novice workers acquire knowledge from experienced workers, they should not only learn a single form of knowledge, but also understand the interrelationships among these four knowledge sources. In this study, we look at a case study involving a snow control plan (SCP) for highways. This study presents a case in which KTS is being implemented well.

The rest of this paper is organized as follows: Section 2 deals with related works. In Section 3 we provide a system architecture overview. Subsequently, we provide details about knowledge sources. In Section 4 we provide an overview of how to use KTS. In Section 5 we look at a case study, and in Section 6 we conclude our study.

---

[*] Corresponding authors.

R. Buchmann et al. (Eds.): KSEM 2014, LNAI 8793, pp. 357–367, 2014.

2   Related Works

In knowledge management, Nonaka proposed the SECI model for organizational knowledge creation [1]. The SECI model shows how organizational knowledge is created by a synthesis of explicit and implicit knowledge: Socialization, Externalization, Combination, and Internalization. Nonaka also proposed "ba". "Ba" is a place where these syntheses are conducted [2]. Hijikata proposed a computerized "ba" where two domain experts externalize and combine their tacit knowledge efficiently with the help of a computer that identifies inconsistencies in the tacit knowledge of the two domain experts and the knowledge created by inductive case learning [3]. The SECI model focuses on organizational knowledge creation. Our proposal, on the other hand, mainly focuses on knowledge transfer to next-generation engineers, and emphasizes that novice workers can perform tasks by making use of transferred knowledge under different situations. We focus on the "Combination" aspect of the SECI model and discuss how to interconnect various knowledge sources.

In ontology engineering, Liana Razmerita proposed an ontology-based user modeling architecture for a knowledge-management system [6]. In the business process, Norbert Gronau said the knowledge process and the business process should link together [7]. In the business process and goal, David Martindho used workflow and a goal-driven strategy to obtain tacit knowledge [12]. In knowledge integration, Okabe Masao and Kobayashi Keido integrated three knowledge resources: ontology, rule base, and rule ontology [8, 9]. Ning Huang proposed ontology-based integration with workflow and rule base [10]. Yinglin Wang put forward a knowledge-management system framework integrating workflow, ontology and rule base [11]. Yinglin Wang used ontology for the reusability of a knowledge-management system. Our proposal deals mainly with how to integrate four knowledge sources for knowledge transfer.

3   System Architecture

3.1   System Overview

In this section, we construct the four knowledge sources to manage 3W1H questions. First, we show an overview of the KTS in Fig. 1. We then explain the four knowledge sources in Section 3.2.

Figure 1 shows a system architecture overview for a KTS. This system has four knowledge sources: workflow, rule base, goal tree, and domain ontologies. All links are defined with ontology. Novice workers can learn by applying these sources. The learning scenario is as follows: 1. Workflow, 2. Rules, and 3. What novice workers are interested in (e.g. meaning of terms, related work, and reason of rules). Objects drawn with dotted lines means they are essential objects for a KTS, but they were not create with syntax.

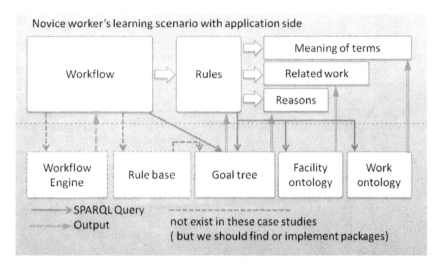

Fig. 1. Knowledge Transfer System Overview

3.2 Knowledge Sources

Workflow. Generally, it takes a great deal of time for novice workers to understand both the overall workflow and the details contained in the workflow. In addition, it is very difficult for novice workers to understand the workflow hierarchical structure from coarse grain size to fine grain size. To understand the workflow hierarchical structure, a workflow hierarchical structure should be explicit, see Fig. 2. The structure consists of three levels of workflow: Top, Middle, and Low. The top level workflow shows us how the workflow proceeds in coarse granularity. the middle level workflow is a sequence of work decomposed from the top level workflow. The elements of the sequence all have their own names, and the people involved in an enterprise use these names generally. The low level workflow shows us the details from which people in a workplace can understand what to do. From the viewpoint of knowledge management, this structure helps novice workers distinguish between the overall workflow and the processes. Finally, we describe knowledge related to decision-making in primitive work in the rule base section. We take the information that is needed for defining primitive works into the work ontology section.

Rule Base. Some tasks include rules by which novice workers can determine what they should do next. Even if novice workers understand the overall workflow and the processes throughout the top level workflow to the low level workflow, they still don't understand how to do the work in concrete terms. Therefore, we use a rule base because it aids making decisions in a specific situation. Because a task includes rules, the rule base should be related to the work in a workflow. We deal with rules that are related to two types of judgments: what to command, and how to do the work. Learning using a rule base after learning the workflow supports novice workers in developing a good understanding of a task. Top-down learning helps novice workers understand quickly and deeply. A rule has a rule ID, related work, conditions (If part), conclusion (Then part), and sentences in Fig. 3. Sentences don't include machine-readable structures, and the sentences and rule IDs are written in the goal tree.

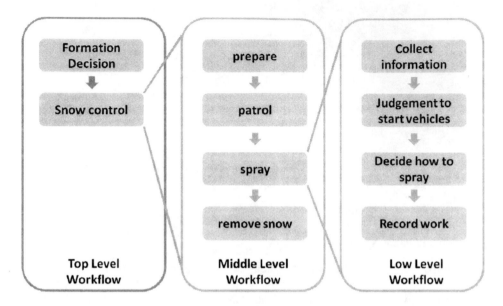

Fig. 2. Three-Layered Workflow Structure

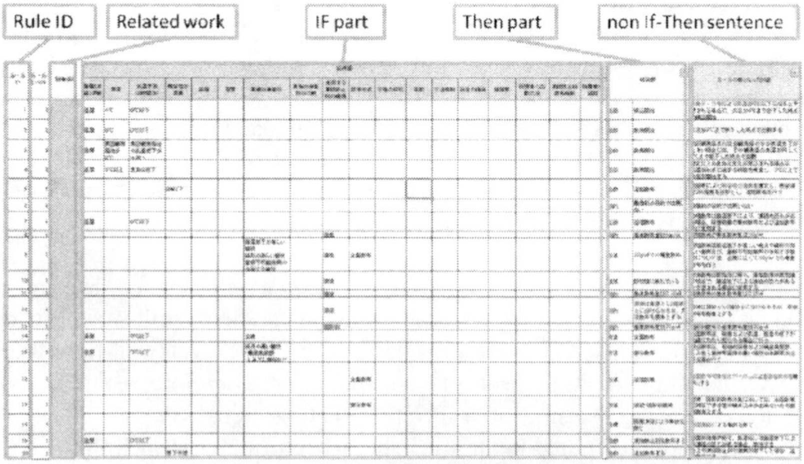

Fig. 3. Rule Base Overview

Domain Ontologies. Domain ontologies have a concept classification hierarchy and semantic networks between facilities and work tasks. Domain ontologies, in particular, help novice workers understand terms in a specific domain. Domain ontologies are separated into facility and work ontology. They help novice workers understand the meaning of terms in a specific domain. We create the ontologies using Protégé [4].

(1) Facility Ontology
We take the property "purpose," whose domain is "facility" in Fig. 4 and whose range is "work" in Fig. 5 as discrimination attributes to build a concept hierarchy structure. Facility instances include pictures that support novice workers in understanding what the instances really were.

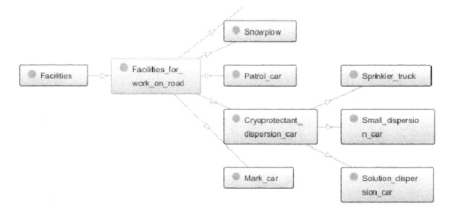

Fig. 4. Facility Ontology

(2) Work Ontology
We take property "has-a" whose domain is "work" and whose range is "work" as a discrimination attribute to build a concept hierarchy structure. Some instances include movies to show how to act in a specific situation.

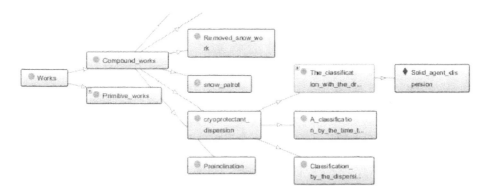

Fig. 5. Work Ontology

Goal Tree. A goal tree is an ontology that deals with a rule as a concept. A goal tree defines the hierarchy of the rules and the network of rules and reasons. In Fig. 6, there are five categories: A is an instance of rule class, and the others are instances of reasons, We call them Rule A, and Reasons B, C, D, and E. Novice workers learn using Rule A. If they follow Reason B from Reason A, they can understand why Rule A is implemented. If they don't understand why Reason B is needed, Reason C can show them the justification or the reason for Reason B. As they follow the reasons and find the last reason (Rule E in Fig. 6), they find the last reason is an instance of the managerial goal. Thus novice workers can understand a managerial goal from a rule used in a scenario. By learning the goal tree, novice workers can understand not only why they should do a task, but the managerial goals. As a result of knowing the reasons they should perform a task, novice workers can acquire the ability to deal with work that is an exception to the norm.

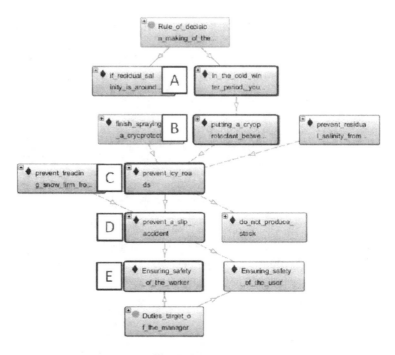

Fig. 6. Goal Tree

4   System Flow

We develop a system by integrating four knowledge sources: workflow, a rule base, domain ontologies, and a goal tree (we developed the system in Japanese, so some of the examples in the figures are in Japanese). The learning process of this system is basically divided into two fields. One field is the workflow side, and the other field is the ontology side, see Fig. 7.

If novice workers use this system, they can start to learn the workflow shown in Fig. 7. Novice workers start to study from the top level workflow down to the low level workflow shown in Fig. 8-A. In Fig. 8-A, each task in a workflow is a button. When a novice worker pushes a button in the top level workflow, the middle level workflow appears. And when they push a button in the middle level workflow, they can see the low level workflow. These workflows are written in HTML and Java-Script. When they then push a button in the low level workflow, the system shows the rules related to the task in Fig.8-B.

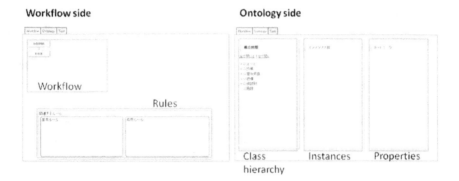

Fig. 7. Workflow and Ontology

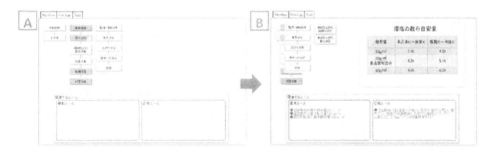

Fig. 8. System Flow

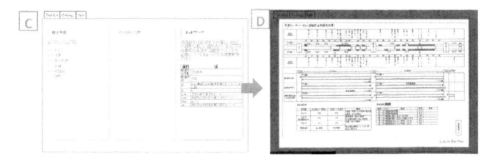

Fig. 9. System Flow

When the novice worker pushes a button in the low level workflow, a SPARQL query is sent to the goal tree to obtain rule instances related to the workflow and to show rule links. If novice workers follow a rule link, the system shows the contents related to the rule link in Fig. 9-C. When novice workers follow a rule link, a SPARQL query is sent to obtain objects related to the rule. Figure 9-C shows an ontology page. While a workflow page has the workflow and the rules related to the workflow currently being viewed by a novice worker, a ontology page contains three items: class hierarchy, instances, and properties. When novice workers click a class in the bottom layer, the system shows instances related to what is being clicked in the Instances area. When an Instance is clicked, the system shows the properties and values. Some rule instances have "picture" properties and users can see what the Instances really are in Fig.9-D.

5   Case Studies and Discussion

Because skilled workers are is short supply in social infrastructure maintenance [5], it is a key issue for social infrastructure companies to develop systems to transfer knowledge from skilled workers to novice workers. Thus we apply our knowledge-transfer system to a SCP for highways, which consists of two tasks: anti-freezing work and snow-removal work. As a SCP is a complex task related to natural environments and is carried out just in the winter season (from December to February/March), it is very difficult to transfer knowledge and skills to novice workers.

We have analyzed SCP manuals, interviewed skilled workers many times, and personally observed a SCP in operation. Thus we have obtained four types of knowledge: workflow, rule base, domain ontologies and goal tree. Table 1 shows the form and specifications of the knowledge sources.

- Workflow: We have a three-layered workflow. The top level workflow has two elements. The middle level workflow has two elements (anti-freezing and snow removal). The low level element has five processes.
- Rule base: Rule LHS (left-hand side) has 18 elements: road condition, temperature, forecast (temperature after two hours), salinity concentration, snow, snow type, workplace, condition of workplace, type of anti-freezing agent, how to spray, pressed-snow condition, season, traffic control, how to remove snow, time interval, dispatch of snow-removal vehicles, timing of spraying, and formation of snow-removal vehicles. The rule base has about 100 rules with the above-mentioned LHS.
- Domain ontologies: The facility ontology has 26 classes, 45 instances, and 200 RDF (Resource Description Framework) triples. We have a class hierarchy with discrimination properties from interviews with skilled workers for SCP. The SCP work ontology has 27 classes, 37 instances, and 200 RDF triples. Based on interviews with skilled workers, we categorized the classes in the following order: primitive/compound work, SCP task type, means, timing of action, and place of work.
- Goal tree: We asked skilled workers the following question: Why do you implement a rule? Then we produced a goal tree where super rules justify sub rules. The goal tree has 98 nodes (rules) with 35 rules from anti-freezing work, 44 rules from snow removal ,and 19 rules from management issues and others.

Table 1. Knowledge Sources for SCP KTS

Knowledge Source	Formality	Specification
domain ontologies	formal	owl file (made with Protégé)
goal tree	semiformal	owl file (made with Protégé)
rule base	semiformal	Text (Only rule ids and sentences are defined with Protégé. But sentences have If-Then part.)
work flow	informal	html + javascript (no syntax)

5.1   Case Study

Thirteen novice workers participated in our case study to evaluate our KTS using the SCP for highways. They learned the SCP from KTS and/or a worker's manual. They are divided into the following three groups:

- Group A learns the SCP using a worker's manual for one hour.
- Group B learns the SCP using KTS for one hour.
- Group C learns the SCP using a worker's manual for half an hour and using KTS for half an hour.

Just after learning the SCP with a worker's manual and/or KTS, they are given a written test and questionnaire. Some questions are related with workflow and others are not. Fifteen questions were devised as follows in collaboration with skilled workers:

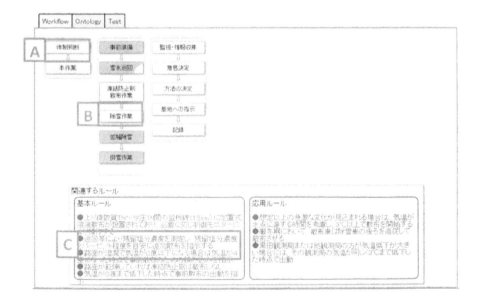

Fig. 10. Written Test Using Personal Computer

- Nine questions related to work (or rules) order such as "What do you do first in Fig. 10-A?", "After spraying an anti-freezing agent, what do you do next in Fig. 10-B?", and "If the percentage of residual salinity becomes 2%, what do you do in Fig. 10-C?")
- Six questions not related to work (or rules) order such as "Generally, when is it easiest to remove pressed snow?"

The answers to all questions are in both KTS and the worker's manual.

5.2   Discussion

Table 2 shows how well each group fared in the written test. Group B's accuracy rate is much better than Group A's and so we conclude KTS works very well. We make SWOT analysis on KTS as follows:

Table 2. Case Studies Results

Subject group	Question Types and Accuracy Rate				Total Accuracy Rate
	When	How	What	Why	
Group A: Manual	30%	50%	30%	50%	32.8%
Group B: KTS	50%	100%	45%	81.3%	62.5%
Group C: Manual & KTS	60%	100%	60%	75%	61.3%

- Strengths: Novice workers have a good understanding of when/how to perform a task and the reason for performing a task through KTS. Domain ontologies with class hierarchy and relationship network support novice workers in understanding technical terms. Based on these finding, we observe that structured knowledge works well so that novice workers can understand SCP work quickly and easily. Although the case study was conducted for just one hour, KTS work better over a longer time.
- Weaknesses: Compared to manual, it is difficult for novice workers to grasp how much to learn. KTS doesn't have user models to learners and so does not provide personalization facilities.
- Opportunities: Because many enterprises in social infrastructure maintenance suffer from a shortage of workers, KTS becomes more and more important. Knowledge integration will support more various novice workers in understanding works well.
- Threats: Because of a shortage of workers, it will take many costs to elicit expertise from experienced workers.

6   Conclusion

We have developed a knowledge-transfer systems (KTS), integrating four kinds of knowledge: workflow to manage When-type questions, a rule base to manage How-type questions, domain ontologies to manage What-type questions, and a goal tree to manage Why-type questions. Applying KTS to a case study involving a snow control plan for highways, KTS was found to work much better than a worker's manual. When skilled workers in another snow area used our KTS, they agreed with the generic knowledge but not with specific knowledge. Specific process and knowledge might differ depending on snow quality. We should identify what knowledge can be reused or not from the point of knowledge-sharing among different organization divisions.

References

1. Nonaka, I., Takeuchi, H.: The Knowledge-Creating Company: How Japanese Companies Create the Dynamics of Innovation. Oxford University Press, Oxford (1995)
2. Nonaka, I., Konnno, N.: The Concept of 'ba': Building a foundation for knowledge creation. California Management Review 40(3), 30–54 (1998)
3. Hijikata, Y., Takenaka, T., Kusumura, Y., Nishida, S.: Interactive knowledge externalization and combination for SECI model. In: Proceedings of the 4th International Conference on Knowledge Capture, Whistler BC, Canada, pp. 151–158 (2007)
4. Protégé, http://protege.stanford.edu/
5. Japan Society of Civil Engineers. Public Works Management Journal (424), 32–36 (2013)
6. Razmerita, L., Angehrn, A., Maedche, A.: Ontology-Based User Modeling for Knowledge Management Systems. In: Brusilovsky, P., Corbett, A.T., de Rosis, F. (eds.) UM 2003. LNCS (LNAI), vol. 2702, pp. 213–217. Springer, Heidelberg (2003)
7. Gronau, N., Weber, E.: Management of Knowledge Intensive Business Processes. In: Desel, J., Pernici, B., Weske, M. (eds.) BPM 2004. LNCS, vol. 3080, pp. 163–178. Springer, Heidelberg (2004)
8. Okabe, M., Yanagisawa, M., Yamazaki, H., Kobayashi, K., Yoshioka, A., Yamaguchi, T.: Organizational Knowledge Transfer of Intelligence Skill Using Ontologies and a Rule-Based System. In: Yamaguchi, T. (ed.) PAKM 2008. LNCS (LNAI), vol. 5345, pp. 207–218. Springer, Heidelberg (2008)
9. Kobayashi, K., Yoshioka, A., Okabe, M., Yanagisawa, M., Yamazaki, H., Yamaguchi, T.: How Much Well Does Organizational Knowledge Transfer Work with Domain and Rule Ontologies? In: Karagiannis, D., Jin, Z. (eds.) KSEM 2009. LNCS (LNAI), vol. 5914, pp. 382–393. Springer, Heidelberg (2009)
10. Ning, H., Shi Han, D.: Ontology-based Enterprise Knowledge Integration: Robotics and Computer-Integrated Manufacturing, pp. 562–571 (2008)
11. Wang, Y., Guo, J., Hu, T., Wang, J.: An Ontology-Based Framework for Building Adaptable Knowledge Management Systems. In: Zhang, Z., Siekmann, J.H. (eds.) KSEM 2007. LNCS (LNAI), vol. 4798, pp. 655–660. Springer, Heidelberg (2007)
12. Martinho, D., Silva, A.R.: A Recommendation Algorithm to Capture End-Users' Tacit Knowledge. In: Barros, A., Gal, A., Kindler, E. (eds.) BPM 2012. LNCS, vol. 7481, pp. 216–222. Springer, Heidelberg (2012)

A Conceptual Reference Model
of Modeling and Verification Concepts
for Hybrid Systems

Andreas Müller[1], Stefan Mitsch[2],
Werner Retschitzegger[1], and Wieland Schwinger[1]

[1] Johannes Kepler University Linz, Altenbergerstr. 69, 4040 Linz, Austria
[2] Computer Science Dept., Carnegie Mellon University, Pittsburgh, PA-15213, USA

Abstract. *Cyber-physical systems* (CPS), which are computerized systems directly interfacing their real-world surroundings, leverage the construction of increasingly autonomous systems. To meet the high safety demands of CPS, verification of their behavior is crucial, which has led to a wide range of tools for modeling and verification of hybrid systems. These tools are often used in combination, because they employ a wide range of different formalisms for modeling, and aim at distinct verification goals and techniques. To manage and exchange knowledge in the verification process and to overcome a lack of a common classification, we unify different terminologies and concepts of a variety of modeling and verification tools in a *conceptual reference model* (CRM). Furthermore, we illustrate how the CRM can support comparing models and propose future extension.

1 Introduction

Systems that exhibit physical behavior by interfacing and interacting directly with their real-world surroundings through sensors and actuators are known as *cyber-physical systems* (CPS). CPS become increasingly autonomous (e. g., autonomous cars); consequently, significant demands are imposed on the *safety* of such a CPS and the knowledge needed to design and implement them correctly. Therefore, the field of *formal verification*, i. e., mathematically proving that a CPS behaves as intended, is key to engineering CPS for safety-critical application domains. The behavior of a CPS can be described using *hybrid system* models [2], which simultaneously capture the continuously evolving real-world behavior and the discrete control decisions of the CPS within one model.

In order to model a CPS and formally verify the desired behavior, the computational and the physical behavior of a CPS need to be considered in conjunction, which introduces unprecedented complexity into verification. Modelers face many modeling and verification tools, which employ a wide range of modeling formalisms (e. g., hybrid automata [18], hybrid programs [36,38]), aim at distinct verification goals (e. g., safety, liveness) and incorporate heterogeneous verification techniques (e. g., theorem proving, reachability analysis). Often, using multiple tools in combination is beneficial because their capabilities differ

R. Buchmann et al. (Eds.): KSEM 2014, LNAI 8793, pp. 368–379, 2014.

strongly. The downside of this diversity are *compatibility* issues, where specifically questions of knowledge management arise, such as: *(i)* which model representation is useful for which aspect of the system? *(ii)* which parts of the system can be formally verified using which tools? *(iii)* what are the trade-offs between modeling and verification (detail vs. automation)? *(iv)* which parts of a system are verified and how should the verification results be composed to a comprehensive correctness argument? A major difficulty for systematically analyzing modeling and verification and managing knowledge in the verification process of hybrid systems is a lack of a common classification of *hybrid system modeling and verification* concepts.

To overcome this lack of a common classification, we unify different terminologies and concepts of a variety of modeling and verification tools in a *conceptual reference model* (CRM), methodologically adhering to our previous work (e. g., [49]). We illustrate how the CRM can assist in classifying the capabilities of modeling formalisms and tools. Additionally, we identify future extensions and enhancements for modeling and verification tools for CPS.

2   Related Work

To the best of our knowledge, no other CRM for hybrid system modeling and verification concepts has been proposed so far. Nevertheless, prior surveys on CPS and hybrid system modeling and verification provide classification fragments, which will be discussed below.

Broman et al. [8] introduce a coarse-grained model for categorizing hybrid systems. Their framework comprises *Viewpoints* (of stakeholders and their concerns), *Formalisms* (modeling formalisms for hybrid systems) and *Languages and Tools* (which implement formalisms). They conclude that their framework serves as a basis for assisting CPS designers in the modeling process. Their framework reviews tools primarily based on the requirements of stakeholders, whereas we focus on the engineering and knowledge representation aspects of hybrid systems design. Alur [1] reviews formal verification approaches, but not modeling and tool support. Carloni et al. [9] analyze the syntax and semantics for hybrid systems modeling w.r.t. verification and simulation. We, in addition, discuss modeling and tool support.

A large body of research addresses solely the systematic modeling and specification of CPS, but does not address verification: Giese et al. [15] survey visual model-driven development of software-intensive systems. Shi et al. [45] provide a short overview and further research challenges of CPS. Sanislav et al. [43] focus their work on challenges, concepts and research goals in the area of CPS. Wan et al. [48] investigate the applicability of different composition mechanisms for cyber-physical applications. Kim et al. [23] provide a broad overview of CPS research and Lee [27] examines the challenges in designing CPS. Finally, the Open Model Community's[1] formal definition of modeling methods [13] may act as an alternative to UML for representing our CRM.

---

[1] http://www.omilab.org

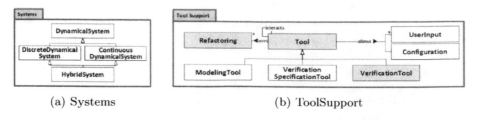

(a) Systems   (b) ToolSupport

Fig. 1. Systems and tool support concepts of the CRM

3   Conceptual Reference Model

In this section, we present a CRM of modeling and verification concepts for hybrid systems. In principle we followed a top-down approach for constructing the CRM, meaning that several concepts of the CRM have been adopted from existing other surveys in this area as referred to in Section 2. We supplemented our CRM in a bottom-up way with concepts prevalent in existing tools. Finally, we structured our CRM into four packages: (i) the `Systems` package describes the real world systems; (ii) the `Modeling` package abstracts from real-world systems to models of their behavior and specifications of important properties; (iii) the `Verification` package aims at verifying the modeled systems; (iv) the `ToolSupport` package contains tool related aspects. In the following sections, the concepts of the CRM are described along these four packages (see `http://cis.jku.at` for complete CRM).

We express the CRM as Unified Modeling Language (UML) classes, since UML is the prevailing standard in object-oriented modeling[2] and expose the basic components of hybrid systems and the interrelations between them. Naturally, the CRM thus serves also as a *framework*, which can be extended by means of sub-classing if further hybrid system concepts need to be captured.

3.1   Systems

The classes in the systems package (cf. Fig. 1a) describe a high-level systems perspective to anchor modeling and verification tools. We follow Teschl [46], and distinguish `DynamicalSystems` into `DiscreteDynamicalSystems` (state space is $\mathbb{N}/\mathbb{Z}$) and `ContinuousDynamicalSystems` (state space is $\mathbb{R}$); systems that have both characteristics are `HybridSystems` [18], focused on in this paper. Specifically, the dimensions of space and time are important characteristics for many systems. A difference in handling those in a discrete or continuous manner indicates a potentially fundamental conflict between modeling concepts and tools.

---

[2] UML meta-model as included in the OMG "Unified Modeling Language: Superstructure" version 2.4.1, available at
`http://www.omg.org/spec/UML/2.4.1/Superstructure/PDF/`

3.2   Modeling

For the modeling package (cf. Fig. 2a) we follow Gupta [16] to distinguish a model and its correctness specification (implementation and specification).

Model. A **Model** captures the relevant features of a dynamic system. It is expressed in a **ModelingFormalism** and can be constrained by **Conditions**.

Verification Specification. A **VerificationSpecification** describes a model's expected behavior utilizing a **SpecificationFormalism** [11], [16]. A verification specification, being a logical formula in many approaches (e. g., [26], [36]), consists of a **StartCondition** that specifies the initial conditions under which we want a system to be safe to start, and a **CorrectnessCriterion** that we want a system to fulfill (e. g., throughout, after all, or after at least one of its executions). Furthermore, it is often possible to annotate models with **Hints/Strategies**, that guide a verification tool but do not influence the behavior of a model directly (e. g., *invariants* in KeYmaera [40] and UPPAAL [26]).

Formalism. A major part of the Modeling package is the **Formalism** sub-package, which is divided into modeling formalisms for creating models and specification formalisms for creating verification specifications (these may reference the created models). The included formalisms are the most commonly used in literature, namely **Automata** and **Programs** [16] for modeling of discrete systems, **Differential Equation** for modeling of continuous systems [10], as well as their combination in the form of **HybridAutomata/HybridPrograms** [18], [36]. In order to constrain a model to realistic behavior (e. g., the "bouncing ball" can never fall through the floor), the CRM introduces conditions. Following Meyer [28], these conditions are further subdivided into **PreConditions**, **PostConditions** and **Invariants**. Moreover, a modeling formalism can have multiple characteristics further describing its capabilities. We include subclasses of **Characteristic** in the CRM to handle **Compositionality** (compositional models, for instance, through **Parallelism** [19], **Urgency** [5], **Synchronization** or *sequences* [44]) and **Non-Determinism** (e. g., *non-deterministic choice* in d$\mathcal{L}$ [36]).

Automaton and Program. An automaton comprises a set of **States** and a set of **Transitions**, and can be visualized as a directed graph [20]. A condition, when attached to an **AutomatonElement**, restricts or details the behavior of the automaton. Another modeling formalism are programs, representing a sequence of instructions. Although they are often interchangeable we separate these formalisms, because their structural differences can be utilized by verification tools.

Differential Equations. Following [51] we classify differential equations (DE) into **PartialDE** (PDE) and **OrdinaryDE** (ODE). Both can have special restricted constant and linear forms (i. e., **LinearPDE**, **LinearODE**). In accordance with [37], we allow conditions to restrict a DE to remain within a particular region, transforming DEs into **DifferentialAlgebraicEquations** (DAE). DAEs can further be equivalently transformed into differential inclusion, which is useful to express disturbance in continuous dynamics.

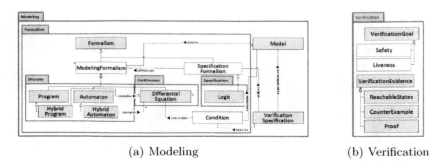

(a) Modeling   (b) Verification

Fig. 2. Modeling and Verification Concepts of the CRM

HybridAutomaton (HA) and HybridProgram (HP). The generic concept of HA is the basis for numerous hybrid formalisms with different levels of expressiveness and detail, such as *HybridUML* [6] or *Hybrid Petri Nets* [12], [34]. For HA, we introduce a `ContinuousState` that references a set of DEs. These differential equations represent the continuous behavior of a system while their respective state is active. `TimedAutomata` are restricted to modeling real-time systems and correspond to HA with only clocks [4]. Like above, we introduce HPs [36], [38] as sub-class of programs which allows differential equations as instruction. As already mentioned, although HA can be encoded as HPs [36], they differ in structural aspects that can be exploited by hybrid system verification tools.

Logic. A verification specification is expressed in terms of a `Logic` [16], such as Temporal Logic (TL) [41] or differential dynamic logic ($d\mathcal{L}$) [36]. These logics differ in terms of capabilities and expressiveness, and support various kinds of quantifiers and modalities (e.g., []-safety or <>-liveness modalities in $d\mathcal{L}$, A or E path quantifiers in CTL). Logics currently included in the CRM are `CTL`, `LTL`, and their common superset `CTL*`, `dL` used together with HP for deductive verification and `TCTL` used in model checking of timed systems.

3.3 Verification

Verification Goal. *Verification* is usually defined to check the behavior of a system w.r.t. its intent [24]. Kern et al. [22] conceptually distinguish between *property-* and *implementation verification*. Property verification is concerned with specifying properties that are desired for a design (equivalent to what we consider *verification* in our CRM), while implementation verification deals with the relationship between high-level models and the implementation. In the CRM (cf. Fig. 2b) we focus on property verification, as implementation verification is rather a topic of model transformation which is not dealt with here. For ensuring such an intended behavior, formal verification methods are typically distinguished into *Model Checking* and *Verification by Deduction* [24]. Such methods provide rigorous evidence (e.g., a proof) that a specification aiming at a `VerificationGoal` is correct. Common verification goals include `Safety` (i.e., something will *not* happen) or **Liveness** (i. e., something *must* happen) [25], **Controllability/Reactivity** [38], **Fairness** [47] and **Deadlock Freedom** [7].

Verification Evidence. **VerificationEvidence** witnesses the correctness of a model w.r.t. a verification specification. What is considered a witness typically depends on the employed formal verification method (e. g., reachable states for model checking, a formal proof for deductive verification or counterexamples in both methods to witness correctness violation).

Proof. A **Proof** consists of arbitrary many proof steps and aims at verifying that the model is correct w.r.t. the specification. Entire proofs or parts of a proof might be transferred to other users using either the same tool or other tools to make progress or even close the entire proof. An implementation of the CRM should support composition and decomposition of proofs, as well as exchanging proofs, partial proofs, and lemmas. An important aspect arises from exchanging evidence: how can the correctness of exchanged artifacts be substantiated (e. g., certificates or by providing an exact listing of all proof steps)?

CounterExample. A specification for a selected technique can be refuted by a **CounterExample**, which is mutually exclusive to a successful proof. Multiple counter examples might be found for each open proof step. As soon as one is found, the refutation of the entire verification specification is inferred.

ReachableStates. Another possible output is the set of **ReachableStates**, if a reachability analysis was performed.

3.4 Tool Support

Because of the large number of tools available for modeling and verification of CPS (cf. Fig. 1b), in this section we restrict the discussion to classes related to modeling and proof collaboration, as motivated by our previous work [30]. However, as usual, the package can be extended to fit other tools as required.

Tool. The ToolSupport package includes concepts for **Tools**: (i) a **ModelingTool** supports users in creating a model using one of the respective formalism, (ii) a **VerificationSpecificationTool** allows formulation of a verification specification about a model, and (iii) the **VerificationTool** takes the model and its specification and produces a verification result.

A tool manipulates various **Artifacts**; it uses input (e. g., a model) to produce a corresponding output (e. g., a verification specification). Tools might require **UserInput** at run-time and additional prior **Configuration** (i. e., meta information required to run the tool, e. g., library paths). For example, verification tools often make a trade-off between expressiveness to achieve full automation (e. g., affine linear dynamics in SpaceEx [14]) and user interaction to handle undecidability (e. g., KeYmaera [40]). Furthermore, tools might interact with each other to enable **Collaboration** between instances of a single tool (e. g., multiple users might collaboratively produce a complex model) or different kinds of tools (e. g., different verification tools can be used to verify a single specification).

Refactorings. Artifacts can influence other artifacts (e. g., a counterexample may lead to revision of the initial model), since hybrid systems are often developed incrementally by model refinement. `Refactorings` support the refinement process through automated restructuring of artifacts. In case a model is accompanied with a correctness specification and a proof, it is important to spare full re-verification when the behavior of the model is refactored [31].

Another benefit of refactoring is to reduce verification complexity. For example, concurrent transition systems are exponentially harder to verify than sequential ones [17]; so concurrent real-world components that are independent in their read-write variables can be modeled sequentially in arbitrary order to reduce verification complexity. We can therefore further distinguish refactorings that change the behavior of a model into those increasing modeling detail (`Refinement Refactorings`) and those reducing details (`Abstraction Refactorings`).

Verification Tools. In our CRM we include two types of verification tools corresponding to the two major verification techniques: *model checking* and *theorem proving* [33]. For theorem proving, we introduce the class `DeductiveVerificationTool`. These tools are based on some logic and use inference rules (i. e., proof rules) to transform formulas until they yield axioms or logical tautologies. Kern et al. [22] refer to these techniques as *Deductive Methods*, while Clarke et al. [11] call them *Theorem Proving*. Model checkers calculate the states that can be reached by a model and check if all are desirable. Since the term model checking, however, refers to a technique mainly used for verification of purely discrete systems, we choose *reachability analysis* as the equivalent of model checking for hybrid systems. A `ReachabilityAnalyzer` calculates reachable states either in an exact manner by limiting the continuous dynamics to simple abstractions or in an approximate manner by over-approximating the set of reachable states [3]. Many of the techniques that work with (over-)approximations have to deal with floating point issues. The exactness of a verification technique (e. g., floating point variables do not store exact real valued numbers, but might result in rounding errors) has to be exchanged when different tools interact.

4   Knowledge Representation in the CRM

In this section we use the CRM to represent knowledge about hybrid systems.In order to simplify comparison of verification and modeling tools with the help of our CRM, we deduce a set of criteria that allow to classify and evaluate hybrid systems modeling and verification tools.

Detailed criteria are analogous to the concepts of the CRM and reveal fundamental differences between the hybrid systems modeling and verification tools as discussed below. As sample tools, we chose verification tools for hybrid systems (SpaceEx [14], and KeYmaera [40]) and for timed systems (UPPAAL [26]). The tools have different capabilities (cf. Fig. 3a): SpaceEx is a verification platform

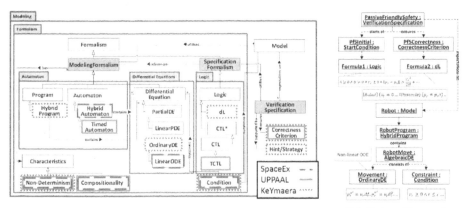

(a) Knowledge about SpaceEx, UPPAAL and KeYmaera   (b) Robot Example

Fig. 3. Knowledge representation in the CRM

for hybrid systems by performing reachability analysis, UPPAAL is an environment for verification of timed automata through model-checking and reachability analysis, and KeYmaera is a theorem prover for hybrid systems.

Modeling Formalism. Each of the tools uses a slightly different modeling formalism. SpaceEx and UPPAAL both work with networks of automata, where SpaceEx accepts HA and UPPAAL is restricted to timed automata. KeYmaera chooses a different approach and uses HPs. As for *continuous dynamics*, UPPAAL is restricted to the use of synchronous clocks, while the other tools can handle differential equations. However, SpaceEx is limited to affine linear dynamics while KeYmaera can also handle non-linear ODEs, algebraic DEs and differential inclusion. Additionally, SpaceEx and UPPAAL—as they use networks of automata—support compositionality (e. g., by parallel composition).

Timed automata can be translated into HA, but not necessarily the other way around, as modeled through timed automata being a subclass of HA. This in turn means that transformations from SpaceEx models containing differential equations will fail when translated to UPPAAL. Similarly, specifications provided in UPPAAL (e. g., safety properties) in terms of TCTL, will be lost when translating a model to SpaceEx. Furthermore, both SpaceEx and UPPAAL support guards and invariants (highlighted in the CRM by means of the marked Condition class), allowing loss-less translation of conditions.

Additionally, the CRM can be implemented as an object-oriented *knowledge representation scheme* to provide automated support for model selection and compatibility checks. For example, the *modeling formalism* for UPPAAL is timed automaton while it is HA for SpaceEx, and HP for KeYmaera. For a concrete model using a timed automaton as modeling formalism a knowledge representation query would return tools that can handle timed automata directly (SpaceEx and UPPAAL), and those tools that can handle the model with some transformation (KeYmaera). In the latter case, the models resulting from the transformation, however, are likely to not benefit from KeYmaera in the fullest possible extent, because the structure of the source model is different from the expected program structure. When comparing SpaceEx and UPPAAL, the knowledge representation would also provide that an instance of a SpaceEx model (containing a HA) cannot necessarily be assigned to UPPAAL, as the modeling formalism of UPPAAL cannot handle HA in general.

Specification Formalism. KeYmaera and UPPAAL use logical statements to define desired properties about a system, whereas SpaceEx computes a set of reachable states that can be compared either for intersection with the set of unsafe states (safety) or with the set of goal states (liveness).

The CRM also supports comparison of specification formalisms, as a similar class hierarchy as described above, also exists for logics. While UPPAAL uses a subset of TCTL for its specifications, KeYmaera uses d$\mathcal{L}$.

Verification Specification. For conditions, all tools support guards and invariants to restrict the behavior of the models. While SpaceEx returns a set of reachable states which has then to be analyzed for intersections with desirable or undesirable sets of states, UPPAAL supports the use of path formulae (further classified into reachability, safety and liveness) and KeYmaera allows arbitrary d$\mathcal{L}$ fomulae to specify safety and liveness properties. KeYmaera furthermore allows annotating its models with additional conditions (e.g., variants and invariants), to support the tool during verification of the models.

The CRM supports comparing the available kinds of correctness criteria specifiable within different tools. When translating from SpaceEx to UPPAAL, additional correctness criteria must be specified after the translation so that UPPAAL can verify the model. From UPPAAL to KeYmaera we may want to enrich the model with annotations to guide the proof search.

Applying the CRM. We use a robot collision avoidance model [29] to illustrate how hybrid systems models can be captured in the CRM. The collision avoidance model was designed using *hybrid programs* as a modeling formalism; its correctness requirements are defined by a specification. Since the robot moves along sequences of circular arcs and must not drive backwards, the hybrid program contains non-linear differential-algebraic equations. The starting condition *PfSInitial* of the system is a logical formula, which describes conditions on acceleration A, braking B, robot speed v_r, robot direction d_r, robot positions p_r, and obstacle positions p_o, under which it is safe to start the robot. The correctness criterion *PfSCorrectness* expresses that all possible behavior of the robot has to avoid collision, and furthermore has to retain at least one maneuver for obstacles to avoid collision as well, which means that it uses nested modality operators and, hence, can only be expressed in d$\mathcal{L}$. From these instances we immediately see that KeYmaera is the tool which fits best for verification of this model, since d$\mathcal{L}$ and non-linear DE are not supported by SpaceEx or UPPAAL (cf. Fig. 3b).

5 Conclusion and Future Work

In this paper, we introduced a conceptual reference model that can be used to analyze the properties of hybrid system modeling and verification tools and classify them accordingly. The resulting CRM can be used to (i) unify the used terminology, (ii) compare the capabilities of modeling and verification methods, and (iii) represent and exchange knowledge about those methods and about models. Furthermore, we see several promising application areas for such a CRM, spanning from education purposes in order to provide students with a road-map for CPS development and verification, to an object-oriented knowledge representation system, which can be used to automatically check the compatibility between models and tools, and exchange information between different tools as emphasized for interchange formats (e. g., [5], [32], [35]).

The CRM is a first step towards a comprehensive classification framework of hybrid system modeling and verification concepts and should be extended with further details: For future work, we plan to extend the CRM with *stochastic* or *probabilistic hybrid systems* (e. g., [21], [39], [50]), hybrid games (e. g., [42]), component-based modeling, and transformation. Finally, we plan to conduct a survey of hybrid system modeling and verification approaches using a criteria catalog derived from the CRM.

Acknowledgements. This work was funded by the Austrian Federal Ministry of Transport, Innovation and Technology (BMVIT) grant FFG FIT-IT 829598, FFG BRIDGE 838526 and FFG Basisprogramm 838181, and by PIOF-GA-2012-328378. The authors thank Andre Platzer for fruitful discussions and feedback.

References

1. Alur, R.: Formal verification of hybrid systems. In: Proc. of the 9th ACM Intl. Conf. on Embedded Software, EMSOFT 2011, pp. 273–278. ACM, NY (2011)
2. Alur, R., Courcoubetis, C., Halbwachs, N., Henzinger, T.A., Ho, P.H., Nicollin, X., Olivero, A., Sifakis, J., Yovine, S.: The algorithmic analysis of hybrid systems. Theor. Comput. Sci. 138(1), 3–34 (1995)
3. Alur, R., Dang, T., Ivančić, F.: Reachability Analysis of Hybrid Systems via Predicate Abstraction. In: Tomlin, C.J., Greenstreet, M.R. (eds.) HSCC 2002. LNCS, vol. 2289, pp. 35–48. Springer, Heidelberg (2002)
4. Alur, R., Dill, D.: The theory of timed automata. In: Huizing, C., de Bakker, J.W., Rozenberg, G., de Roever, W.P. (eds.) REX 1991. LNCS, vol. 600, pp. 45–73. Springer, Heidelberg (1992)
5. van Beek, D.A., Reniers, M.A., Schiffelers, R.R.H., Rooda, J.E.: Foundations of a Compositional Interchange Format for Hybrid Systems. In: Bemporad, A., Bicchi, A., Buttazzo, G. (eds.) HSCC 2007. LNCS, vol. 4416, pp. 587–600. Springer, Heidelberg (2007)
6. Berkenkötter, K., Bisanz, S., Hannemann, U., Peleska, J.: The HybridUML profile for UML 2.0. J. on Software Tools for Technology Transfer 8(2), 167–176 (2006)
7. Bingham, B.D., Greenstreet, M.R., Bingham, J.D.: Parameterized verification of deadlock freedom in symmetric cache coherence protocols. In: Formal Methods in Computer-Aided Design (FMCAD 2011), pp. 186–195 (2011)

8. Broman, D., Lee, E.A., Tripakis, S., Törngren, M.: Viewpoints, Formalisms, Languages, and Tools for Cyber-Physical Systems. In: Proc. of the 6th Intl. Workshop on Multi-Paradigm Modeling (MPM 2012) (2012) (preprint)
9. Carloni, L.P., Passerone, R., Pinto, A., Sangiovanni-Vincentelli, A.L.: Languages and Tools for Hybrid Systems Design. Foundations and Trends in Electronic Design Automation 1(1), 1–193 (2006)
10. Cellier, F.: Continuous System Modeling. Springer (1991)
11. Clarke, E.M., Wing, J.M.: Formal Methods: State of the Art and Future Directions. ACM Comput. Surv. 28(4), 626–643 (1996)
12. David, R., Alla, H.: On Hybrid Petri Nets. DEDS 11(1-2), 9–40 (2001)
13. Fill, H.-G., Redmond, T., Karagiannis, D.: Formalizing Meta Models with FDMM: The ADOxx Case. In: Cordeiro, J., Maciaszek, L.A., Filipe, J. (eds.) ICEIS 2012. LNBIP, vol. 141, pp. 429–451. Springer, Heidelberg (2013)
14. Frehse, G., et al.: SpaceEx: Scalable Verification of Hybrid Systems. In: Gopalakrishnan, G., Qadeer, S. (eds.) CAV 2011. LNCS, vol. 6806, pp. 379–395. Springer, Heidelberg (2011)
15. Giese, H., Henkler, S.: A survey of approaches for the visual model-driven development of next generation software-intensive systems. Journal of Visual Languages & Computing 17(6), 528–550 (2006)
16. Gupta, A.: Formal Hardware Verification Methods: A Survey. In: Kurshan, R. (ed.) Computer-Aided Verification, pp. 5–92. Springer (1993)
17. Harel, D., Kupferman, O., Vardi, M.: On the complexity of verifying concurrent transition systems. In: Mazurkiewicz, A., Winkowski, J. (eds.) CONCUR 1997. LNCS, vol. 1243, pp. 258–272. Springer, Heidelberg (1997)
18. Henzinger, T.A.: The Theory of Hybrid Automata. In: LICS, pp. 278–292. IEEE Computer Society Press (1996)
19. Hoare, C.A.R.: Communicating sequential processes, vol. 178. Prentice-Hall, Englewood Cliffs (1985)
20. Hopcroft, J.E., Motwani, R., Ullman, J.D.: Introduction to Automata Theory, Languages, and Computation, 3rd edn. Addison-Wesley, Boston (2006)
21. Hu, J., Lygeros, J., Sastry, S.: Towards a Theory of Stochastic Hybrid Systems. In: Lynch, N.A., Krogh, B.H. (eds.) HSCC 2000. LNCS, vol. 1790, pp. 160–173. Springer, Heidelberg (2000)
22. Kern, C., Greenstreet, M.R.: Formal Verification in Hardware Design: A Survey. ACM Trans. Des. Autom. Electron. Syst. 4(2), 123–193 (1999)
23. Kim, K.D., Kumar, P.: Cyber-Physical Systems: A Perspective at the Centennial. Proc. of the IEEE 100(special centennial issue), 1287–1308 (2012)
24. Kreiker, J., Tarlecki, A., Vardi, M.Y.: Reinhard Wilhelm: Modeling, Analysis, and Verification - The Formal Methods Manifesto 2010 (Dagstuhl Perspectives Workshop 10482). Dagstuhl Manifestos 1(1), 21–40 (2011)
25. Lamport, L.: Proving the Correctness of Multiprocess Programs. IEEE Transactions on Software Engineering 3(2), 125–143 (1977)
26. Larsen, K.G., Pettersson, P., Yi, W.: Uppaal in a nutshell. Intl. Journal on Software Tools for Technology Transfer 1(1-2), 134–152 (1997)
27. Lee, E.: Cyber Physical Systems: Design Challenges. In: 11th IEEE Intl. Sym. on Object Oriented Real-Time Distributed Computing, pp. 363–369 (2008)
28. Meyer, B.: Applying Design by Contract. Computer 25(10), 40–51 (1992)
29. Mitsch, S., Ghorbal, K., Platzer, A.: On Provably Safe Obstacle Avoidance for Autonomous Robotic Ground Vehicles. In: Robotics: Science and Systems (2013)
30. Mitsch, S., Passmore, G.O., Platzer, A.: Collaborative verification-driven engineering of hybrid systems. Mathematics in Computer Science 8(1), 71–97 (2014)

31. Mitsch, S., Quesel, J.D., Platzer, A.: Refactoring, refinement, and reasoning: A logical characterization for hybrid systems. In: Jones, C.B., Pihlajasaari, P., Sun, J. (eds.) FM (2014)
32. MoBIES team: HSIF semantics (version 3): Technical Report (2002)
33. Ouimet, M., Lundqvist, K.: Formal Software Verification: Model Checking and Theorem Proving (2007)
34. Pettersson, S., Lennartson, B.: Hybrid Modelling focused on Hybrid Petri Nets. In: 2nd European Workshop on Real-time and Hybrid Systems, pp. 303–309 (1995)
35. Pinto, A., Sangiovanni-Vincentelli, A.L., Carloni, L.P., Passerone, R.: Interchange formats for hybrid systems: Review and proposal. In: Morari, M., Thiele, L. (eds.) HSCC 2005. LNCS, vol. 3414, pp. 526–541. Springer, Heidelberg (2005)
36. Platzer, A.: Differential Dynamic Logic for Hybrid Systems. J. Automated Reasoning 41(2), 143–189 (2008)
37. Platzer, A.: Differential-algebraic Dynamic Logic for Differential-algebraic Programs. J. Log. Comput. 20(1), 309–352 (2010)
38. Platzer, A.: Logic and Compositional Verification of Hybrid Systems (Invited Tutorial). In: Gopalakrishnan, G., Qadeer, S. (eds.) CAV 2011. LNCS, vol. 6806, pp. 28–43. Springer, Heidelberg (2011)
39. Platzer, A.: Stochastic differential dynamic logic for stochastic hybrid programs. In: Bjørner, N., Sofronie-Stokkermans, V. (eds.) CADE 2011. LNCS (LNAI), vol. 6803, pp. 446–460. Springer, Heidelberg (2011)
40. Platzer, A., Quesel, J.-D.: KeYmaera: A Hybrid Theorem Prover for Hybrid Systems. In: Armando, A., Baumgartner, P., Dowek, G. (eds.) IJCAR 2008. LNCS, vol. 5195, pp. 171–178. Springer, Heidelberg (2008)
41. Pnueli, A.: The temporal logic of programs. In: Proc. of the 18th Annual Symposium on Foundations of Computer Science, SFCS 1977, pp. 46–57. IEEE Computer Society, Washington, DC (1977)
42. Quesel, J.-D., Platzer, A.: Playing hybrid games with keymaera. In: Gramlich, B., Miller, D., Sattler, U. (eds.) IJCAR 2012. LNCS (LNAI), vol. 7364, pp. 439–453. Springer, Heidelberg (2012)
43. Sanislav, T., Miclea, L.: Cyber-Physical Systems - Concept, Challenges and Research Areas. Journal of Control Engineering and Applied Informatics 14(2) (2012)
44. Schmidt, D.C., Buschmann, F., Henney, K.: Pattern-oriented software architecture. Wiley series in software design patterns. Wiley, Chichester (2000)
45. Shi, J., Wan, J., Yan, H., Suo, H.: A survey of Cyber-Physical Systems. In: Intl. Conf. on Wireless Communications and Signal Processing, pp. 1–6 (2011)
46. Teschl, G.: Ordinary differential equations and dynamical systems, Graduate studies in mathematics, vol. 140. American Mathematical Society (2012)
47. Völzer, H., Varacca, D.: Defining Fairness in Reactive and Concurrent Systems. Journal of the ACM (JACM) 59(3), 13:1–13:37 (2012)
48. Wan, K., Hughes, D., Man, K.L., Krilavicius, T., Zou, S.: Investigation on Composition Mechanisms for Cyber Physical Systems. Intl. Journal of Design, Analysis and Tools for Integrated Circuits and Systems 2(1), 30–40 (2011)
49. Wimmer, M., Schauerhuber, A., Kappel, G., Retschitzegger, W., Schwinger, W., Kapsammer, E.: A survey on UML-based aspect-oriented design modeling. ACM Computing Surveys 43(4), 28:1–28:33 (2011)
50. Zhang, L., She, Z., Ratschan, S., Hermanns, H., Hahn, E.M.: Safety Verification for Probabilistic Hybrid Systems. In: Touili, T., Cook, B., Jackson, P. (eds.) CAV 2010. LNCS, vol. 6174, pp. 196–211. Springer, Heidelberg (2010)
51. Zwillinger, D.: Handbook of differential equations. Academic Press Inc. (1998)

A Semantic-Based EMRs Integration Framework for Diagnosis Decision-Making

Huili Jiang, Zili Zhang, and Li Tao

Key Laboratory of Intelligent Software and Software Engineering,
Southwest University, Chongqing 400715, China
{jianghuili,zhangzl,tli}@swu.edu.cn

Abstract. Discovering latent information from Electronic Medical Records (EMRs) for guiding diagnosis decision making is a hot issue in the era of big data. An EMR composes of various data (e.g., patient information, medical history, diagnosis, treatments, symptoms), but most of them are stored in the relational database. It is difficult to integrate the data and infer new knowledge based on existing data structures. Semantic technology (ST) is a flexible and scalable method for integrating heterogeneous, distributed information from big data. Taking advantage of these features, this paper proposes a framework that leverages ontology to improve EMRs decision-making. A case study shows that this framework is feasible to integrate information, and can provide specific and personalized information services for facilitating medical diagnosis.

Keywords: Semantic Technology, Integration, EMRs, Ontology, Diagnosis Decision-making.

1 Introduction

Electronic Medical Records (EMRs) have been carried out in-depth study in Europen and American. It is still one of the hot topics especially in the era of big data. An EMR document contains massive important information, such as basic patient information, medical history, diagnosis, adverse drug history, treatments, symptoms and so on. It plays a vital role in discovering latent information to guide doctors, community physicians and other individuals for diagnosis. There is a big challenge in effectively utilizing this massive information in the era of big data, especially in uncovering the hidden information to guide decision making [1].

However, the format of medical data is unstructured or semi-structured, such as images, office documents, and XML documents. It is so diverse that hardly to be understood by computer automatically. Meanwhile, nearly all medical data is stored in the relational database. It is difficult to infer new knowledge to provide specific and personalized information services to facilitate medical diagnosis. It is a big challenge to reuse and share these information. The use of ontology for the explication of implicit and hidden knowledge to integrate this information is a possible approach to overcome the problem of semantic heterogeneity. Only

R. Buchmann et al. (Eds.): KSEM 2014, LNAI 8793, pp. 380–387, 2014.
© Springer International Publishing Switzerland 2014

integration of these medical information, can make them more valuable. Our work is to integrate medical information which is semantic heterogeneity.

In 1998, the concept and architecture of Semantic Web [2–4] were proposed by T. Berners-Lee and J. Hendler. Many researchers and organizations have contributed great efforts to promote the Semantic Technology (ST). There are three features of ST: Uniform Resource Identifier (URI), the Resource Description Format (RDF), and Ontology. The use of ontology for the explication of implicit and hidden knowledge is a possible approach to overcome the problem of semantic heterogeneity.

In this paper, we present a new framework based on ST for integrating heterogeneous and distributed information from big data. One aspect of our work is creating the ontology that is focused on the current medical development in China. Based on the EMR ontology, specific and personalized information services are provided to facilitate medical diagnosis.

The remainder of the paper is organized as follows. Section 2 gives the framework of our work. Section 3 provides some cases by using the EMRs ontology to make decisions. Section 4 presents the conclusions and future work.

2 A Semantic-Based EMR Framework for Decision-Making

As ontology can explicitly describe relationships between concepts and entities [5], the basis for EMRs decision-making is integrating these documents in the same standard through building EMRs ontology. The aim of our work is to provide a framework to integrate the distributed, diversity information and support a decision-making for doctors, community physicians and other individuals. The framework is demonstrated in Fig.1 which we adopt three data sources to create this ontology.

2.1 Basic Ontology

With the help of ontology engineers and experts in medicine, we create the basic ontology which is the experienced-driven knowledge warehouse. The raw data that consists of disease names, trials, symptoms and so on are mainly from the Chinese medicine experts.

2.2 Extract Information from Websites

Besides the experienced-based knowledge, there is some valuable information in the websites. To make the information more comprehensive, we extract knowledge into the aforementioned ontology from the famous healthcare websites such as Sina and Sohu. Data from these websites is mainly some prevention tips, diet tips, and exercise. It is exactly complementary with the knowledge from medicine experts.

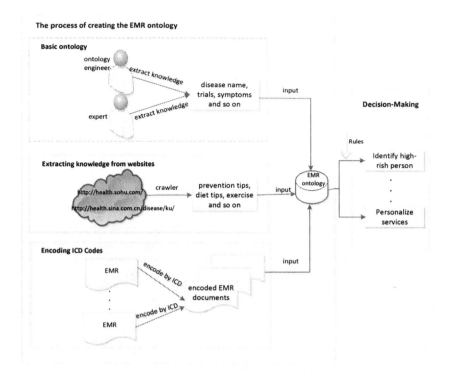

Fig. 1. The Framework of our work

2.3   Encoding ICD Codes

To integrate EMRs knowledge, only aforementioned ontology is not enough because descriptions or records of a disease maybe not represented as the same
symbol. To avoid this ambiguous, a standard encoding is necessary. In our work,
we use the International Classification of Diseases (ICD) code which is widely
adopted in the developed countries.

2.4   Decision-Making in the Diagnosis

Using ontology which leverages knowledge and web intelligence, specific and
personalized information services are provided in the process of decision-making
for doctors, physicians, and individuals. In our work ,we focus on the personalized
services based on integrating information of one person and identifying high-risk
person. More details will be presented in section 3.2.

3   An Application of the Semantic-Based Framework

ST has explicit semantic and is easy to infer new knowledge based on existing data. The core and basis of ST are ontology database and inference based

on ontology. Section 3.1 describes the process of creating the ontology from three data sources in detail. Using the ontology to provide specific and personalized information services to facilitate medical diagnosis will be presented in section 3.2.

3.1 Three Data Sources to Create Ontology

Domain ontology based on the specific domain concepts and accurate relations description lays the foundation of the knowledge base. What is more, well developed and widely adopted ontologies are an important cornerstone, providing some basic term relationships and guiding the further development of the domain knowledge base [6]. In this section, we uses knowledge and web intelligence to create ontology.

3.1.1 Creating the Basic Ontology

The purpose of establishing the EMR ontology is to integrate the massive electronic medical records, to facilitate online retrieval for doctors, patient, to promote the ability and knowledge of doctors, and then to reduce health care costs by improving the efficiency of medical institutions. The main classes of EMR ontology are shown in Table.1. In this paper, we adopt the top-down approach to define the class hierarchy. After defining classes, relevant attributes of the class are defined. Fig.2 is the RDF triples topographies of Coronary heart disease by TopBraid Composer [7] that is an enterprise-class modeling environment for developing Semantic Web ontologies and building semantic applications.

Table 1. The main class of the EMR ontology

Class Name	Explanatory Note
Disease	Name of a disease
Risk_factors	Predisposing factors, including genetic, environmental, and lifestyle risk factors.
Symptoms	The things that show that someone has it
Trials	Medical trials targeting the detection and/or treatment of the disease
Prevention_ factors	Ways of preventing a disease, including vaccines and precautions such as hand washing
Treatment	Medical treatments, such as drugs and procedures to target the illness.
Affiliated_disease	Diseases that often go hand in hand with the disease, such as those can cause or be caused by, or have a high correlation with ,this disease.

3.1.2 Extracting Information from Websites

Meglic et al. [8] evaluated network intervention method. They indicated that the network community and the professional health websites are well known by patients and medical staff. Healthcare knowledge on the network plays a

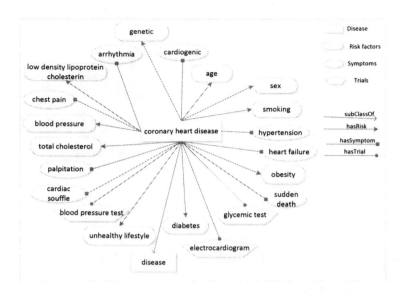

Fig. 2. The RDF triples topographies of Coronary heart disease

huge role in people's life. In the real world, besides the sophisticated doctors knowledge, some health care information is also valuable in the websites [9] such as Sina and Sohu. So extracting knowledge of disease prevention is a necessary assignment. We find that it is feasible by analyzing the source codes of websites.

In our experiments, we use open source Heritrix to crawl the corresponding data, parse prevention and health care knowledge and write them into ontology repository. Fig.3 is the source codes from one website. It is clear that they are all semi-structure data. Besides the disease name, we mainly extract the knowledge of prevention and diet care into our ontology. As shown in Fig.3, we extract prevention and put them into the term prevention.

```
<ul>
  <li><span>Disease name:</span>coronary heart disease</li>
  <li><span>Location:</span> <a href=" /body-1017.shtml" >chest</li>
  <li><span>Department:</span><a href=" /dept-2002.shtml" >cardiovascular medicine</li>
  <li><span>Symptom:</span>Sudden death|sleep-disordered breathing|chest pain|cardiac
murmur|cardiogenic|Arrhythmia    |hypertension|palpitation|</li>
  <li><span>intro</span> coronary artery coronary heart disease (CHD) is a kind of organic (atherosclerosis or
dynamic vasospasm) stenosis or occlusion of myocardial ischemia caused by lack of oxygen (angina) or myocardial
necrosis (myocardial infarction), heart disease, also known as ischemic heart disease···   </li>
<ul>
<h3><span>Coronary heart disease</span>prevention</h3>
<p>Control and prevention against three levels, primary prevention means there is no disease prevention,
secondary prevention is<a href="/disease-3474_5.shtml" target="_blank">[more]</a>
</p>
```

Fig. 3. The source codes of one website

3.1.3   Manual Coding in ICD

A unified coding is the core of data integration. ICD codes are the standard diagnostic tool for epidemiology, health management and clinical purposes. However, there exists some similar ICD codes about relevant diseases. As far as we know, traditional machine learning and Natural Language Processing can not achieve high accuracy even in the best situations.

Considering above factors, this paper translate ICD code manually. A concrete instance about Intra Uterine Contraceptive Device(IUCD) is shown in Table 2. It is no doubt that there are similar codes for different diseases which are too hard to distinguish them literally. IUCD codes are so complicated, not to mention to all the ICD codes. We encode the electronic medical records into ICD through the diagnoses in the documents.

Table 2. The similar ICD codes similar disease

ICD code	Disease name	Mnemonic in Chinese
O35.701	IUCD pregnancy	zgnbyqrs
O35.702	IUCD failure	zgnbyqsb
T83.301	IUCD incarcerate	zgnbyqqd
T83.305	IUCD breakage	zgnbyqdl
Z30.501	Remove IUCD	qczgnbyq

3.2   EMRs Ontology-Based Inferring

Creating EMRs ontology and making decision based on it are two major works. The EMR ontology provides more comprehensive and accurate health information. Section 3.1 has discussed the details to create EMRs ontology from three data sources. On the other side, in the process of intelligent reasoning, personalized services for users are provided. Section 3.2.1 applies EMRs ontology to provide healthcare information. Section 3.2.2 provides an example of using rules to identify the high-risk person.

3.2.1   Using ICD EMRs to Provide Healthcare Information

Traditional data mining provides services for all patients or a particular group. However, it is unable to provide suggestions for a single person. In the era of personalized therapy that people need specific suggestions based on their individual patient data and want some specific problems to be solved before the disease appears. In this case, only the data mining alone is not enough.

The steps of using EMRs ontology to provide personalize offerings are as follows: firstly, annotate personal information and symptoms, then put these information into the ontology. From each individual's perspective, the information from different hospitals can easily integrate. Meanwhile, the related preventions, diets, sport tips will be offered for the patients. Based on making personal documents into a semantic graph, the information from different hospitals is easily integrated, the diagnosis of the person will be provided on the EMRs ontology.

Zheng, female, 49, healthcare, shanghai hospital, bed number :0903, outpatient number: 06289170, hospital number: 00367285, director of the doctor: Liang, doctor: Shen.

Hospital diagnosis: patients after admission to perfect the related inspection, with interest rates and expanding cervix, hysteroscopy examination in 2011.1.13, positioning for ring. Under the HSC see: endometrial hyperplasia of mild hypertrophy, the luster is normal. Palace, a ring at the bottom of the shade part embedded in the endometrium. On both sides of the normal fallopian tube openings. Positioning took out a ring of integrity. Smooth operation, less intraoperative bleeding. Anti-inflammatory therapy after surgery. , discharge conditions: a small amount of vaginal bleeding, no abdominal pain, abdominal distension. Temperature is normal. Hospital, the doctor's advice: 1. The outpatient follow-up.

(a) The Raw EMR from hospital　　　　　　　**(b) Result with ICD code of using ontology**

Fig. 4. The graph of one person's EMR document

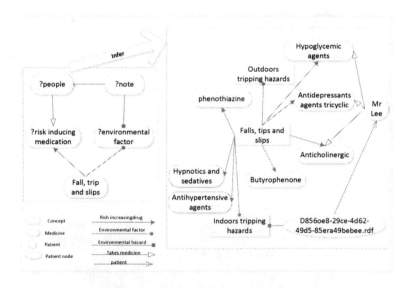

Fig. 5. A case to use knowledge to infer wrestle

3.2.2　Using Rules to Identify High-Risk Persons

Predicting diseases based on symptoms is one of goals we try to achieve. Warning some high-risk persons before real diseases come is the another one. This section introduces the process identify high-risk people who will wrestle. The steps of using rules to identify high-risk person are as follows: 1) extracting information from EMRs written in Chinese, 2) establishing rules of risk in terms of falls, tripping, slipping, 3) using the integrated EMRs ontology, screen out these patients who have this risk. More detailed are presented in Fig.5.

4   Conclusion

We propose a semantic-based EMRs integration framework which uses ontology to predict latent diseases and warn some high-risk persons. A case study shows that this framework is feasible and good scalability. In the course of creating EMRs ontology, we adopt knowledge and web intelligence to make this ontology more valuable. We use medicine, diet tips, exercise as a treatment in this paper. In the future, we will optimize our ontology by adding feedback and introducing other ontologies, such as SNOMED CT.

Acknowledgement. This Project was supported by the national science and technology support program (2012BAD35B08). The authors would like to thank Dr. Chao Gao for his great help in this paper.

References

1. Lopez, D.M., Blobel, B.: A development framework for semantically interoperable health information systems. International Journal of Medical Informatics 78(2), 83–103 (2009)
2. Berners-Lee, T., Hendler, J., Lassila, O., et al.: The semantic web. Scientific American 284(5), 28–37 (2001)
3. Berners-Lee, T., Hendler, J.: Publishing on the semantic web. Nature 410(6832), 1023–1024 (2001)
4. Shadbolt, N., Hall, W., Berners-Lee, T.: The semantic web revisited. Intelligent Systems 21(3), 96–101 (2006)
5. Gruber, T.R.: A translation approach to portable ontology specifications. Knowledge Acquisition 5(2), 199–220 (1993)
6. Jain, H., Thao, C., Zhao, H.: Enhancing electronic medical record retrieval through semantic query expansion. Information Systems and e-Business Management 10(2), 165–181 (2012)
7. Weiten, M.: Ontostudio® as a ontology engineering environment. In: Semantic Knowledge Management, pp. 51–60. Springer (2009)
8. Meglic, M., Furlan, M., Kuzmanic, M., et al.: Feasibility of an eHealth service to support collaborative depression care: results of a pilot study. Journal of Medical Internet Research 12(5) (2010)
9. Chang, C.H., Kayed, M., Girgis, M.R., et al.: A survey of web information extraction systems. Knowledge and Data Engineering. Knowledge and Data Engineering. IEEE Transactions on Knowledge and Data Engineering 18(10), 1411–1428 (2006)

Author Index